Conceptual Foundations of Materials

A Standard Model for Ground- and Excited-State Properties

Series: Contemporary Concepts of Condensed Matter Science
Series Editors: E. Burstein, M.L. Cohen, D.L. Mills and P.J. Stiles

Conceptual Foundations of Materials

A Standard Model for Ground- and Excited-State Properties

Steven G. Louie and Marvin L. Cohen
Department of Physics
University of California
Berkeley, CA, USA

ELSEVIER
Amsterdam – Boston – Heidelberg – London – New York – Oxford
Paris – San Diego – San Francisco – Singapore – Sydney – Tokyo

Elsevier
Radarweg 29, PO Box 211, 1000 AE Amsterdam, The Netherlands
The Boulevard, Langford Lane, Kidlington, Oxford OX5 1GB, UK

First edition 2006

Library of Congress Cataloging-in-Publication Data
A catalog record for this book is available from the Library of Congress

British Library Cataloguing in Publication Data
A catalogue record for this book is available from the British Library

ISBN-13: 978-0-444-50976-5
ISBN-10: 0-444-50976-3
ISSN: 1572-0934

For information on all Elsevier publications
visit our website at books.elsevier.com

Printed and bound in The Netherlands

06 07 08 09 10 10 9 8 7 6 5 4 3 2 1

CONTENTS

LIST OF CONTRIBUTORS

P. B. Allen	Department of Physics and Astronomy, State University of New York, Stony Brook, NY 11794-3800, USA
R. Car	Department of Chemistry, Princeton University, Princeton, NJ 08544, USA
J. R. Chelikowsky	Departments of Physics and Chemical Engineering, Institute for Computational Engineering and Sciences, University of Texas, Austin, TX 78712, USA
M. L. Cohen	Department of Physics, University of California at Berkeley and Materials Sciences Division, Lawrence Berkeley National Laboratory, Berkeley, CA 94720, USA
S. G. Louie	Department of Physics, University of California at Berkeley and Materials Sciences Division, Lawrence Berkeley National Laboratory, Berkeley, CA 94720, USA
R. Resta	INFM-DEMOCRITOS National Simulation Center, Via Beirut 4, I-34014 Trieste, Italy and Dipartimento di Fisica Teorica, Università di Trieste, Strada Costiera 11, I-34014 Trieste, Italy
D. Vanderbilt	Department of Physics and Astronomy, Rutgers University, Piscataway, NJ 08854-8019, USA

PREFACE

The fundamental interactions of importance in condensed matter are the Coulomb attraction between the electrons and the nuclei and the Coulomb repulsion among the electrons and among the nuclei. The effects of many electrons under these interactions, obeying the laws of quantum mechanics, ultimately determine all the observed phenomena and properties of condensed matter systems. Thus, electronic structure theory has played a central role in our understanding of atoms, molecules, and solids since the dawn of quantum mechanics. Electronic structure theory is now a pillar of condensed matter science and is perhaps the largest theoretical sub-area in condensed matter physics.

The goal of this volume in the series on "Conceptual Foundations of Material Properties: A Standard Model for Ground- and Excited-state Properties" is to present the fundamentals of electronic structure theory that are central to the understanding and prediction of materials phenomena and properties. The emphasis will be on foundations and concepts. The chapters are designed to offer a broad and comprehensive perspective of the field. They cover the basic aspects of modern electronic structure approaches and highlight their applications to the structural (ground state, vibrational, dynamic and thermodynamic, etc.) and electronic (spectroscopic, dielectric, magnetic, transport, etc.) properties of real materials including solids, clusters, liquids, and nanostructure materials. This framework also forms a basis for studies of emergent properties arising from low-energy electron correlations and interactions such as the quantum Hall effects, superconductivity, and other cooperative phenomena.

Although some of the basics and models for solids were developed in the early part of the last century by figures such as Bloch, Pauli, Fermi and Slater, the field of electronic structure theory went through a phenomenal growth during the past two decades, leading to new concepts, understandings and predictive capabilities for determining the ground- and excited-state properties of real, complex materials from first principles. For example, theory can now be used to predict the existence and properties of materials not previously realized in nature or in the laboratory. Computer experiments can be performed to examine the behavior of individual atoms in a particular process, to analyze the importance of different mechanisms, or just to see what happens if one varies the interactions and parameters in the simulation. Also, with *ab initio* calculations, one can determine from first principles important interaction parameters which are needed in model studies of complex processes or highly correlated systems. Each time a new material or a novel form of a material is discovered, electronic structure theory inevitably plays a fundamental role in unraveling its properties.

This volume is organized as follows. Chapter 1 gives a brief overview of the field. Chapter 2 is devoted to the foundations of the theory of ground- and excited-state properties of matter. Important concepts such as pseudopotentials, density functional theory, electron correlations, total energy and forces in ground-state properties, and the effects of quasiparticle excitations and electron–hole interactions in spectroscopic properties are discussed. Chapter 3 describes an *ab initio* molecular dynamics approach to finite temperature effects and thermodynamic and dynamic properties. The concepts and techniques of combining forces from electronic structure theory with molecular dynamics methods to form a unified approach for calculating the properties of solids and liquids are developed. Chapter 4 presents applications of electronic structure methods to the understanding and prediction of the structure and electronic properties of complex materials including clusters, nanocrystals, and liquids. The theory of electric field effects in insulators is discussed in Chapter 5. In particular, modern concepts of solids in electric fields involving Berry's phase and Wannier functions are presented. Finally, the fundamentals of electronic transport in solids are presented in Chapter 6. Developments from a semi-classical Boltzmann equation treatment to the Landauer formalism for quantum conductance are discussed.

We thank Jack Deslippe for assistance with the technical layout and proof reading of this volume.

Steven G. Louie and Marvin L. Cohen
Department of Physics,
University of California,
Berkeley, CA, USA

Chapter 1
OVERVIEW: A STANDARD MODEL OF SOLIDS

M. L. Cohen

1. BACKGROUND

Understanding the properties of matter is one of the oldest branches of science. Today, condensed matter or solid-state physics is the largest branch of physics, and the quest for a theoretical framework to explain and predict the properties of solids represents a significant fraction of theoretical physics in general. So, like the ancients, we are still trying to understand the properties of matter, but now we have powerful and useful conceptual models.

A significant breakthrough has been the concept that solids are composed of atoms. In the time of the alchemists, the properties of solids were attributed to a veneer superimposed on base matter. The shining properties of gold were attributed to the modifications of the gray base solid. So it is not so surprising that alchemists would have the goal of transforming lead into gold. We now know that without nuclear physics one cannot cause or understand this transformation. Once the atomic view is adopted then the problem is changed. The atoms are a "given" with no transformations between elements when considering low-energy processes. The remaining problem is how to describe the properties of solids knowing that they are composed of strongly interacting atoms.

Before going directly to the paths leading to successful models and methods, it is worthwhile discussing the two dominant current "pictures" or "philosophies" held consciously or unconsciously by researchers in this field.

Many in physics rely heavily on reductionism. In constructing a model or picture of a solid, this view would propose that, since a solid is made of interacting atoms

Contemporary Concepts of Condensed Matter Science
Conceptual Foundations of Materials: A Standard Model for Ground- and Excited-State Properties

ISSN: 1572-0934/doi:10.1016/S1572-0934(06)02001-4

and because of the length and energy scales involved and the fact that the dominant interaction is electromagnetism, then to determine the electronic structure of solids, we should start with free atomic states and then determine the effects arising from atom–atom interactions. This model is particularly suited to systems where the atom–atom interactions are weak and can be treated quantum mechanically with perturbation theory. A specific model currently in use, which derives from these ideas, is the tight-binding model (TBM) which is appropriate when the overlapping of electron wave functions of different atoms is small. The TBM can be extended to cases where there is considerable overlap; but, for enough overlap, it is best to consider the valence electrons as liberated from individual atoms. This nearly free electron model (NFEM) is more appropriate for a large class of solids.

Another view, when thinking of a model of solids, is to take a more "emergent" rather than "reductionist" view. We determine properties of solids by probing them and measuring their responses to the probes through response functions. For electromagnetic probes, dielectric functions and magnetic susceptibilities are the response functions, while for temperature changes, the heat capacity is the response function. One can model the behavior of response functions in terms of elementary excitations of solids. For example, temperature changes affect lattice vibrations, and this can be viewed as excitations of phonons. By viewing the collective excitations of vibrating atoms in terms of phonons, and collective excitations of electrons and spins as plasmons and magnons, we introduce fictitious particles to model the responses of a solid and through the use of response functions. For collective excitations these elementary excitations are bosons. For electrons, the elementary excitations corresponding to individual particle behavior are quasiparticles. The quasiparticles are fermions. They resemble the electrons that went into the solid, but they are modified. For example, a polaron is a quasiparticle that has the charge of an electron, but an enhanced mass because of electron lattice interactions. For the fractional quantum Hall effect, the quasiparticles can have fractional charges. A hole, which represents the absence of an electron, is a quasiparticle with positive charge. We do not expect to produce a beam of real holes propagating in a vacuum, but this mental construction is very useful for understanding the properties of solids.

So, when asked whether, in this branch of physics, the theorists are reductionists or proponents of emergent philosophies, the answer is yes. Calculating the properties of matter requires a dual approach in the same spirit as the need for a particle-wave approach is necessary in many areas of physics. Also, properties "emerge." Given all the particles and their fundamental interactions, it is still unlikely that anyone would think of properties such as superconductivity or the fractional quantum Hall effect before they are discovered experimentally. So for now, the theorists in this field follow experimentalists, but they have some space for prediction. Theoretically, predicting dramatically different states or properties of matter are possible in the future, but for now discoveries of this magnitude are in the domain of the experimentalists. Also, the final decision on whether a theory is correct is made by experiment.

2. THE HAMILTONIAN

For most solid-state effects, the atomic core electrons of a solid will not dominate the bonding and other electronic properties of a solid, hence it is reasonable to follow the suggestion of Fermi [1] for studying highly excited atoms. He introduced a pseudopotential to determine the interactions between valence electrons and cores and to focus on the parts of the wave functions away from the regions near the cores. The pseudopotential has been rediscovered several times and other very productive views of its properties [2] have contributed immensely to our understanding and use of this approach. If we restrict ourselves to this pseudopotential model and employ the Born–Oppenheimer approximation to separate electronic and core motions, one can write a Hamiltonian describing a solid made of "pseudoatoms" if the (valence) electron–electron interactions and core–core interactions are included. The core–core part can be approximated using Coulomb interactions and standard Madelung sums to get this contribution to the total energy of a solid.

The electron–electron problem is more formidable because here we are dealing with itinerant fermions. Exchange and correlation effects are important, and a simple Hartree model where one electron moves in the mean or average field of all the other electrons does not suffice for explaining the detailed properties of real materials. For many of the applications of the pseudopotential model, the density functional theory (DFT) is used [3] for computing the electron–electron contributions to the total energy. When the local density approximation (LDA) within DFT is used, ground-state properties of solids can be evaluated using functionals of the electron density. The resulting Hamiltonian is relatively simple, and it will be discussed in later chapters with more justification for the contributions than presented here. This is essentially the standard model [4] used currently for explaining and predicting ground-state properties. For excited states and the determination of quasiparticle properties, extensions to the model such as the "GW" approximation [5] allow determinations of response functions to explain and interpret properties, such as photoemission and optical spectra. These modifications for describing excited states will be described in other chapters.

3. EMPIRICAL MODELS LEAD THE WAY

The NFEM and TBM mentioned earlier were precursors to some of the modern approaches. Even the completely free electron model (FEM) provided, and still provides, a good starting point for many calculations [6]. In particular, if the structural positions of the cores are ignored and the positive cores are smeared out into a structureless jelly, one still can gain insight for some problems, particularly if confinement effects on electrons are of interest. This jellium model is often the starting point for treating the electron liquid and for determining the properties of simple metals. In some sense, the jellium model is a quantum extension of J.J. Thompson's plum pudding model of a solid. In Thompson's model, the positive background was viewed as a pudding, while the electrons were plums in the pudding. In the FEM, the pudding is the positive jellium background smeared out throughout the solid, while the electrons are allowed to move freely through the solid. However, unlike the plum model of Thompson, the electrons are not localized and they are treated quantum mechanically. For this model, using the wave nature of the electrons confined to a box, one can solve many of the mysteries that arose in classical physics [6] such as the puzzle of the heat capacity of metals. In fact, about a fourth of the periodic table can be modeled using a FEM. Basically it is the effects of confinement to a solid (box) and the Pauli Principle that determine the quantum effects. Much of this is discussed in standard textbooks [6].

A modern application is the study of small structures such as clusters or wires made of free electrons. As an example, studies of jellium spheres explain many of the properties of sodium clusters including the "magic number" associated with the abundances of specific cluster sizes emitted from ovens [7]. Confinement effects on the nanoscale can also be explored with the FEM, but here the reduced dimensionality and symmetry effects associated with structure often dominate.

In moving across a row of the periodic table such as Na, Mg, Al, and Si, one can argue for a jellium model for some properties of the first three elements, but for Si the structure and core potential effects dominate. Bonds are formed and the electrons are no longer completely free. It is almost as if we have passed from Thompson's model of an atom to Newton's model where atoms have hooks to form bonds. Here, we can use the NFEM with a core potential and input the structure using reciprocal lattice vectors **G** [6]. To simplify the problem even more, one can assume that there is a periodic pseudopotential containing both electron–core and electron–electron contributions. Since the pseudopotential $V(\mathbf{r})$ is periodic, it can be expanded as a sum with form factors or coefficients $V(\mathbf{G})$

$$V(\mathbf{r}) = \sum_{\mathbf{G}} S(\mathbf{G})V(\mathbf{G})e^{i\mathbf{G}\cdot\mathbf{r}} \tag{1}$$

where $S(\mathbf{G})$ is the structure factor that places atomic cores in their correct structural positions

$$S(\mathbf{G}) = \frac{1}{w}\sum_{\tau} e^{i\mathbf{G}\cdot\mathbf{r}} \tag{2}$$

where τ is the basis vector to each of the w atoms in the unit cell. For the diamond structure,

$$S(\mathbf{G}) = \cos \mathbf{G}\cdot\tau \tag{3}$$

where $\tau = (\frac{1}{8}, \frac{1}{8}, \frac{1}{8})a$ and a is the lattice constant.

Because the pseudopotential is designed to reproduce the outer parts of the electron wave function and not the strong oscillations in the core, it is relatively weak. This means that the Fourier sum in Eq. (1) can be cut off after a few $V(\mathbf{G})$s. For Si or Ge, only the first three $V(111)$, $V(220)$, and $V(311)$ are needed. In fact, for most of the standard *III–V* and *II–VI* zincblende semiconductors only three coefficients per atom suffice to produce accurate band structures.

In the 1960s and 1970s, these coefficients were fit to reproduce optical data. This empirical pseudopotential method (EPM) produced dozens of highly accurate band structures that explained photoemission spectra, visible and UV optical properties, and many other electronic properties of solids [8,9]. It is probably fair to say that the EPM solved the problem of explaining and interpreting the optical properties of common semiconductors.

An important feature of the EPM that led to future development was the realization that the pseudopotentials determined for different semiconductors were transferable from one semiconductor to another and from one structure to another. Another feature was the fact that electron densities, obtained from the wave functions determined using the EPM, correctly predicted the measured electron densities and bond charges that were later determined using X-ray spectroscopy [9].

Once it was realized that the electron density could be computed accurately, it became possible to extend the EPM to study localized configurations and surfaces [10] and to eventually use DFT. Self-consistency was the key concept. When electronic charge rearranges, the electron–electron potential changes; and this in turn changes the density. Therefore it is necessary to self-consistently feed changes back into the potential. Using this method, one could account for the rearrangement of charge at a surface. Another modification of the EPM was necessary for surfaces. Since the potential expressed in Eq. (1) assumes an infinite solid with translational symmetry, one has to break this symmetry to consider surfaces. The method introduced [10] to do this was called the supercell approach where the structure factor was used to construct a cell that contained slabs of atoms with vacuum regions between the slabs. By reproducing these slabs infinitely as in the EPM, one could represent surfaces using the surface of a slab. Then using self-consistency, the charge rearrangements of electron charge could be incorporated into the potentials. Interfaces could also be studied in this way.

4. TOWARD *AB INITIO* CALCULATIONS

The incorporation of DFT into the pseudopotential scheme became straightforward since the functionals for exchange and correlation could be computed from the electronic charge density. At first, exchange and correlation were treated using approximations by Slater and Wigner. After DFT was incorporated to make the entire scheme parameter free, *ab initio* potentials were developed and used. At this point, the input needed is the crystal structure, the lattice constants, and the atomic numbers to generate the pseudopotentials.

The next step was to directly compute the distances between the atoms and to free the theory from all experimental input except the structure and the atomic numbers and masses of the atoms forming the solid. This was done by utilizing a scheme to calculate the total energy of the solid [4]. For a given structure, the total energy E_t could be evaluated as a function of the lattice constants or volume

$$E_t = E(core-core) + E(el-core) + KE(el) + E(el-el) \tag{4}$$

where $E(core-core)$ can be evaluated as described earlier in terms of Madelung sums. The pseudopotentials for each atom give the $E(el-core)$ terms, while the kinetic energy of the electrons, $KE(el)$, can be determined once the wave function is calculated. The electron–electron term $E(el-el)$ has a Hartree or mean field part and an exchange-correlation part. These are computed using DFT.

Since E_t could be calculated as a function of volume V, one could determine the minimum $E_t(V)$ for a given structure. This would fix the lattice constants. If the curvature of $E_t(V)$ is computed at its minimum energy, the bulk modulus or compressibility could also be determined. When compared with experiment, typical accuracies are less than 1% for the lattice constants and around 5% for the bulk moduli. These calculations were extended considerably. For example, by reducing the lattice constants to simulate the effects of pressure, new high-pressure structures were successfully predicted. Transition volumes and transition pressures for the transformations were determined, and the predictions for these properties were generally within 1% of the measured values. Surface reconstructions and mechanical properties were also explored using this approach.

One very interesting application which added a great deal of credibility for this method was the determination of vibrational (phonon) spectra, electron–phonon interactions, and the successful prediction of superconductivity in high-pressure metallic forms of Si. The phonon spectrum is determined by moving the atoms in the crystal to represent the phonon of interest, calculating the increased energy, and relating this to a "frozen in" phonon. There are variations on this approach where Hellman–Feynman forces are calculated as well as "spring constants." In addition, the electron–phonon coupling can be obtained. Hence, using only the atomic numbers of the constituent atoms to generate the appropriate pseudopotentials, the atomic mass for determining the vibrational properties, and the crystal structure, one is in a position to calculate a wide variety of properties.

A dramatic demonstration of the appropriateness of the plane wave pseudopotential model (PWPM) or standard model is the successful prediction of two high-pressure phases of Si. These were the primitive hexagonal and hexagonal close-packed phases. The transition pressures, lattice constants, electronic properties, vibrational spectra, and superconducting properties were successfully predicted [11]. In fact, until the recent *ab initio* calculation of the superconducting properties of MgB_2 were done [12], this work on Si was the most complete *ab initio* superconducting transition temperature study where minimal information about the material was used as input. Other high-pressure forms of Si were also predicted. All or almost all have been found experimentally.

5. OTHER CHAPTERS IN THIS VOLUME

Although the EPM gave good optical spectra and hence some information about excited states, once more *ab initio* approaches were introduced, problems arose in this area. In particular, when the LDA is used for ground state properties excellent results are obtained; however, the LDA is not designed to give quasiparticle energies. Straightforward application of the LDA leads to a serious underestimation of semiconductor band gaps. Modification [5] of the DFT and the Standard Model to explore this problem and to include excitonic effects as well has been a major challenge to theorists. There has been dramatic progress in this area.

The field of *ab initio* molecular dynamics [13] is having a major influence on this area of research. Allowing atoms to move to calculate thermodynamical properties is an important aspect of the quest to calculate electronic and structural properties of solids. Significant progress has been made in this area too.

Much of the early work on solids was done on structurally simple systems like Si. Models of unit cells or supercells contained an order of 10–50 atoms for simulating complex materials. However, other approaches, including real space methods [14], have been introduced to study clusters, liquids, and more complex solids. This active area bears on condensed matter physics, chemistry, and materials science.

Interesting features and problems arise when one considers the polarizability of solids. Electric fields affect itinerant electrons in complex ways. These effects are sometimes best understood by using a localized basis such as Wannier functions instead of Bloch functions. Problems like the ferroelectric nature of some solids can be addressed, and concepts such as Berry's phase can be used to explore the puzzles related to the properties of electrons in electric fields [15].

One of the oldest problems in the study of solids is electronic transport [6]. Using band theory, it is possible to produce an explanation of the existence of insulators, semiconductors, semimetals, and metals. However, the details of transport in these materials, particularly when dealing with reduced sizes, non-periodic materials, and lower dimensionality are an active area of study.

The standard model, and in particular the version based on the plane wave pseudopotential method, has served us well for explaining and predicting properties of bulk solids, surfaces, interfaces, clusters, and other structural forms of solids. Its extension to the study of nanostructures and lower dimensional systems has been generally quite successful.

Future applications will likely focus on extensions and attempts for wider applicability to more complex solids, correlated electrons, and novel structures. However, even at this point, it is fair to say that we have had considerable success in answering Dirac's 1929 challenge after the development of quantum mechanics: "The underlying physical laws necessary for a large part of physics and the whole of chemistry are thus completely know, and the difficulty is only that the exact application of these laws leads to equations much too complicated to be soluble."

ACKNOWLEDGMENTS

This work was supported by National Science Foundation Grant No. DMR04-39768 and by the director, Office of Science, Office of Basic Energy Sciences, Division of Materials Sciences and Engineering, U.S. Department of Energy under Contract No. DE-AC03-76SF00098.

REFERENCES

[1] E. Fermi, On the pressure shift of the higher levels of a spectral line series, *Nuovo Cimentel* **11**, 157 (1934).

[2] J.C. Phillips and L. Kleinman, New method for calculating wave functions in crystals and molecules, *Phys. Rev.* **116**, 287 (1959).

[3] W. Kohn and L.J. Sham, Self-consistent equations including exchange and correlation effects, *Phys. Rev.* **140**, A1133 (1965).

[4] M.L. Cohen, Pseudopotentials and total energy calculations, *Phys. Scripta* **T1**, 5 (1982).

[5] M.S. Hybertsen and S.G. Louie, First-principles theory of quasiparticles: Calculation of band gap in semiconductors and insulators, *Phys. Rev. Lett.* **55**, 1418 (1985).

[6] C. Kittel, *Introduction to Solid State Physics*, 8th ed. (J. Wiley, Hoboken, NJ, 2005).

[7] W.D. Knight, K. Clemenger, W.A. de Heer, W.A. Saunders, M.Y. Chou and M.L. Cohen, Electronic shell structure and abundances of sodium cluster, *Phys. Rev. Lett.* **52**, 2141 (1984).

[8] M.L. Cohen and T.K. Bergstresser, Band structures and pseudopotential form factors for fourteen semiconductors of the diamond and zincblende structures, *Phys. Rev.* **141**, 789 (1966).

[9] M.L. Cohen and J.R. Chelikowsky, *Electronic Structure and Optical Properties of Semiconductors* (Springer, New York, 1988).

[10] J.R. Chelikowsky, M.L. Cohen, M. Schluter and S.G. Louie, Self-consistent pseudopotential method for localized configurations: Molecules, *Phys. Rev. B* **12**, 5575 (1975).

[11] K.J. Chang, M.L. Cohen, J.M. Mignot, G. Chouteau and G. Martinez, Superconductivity in high-pressure metallic phases of Si, *Phys. Rev. Lett.* **54**, 2375 (1985).

[12] H.J. Choi, D. Roundy, H. Sun, M.L. Cohen and S.G. Louie, The origin of the anomalous superconducting properties of MgB_2, *Nature* **418**, 758 (2002).

[13] S. Fahy, X.W. Wang and S.G. Louie, Variational quantum Monte Carlo nonlocal pseudopotential approach to solids: Cohesive and structural properties of diamond, *Phys. Rev. Lett.* **61**, 1631 (1988).

[14] J.R. Chelikowsky, N. Troullier and Y. Saad, The finite-difference-pseudopotential method: Electronic structure calculations without a basis, *Phys. Rev. Lett.* **72**, 1240 (1994).

[15] I. Souza, J. Iniguez and D. Vanderbilt, Dynamics of Berry-phase polarization in time-dependent electric fields, *Phys. Rev. B* **69**, 085106 (2004).

Chapter 2
PREDICTING MATERIALS AND PROPERTIES: THEORY OF THE GROUND AND EXCITED STATE

S. G. Louie

1. INTRODUCTION

Tremendous progress has been made in the past two decades in employing *ab initio* theories and computation to explain and predict the various properties, and even the existence, of condensed matter systems. These first-principles studies have been particularly successful in investigating bulk- and reduced-dimensional systems that have moderate electron correlations.

The structure and properties of a condensed matter system are basically dictated by the outer valence electrons of its constituent atoms. The mutual interactions of these electrons and their interactions with the ions determine the electronic structure of the system, which in turn determines the behavior of the material. Understanding materials properties from first principles, then, involves solving an interacting quantum many-body problem with the Hamiltonian (for simplicity of notation, we omit spin and relativistic effects):

$$H_{\text{tot}} = \sum_j \frac{\mathbf{P}_j^2}{2M_j} + \sum_i \frac{\mathbf{p}_i^2}{2m} + \sum_{j<j'} \frac{Z_j Z_{j'} e^2}{|\mathbf{R}_j - \mathbf{R}_{j'}|} + \sum_{i<i'} \frac{e^2}{|\mathbf{r}_i - \mathbf{r}_{i'}|} + \sum_{i,j} \frac{-Z_j e^2}{|\mathbf{R}_j - \mathbf{r}_i|}. \quad (1)$$

An exact numerical solution to this problem is, in general, impractical and in fact often undesirable since the many-body wavefunctions would be so complicated that it would be difficult to achieve physical understanding from the solutions. Here we focus the discussion on solving this many-body problem using different physically motivated approaches to arrive at, in an *ab initio* fashion, an approximate but

Contemporary Concepts of Condensed Matter Science
Conceptual Foundations of Materials: A Standard Model for Ground- and Excited-State Properties

ISSN: 1572-0934/doi:10.1016/S1572-0934(06)02002-6

accurate description of a number of fundamental properties of experimental interest. Other properties not treated in this Chapter, such as those related to finite temperature effects, dynamics, and transport phenomena will be presented in more details in the subsequent Chapters of this volume.

For an interacting many-electron system, it is useful to distinguish the ground-state properties from the electronic excited-state or spectroscopic properties. Properties such as structure, cohesive energies, vibrational properties, and phase stability are ground-state properties because they are intrinsic to a system with all its electrons in the ground state. In general, these properties may be determined from knowing the total energy of the system as a function of the atomic coordinates, and theories such as those based on the density functional formalism [1,2] have been very successful in the *ab initio* calculation of these properties for many materials. On the other hand, spectroscopic properties such as those measured in photoemission, transport, and tunneling experiments involve creating an excited particle (electron or hole) above the ground state. It should be thought of as an $N+1$ particle problem that requires a different theoretical treatment [3–6] from those of the ground-state approaches. Understanding the optical properties from first principles is yet a further challenge because it is an $N+2$ particle problem [7–10]. In this case, we need to include electron–hole interaction, which can be quite important in many systems, particularly in lower-dimensional structures. The excited-state responses or spectroscopic properties are best treated with the concept and formalism of the interacting particle Green's functions [11].

In this Chapter, we discuss the conceptual and theoretical bases for the above approaches to material properties and present some selected applications to illustrate the versatility and power of these different approaches. Emphasis will be placed on a physical understanding of the theoretical advances made in recent years. In Section 2, we introduce density functional theory (DFT). This formalism provides a means to treat electron–electron interactions and transforms the many-electron problem to one of a self-consistent-field one-particle problem for ground-state properties. In Section 3, the concept of pseudopotentials is developed, which allows an efficient treatment of electron–ion interactions and the calculation of a host of material properties. Section 4 is devoted to example calculations of electronic, structural, vibrational, and other related ground-state properties. Electron–phonon interaction and superconductivity are discussed in Section 5.

Section 6 gives a general discussion of excited states and the relationship of spectroscopic properties to the Green's functions. The single-particle Green's function and electron self-energy are introduced in Section 7. A powerful approximation, named the GW approximation, to the electron self-energy operator is presented in Section 8. Section 9 presents some calculations of quasiparticle excitations in condensed matter systems. Section 10 discusses electron–hole excitations and the Bethe–Salpeter equation (BSE). Section 11 is devoted to the optical properties of solids, surfaces, and nanostructures. As an example of the application of the GW–BSE formalism to a novel one-dimensional (1D) system, the spectroscopic properties of nanotubes are discussed in Section 12. Finally, in Section 13, a summary and some perspectives are presented.

2. THE GROUND STATE AND DENSITY FUNCTIONAL FORMALISM

As seen from Eq. (1), the determination of the properties of a material is in general a complex quantum many-body problem. One of the major challenges is the proper treatment of electron–electron interaction. For ground-state properties, it is shown that this many-body problem can be exactly reduced to one of solving a self-consistent-field one-particle problem through a DFT. Unlike other mean-field theories such as the Hartree or Hartree–Fock (HF) approximations, DFT is in principle exact in giving the total energy, electron charge density distribution, and other related ground-state properties (such as the structural and vibrational properties) of an interacting many-electron system. Nowadays, DFT is arguably the most popular approach for first-principles studies of the ground-state properties of condensed matter systems, with applications to many disciplines ranging from physics and chemistry to biology and engineering.

For an interacting many-electron system in an external static potential $V(\mathbf{r})$, Hohenberg and Kohn [1] demonstrated that, instead of the electronic ground-state energy being a functional of the ground-state many-body wavefunction $\Psi(\mathbf{r}_1, \mathbf{r}_2, \ldots, \mathbf{r}_N)$, it can be reformulated as a functional of the charge density $\rho(\mathbf{r})$ alone

$$E_V[\rho] = \int V(\mathbf{r})\rho(\mathbf{r})\,\mathrm{d}\mathbf{r} + T_\mathrm{s}[\rho] + \frac{1}{2}\int \frac{\rho(\mathbf{r})\rho(\mathbf{r}')}{|\mathbf{r}-\mathbf{r}'|}\,\mathrm{d}\mathbf{r}\,\mathrm{d}\mathbf{r}' + E_\mathrm{xc}[\rho]. \tag{2}$$

Here, T_s is the kinetic energy of a non-interacting electron system with density $\rho(\mathbf{r})$, E_xc is a universal functional of the density, and E_V is at its minimum when $\rho(\mathbf{r})$ is the physical density. This formalism is a tremendous conceptual and technical simplification since the $3N$-dimensional many-electron wavefunction is eliminated from the problem. However, the existence of the universal functional was only deduced by a proof by contradiction, and the exact form for E_xc remains unknown although a number of successful approximations, such as the local density approximation (LDA) or the generalized gradient approximation (GGA) have been developed over the years [12,13].

Kohn and Sham [2] further showed that the electron charge density and total energy may be obtained by solving an effective system of non-interacting electrons, making the DFT approach practical. In the Kohn–Sham theoretical construct, two systems are considered – the origin material system of interest with interacting electrons and an associated fictitious non-interacting system with the same number of electrons moving in some effective one-body potential $V_\mathrm{eff}(\mathbf{r})$ (see Fig. 1). The electron density of the non-interacting system is used in evaluating the energy functional of the real system. Making use of the variational principle that the physical density minimizes the total energy functional, we may vary the non-interacting system until the functional is minimized and arrive at the physical charge density and energy of the real interacting electron system. This variation

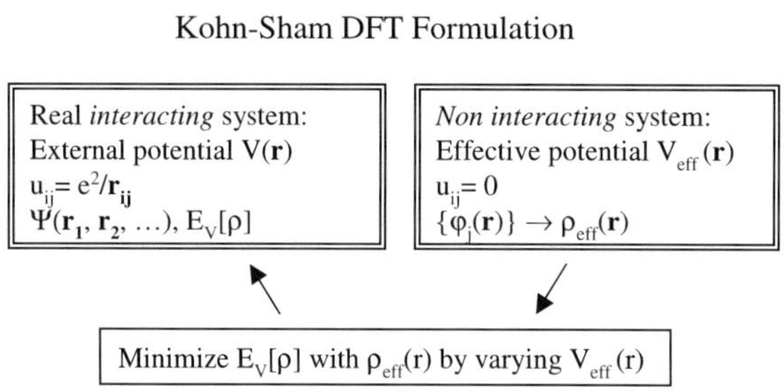

Fig. 1. Kohn–Sham formulation of density functional formalism. The charge density of a non-interacting system is used to minimize the energy functional to arrive at a set of one-electron equations that determine the charge density and total energy of the interacting system.

gives rise to a set of Euler–Lagrange equations that govern the single-particle orbitals and energies of the non-interacting system

$$\left\{\frac{p^2}{2m} + V(\mathbf{r}) + V_{\mathrm{H}}(\mathbf{r}) + V_{\mathrm{xc}}(\mathbf{r})\right\}\varphi_i(\mathbf{r}) = \varepsilon_i\varphi_i(\mathbf{r}) \tag{3}$$

with

$$\rho(\mathbf{r}) = \sum_i^{\mathrm{occ}} |\varphi_i(\mathbf{r})|^2 \tag{4}$$

and

$$V_{\mathrm{xc}}(\mathbf{r}) = \frac{\delta E_{\mathrm{xc}}}{\delta\rho(\mathbf{r})}. \tag{5}$$

This set of equations for the fictitious non-interacting system is known as the Kohn–Sham equations. The effective potential V_{eff} for the non-interacting system has two terms: a Hartree term $V_{\mathrm{H}}(\mathbf{r})$ and an exchange-correlation term $V_{\mathrm{xc}}(\mathbf{r})$ which is given by the functional derivative of E_{xc} with respect to the electron charge density. In principle, if E_{xc} were known, a self-consistent solution to the Kohn–Sham equations would give the exact electron density and ground-state energy of the interacting system as a function of the atomic coordinates and hence also a host of other properties that are related to the ground state. Since E_{xc} is unknown, approximations must be made. The most common form is to assume that

$$E_{\mathrm{xc}} = \int \rho(\mathbf{r})\varepsilon_{\mathrm{xc}}(\mathbf{r})\,\mathrm{d}\mathbf{r} \tag{6}$$

where $\varepsilon_{\mathrm{xc}}(\mathbf{r})$, a local exchange-correlation energy density, is assumed to be a function of the local density $\rho(\mathbf{r})$ in the local density functional approximation (LDA) or a function of $\rho(\mathbf{r})$ and $\nabla\rho(\mathbf{r})$ in the GGA. These approximations, employing data from homogeneous electron gas calculations which are exact in the uniform density limit, have allowed the very accurate *ab initio* computation of many material properties over the past two decades.

We note, however, that the individual Kohn–Sham eigenvalues ε_i and eigenfunctions $\varphi_i(\mathbf{r})$ of Eq. (3) do not correspond physically to the excitation energies and amplitudes of the electrons. Only the electron charge density and total energy are rigorously meaningful. In particular, the Kohn–Sham eigenvalues are only Lagrange multipliers in the Kohn–Sham variational construct, and in general, they are *not* equal to the electron excitation (or quasiparticle) energies of a system even if we know the exact exchange-correlation functional. We can see this from a simple example. Consider the case of the interacting homogeneous electron gas. In this case, because the density is homogenous, both the Hartree potential and the exchange-correlation potential in the Kohn–Sham equation are independent of the coordinates of the electrons regardless of the exact form of $V_{\mathrm{xc}}(\mathbf{r})$. Equation (3) reduces to

$$\left\{\frac{p^2}{2m} + \text{constant}\right\}\varphi_i(\mathbf{r}) = \varepsilon_i\varphi_i(\mathbf{r}). \tag{7}$$

The solution to Eq. (7) is always that of a free-electron dispersion for the Kohn–Sham eigenvalues. That is, if we were to interpret the Kohn–Sham eigenvalues as electron excitation energies, then the band structure and therefore the effective mass m^* and occupied bandwidth would be those of the free electrons, independent of the interaction. Additionally, the lifetime of a quasiparticle created in a particular momentum state would be infinite. This is clearly incorrect. It was this misuse of the Kohn–Sham eigenvalues that led to the famous band gap problem in semiconductors and insulators. Despite this caveat, the Kohn–Sham band structure, nevertheless, often provides a good starting point for the understanding of the electronic structure of materials.

3. *AB INITIO* PSEUDOPOTENTIALS

Since it is essentially the behavior of the outer valence electrons of the constituent atoms that determines the properties of a material, another important ingredient in the theory, the electron–ion interaction, may be described in terms of pseudopotentials. The concept of pseudopotentials explains the apparent weak interaction between the active electrons and the ion cores in solids (e.g., it justifies a nearly free-electron model for the simple metals) and provides great efficiency in the computation of the properties of real materials. Pseudopotentials eliminate the core electrons from the problem and allow the use of significantly simpler basis sets (e.g., planewaves or uniform real-space grids) in the numerical solutions to the self-consistent-field equations.

Although the use of pseudopotentials dates back to the 1930s [14], they are best understood in terms of the Phillips–Kleinman cancellation theorem [15]. Phillips and Kleinman argued that the wavefunctions of the valence and higher-energy electronic states are expected to be smooth away from the atomic sites and oscillatory with atomic character in the core regions (owing to the fact that valence states are orthogonal to the core states). The one-electron wavefunctions (e.g., the Kohn–Sham orbitals in Eq. (3)) may then be written in the form of

$$|\varphi_i(\mathbf{r})\rangle = |\phi_i(\mathbf{r})\rangle + \sum_{\mathrm{c}} |f_{\mathrm{c}}\rangle\langle f_{\mathrm{c}}|\phi_i(\mathbf{r}')\rangle, \tag{8}$$

where $|\phi_i(\mathbf{r})\rangle$ is a smooth pseudowavefunction and $|f_{\mathrm{c}}\rangle$ are appropriately normalized Bloch sums of core states. Solving the one-particle Schrodinger equation

$$\left\{\frac{p^2}{2m} + V\right\}|\varphi_i(\mathbf{r})\rangle = \varepsilon_i|\varphi_i(\mathbf{r})\rangle, \tag{9}$$

where V is a standard one-particle potential, is then equivalent to solving the following equation for the pseudowavefunction $\phi_i(\mathbf{r})$:

$$\left\{\frac{p^2}{2m} + V + V_{\mathrm{R}}\right\}|\phi_i(\mathbf{r})\rangle = \varepsilon_i|\phi_i(\mathbf{r})\rangle. \tag{10}$$

The additional term V_{R}, given by

$$V_{\mathrm{R}} = \sum_{\mathrm{c}} (\varepsilon - \varepsilon_{\mathrm{c}})|f_{\mathrm{c}}(\mathbf{r})\rangle\langle f_{\mathrm{c}}(\mathbf{r}')|, \tag{11}$$

is a non-local, energy-dependent operator. Since the core state energies are in general significantly lower than those of the states interest, V_{R} is effectively a repulsive potential with negligible energy dependence and cancels the strong electron–ion interaction V. The resulting net potential seen by the electrons is then a weak pseudopotential

$$V_{\mathrm{p}} = V + V_{\mathrm{R}}. \tag{12}$$

Within this framework, the solutions to Eq. (10) yield the same eigenvalues as the original Eq. (9) and wavefunctions that are similar outside of the core region.

The above arguments provide the conceptual basis for pseudopotentials and also show that pseudopotentials are not unique since a different choice for the second term on the right-hand side of Eq. (8) would lead to equally valid, but different-looking pseudopotentials. This ambiguity has led to different construction methods to optimize the accuracy and computational efficiency of pseudopotentials [16]. In general, modern constructions of *ab initio* pseudopotentials involve first solving the all-electron problem for a given atom in a particular configuration – that is, solving Eq. (9) within certain approximation (e.g., DFT in the LDA). The resulting valence electron wavefunctions $\varphi_i(\mathbf{r})$ may be used to form a set of pseudowavefunctions $\phi_i(\mathbf{r})$ by joining $\varphi_i(\mathbf{r})$ to a properly chosen smooth, nodeless function for $\mathbf{r}$ less than a certain cut-off radius from the nucleus, as illustrated in Fig. 2 for the case of the 3s and 3p states of sodium. By inverting the Schrodinger equation (for a given $\phi_i(\mathbf{r})$ and ε_i),

$$\left\{\frac{\mathbf{p}^2}{2m} + V_{\mathrm{P}}^i\right\}|\phi_i(\mathbf{r})\rangle = \varepsilon_i|\phi_i(\mathbf{r})\rangle, \tag{13}$$

the corresponding pseudopotential $V_{\mathrm{P}}^i(\mathbf{r})$ is constructed for the ith state of this particular element. The procedure is usually done for a neutral atom in a configuration

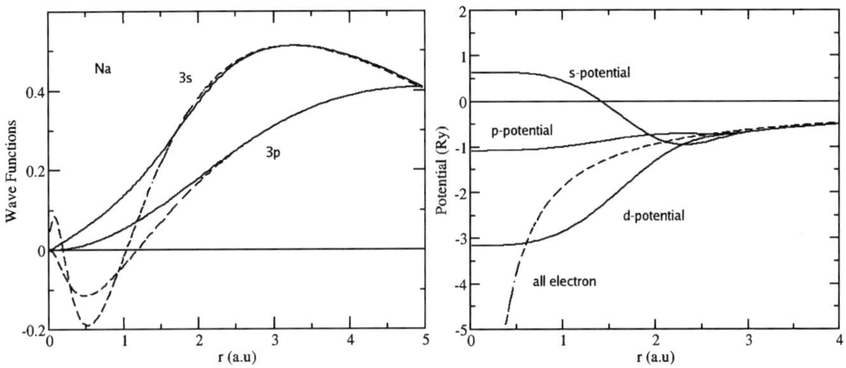

Fig. 2. Left panel: Construction of pseudowavefunctions (solid curves) from atomic wavefunctions (dashed curves) of sodium. Right panel: *Ab initio* ionic pseudopotentials of sodium. (Figure courtesy of J.R. Chelikowsky.)

appropriate in the condensed state. An *ab initio* ionic pseudopotential describing the intrinsic interaction of an electron with an atom stripped of the outer valence electrons may next be obtained by unscreening the neutral pseudopotential, i.e., subtracting off the screening Hartree and exchange-correlation potential resulting from the pseudocharge density of the valence electrons.

Various constraints (such as norm conservation [17]) and different procedures have been developed for the construction of pseudopotentials [16]. The resulting *ab initio* ionic pseudopotentials are in general highly accurate and transferable and have been successfully used in first-principles calculations for different properties and in different environments.

4. ELECTRONIC, STRUCTURAL, VIBRATIONAL, AND OTHER GROUND-STATE PROPERTIES

Investigations of the structure and properties of condensed matter systems have been carried out within different theoretical frameworks and have employed different computational techniques, ranging from empirical to highly accurate first-principles studies. The combination of the density functional formalism and the *ab initio* pseudopotential method is a particularly powerful and versatile approach for the ground-state properties of materials [18]. In this Chapter, we give several selected examples of such calculations.

Figure 3 is taken from the classic work by Yin and Cohen [19], who first showed that structural energies may be accurately determined using the *ab initio* pseudopotential density functional approach. By computing the total energy of the system,

$$E_{\text{total}} = E_{\text{el}} + E_{\text{ion-ion}}, \tag{14}$$

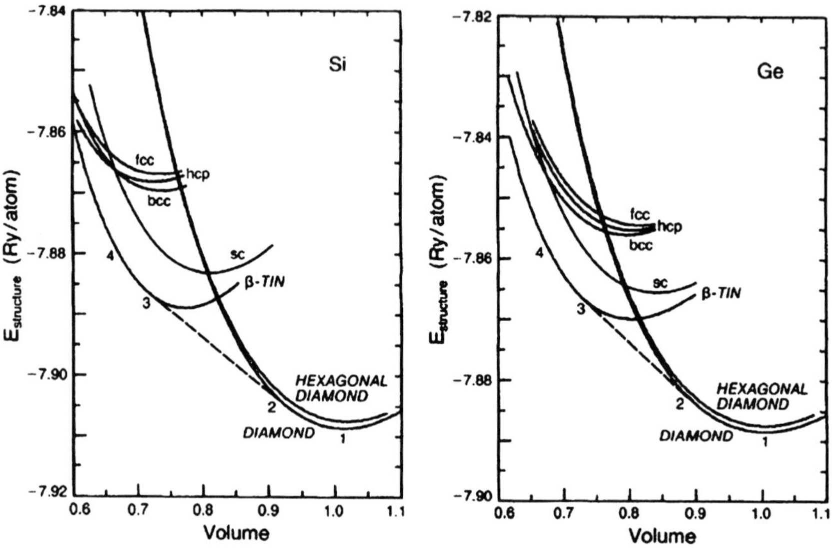

Fig. 3. Total energy per atom of Si (left panel) and Ge (right panel) in different crystal structures. The negative of the common tangent (dashed line) constructed between the diamond and β-tin structure curves gives the critical transition pressure between the two structures. (After Yin and Cohen [19].)

for different structures (where E_{el} is the DFT electronic ground-state energy from Eq. (2) and $E_{\mathrm{ion-ion}}$ is the classical electrostatic interaction energy among the ions), one can determine with high accuracy the stable atomic configuration and other structural parameters such as the cohesive energies, lattice constants, and bulk and elastic moduli. With available *ab initio* exchange-correlation functionals, cohesive energies within a few percent of the experimental values are now obtainable. However, to achieve chemical accuracy, further improvement in the treatment of many-electron effects is required. Relative energies, though, are more accurate, yielding lattice constants and bulk moduli that are typically within 1% and a few percent of experimental values, respectively.

Knowing the equation of state for the various structures allows the investigation of structural phase transitions under pressure by evaluating the enthalpy [20]

$$G = E + PV - TS. \tag{15}$$

In particular, at low temperature, the critical transition pressure between two adjacent structures in Fig. 3 is then given by the negative of the common tangent of the two equations of state. Such analysis has allowed the calculation of not only the critical pressure but also the volume discontinuity at the transition and has provided an *ab initio* understanding and prediction of the high-pressure phases of matter [18].

By minimizing the total energy or the forces on the atoms,

$$F_i = -\frac{\partial E_{\text{total}}}{\partial \mathbf{R}_i}, \tag{16}$$

the approach has been used with success in determining the structural parameters of complex materials, surfaces, defects, clusters, molecules, and nanostructured systems. For non-periodic systems, the standard technique is to carry out the electronic structure calculation employing a supercell scheme [21] in which the surface, defect, or molecule is repeated periodically with sufficient separation in order to mimic the isolated structure. As an illustration, Fig. 4 compares the calculated geometric structure of the Si(111) 2×1 surface with experimental coordinates from low-energy electron diffraction (LEED) measurements [22,23]. This surface is quite interesting both in terms of its structure and its electronic and optical properties. At low temperature, the atoms on the Si(111) surface rearrange themselves from the ideally terminated geometry to form chains of π-bonded atoms, with the position of the atoms on the chains having significant buckling. Figure 4 shows the excellent agreement between the calculated and experimental geometry. Similarly excellent agreement, employing DFT geometry determination, has been achieved for the structural properties of much bigger systems containing up to thousands of atoms in a supercell. Below, we shall also use the Si(111) 2×1 surface as a prototypical system to demonstrate theoretical studies of quasiparticle and optical excitations.

Having the ability to calculate the ground-state total energy and the forces on the atoms also allows us to study from first principles the lattice dynamics of a system within the Born–Oppenheimer approximation (i.e., the electrons remain in the ground state as the ions are moved). The phonon frequencies and eigenvectors are typically obtained from one of two schemes: (i) the frozen phonon method or (ii) the linear response method. Conceptually, in the frozen phonon approach [24] (see Fig. 5), one considers a distortion of the form $\mathbf{u}_i(\mathbf{k}) = \mathbf{u}_0 \cos(\mathbf{k}\mathbf{R}_i + \delta_k)$ frozen into

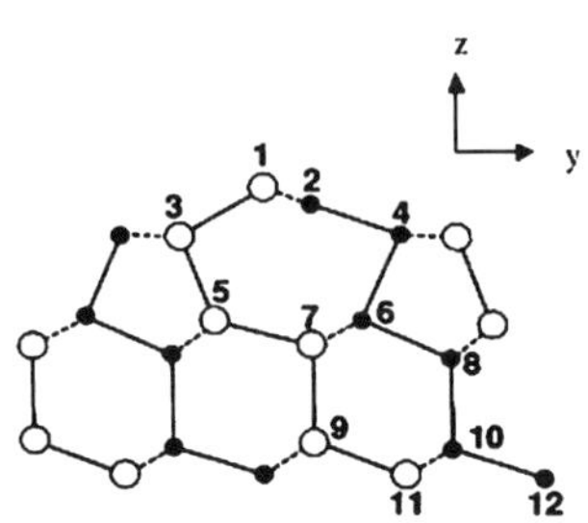

	y-coorinate (Å)		z-coordinate (Å)	
	Theory	Experiment	Theory	Experiment
1	0.0	0.0	0.0	0.0
2	1.11	1.12	-0.47	-0.38
3	-1.99	-2.00	-1.26	-1.26
4	3.30	3.26	-1.21	-1.19
5	-1.14	-1.10	-3.48	-3.46
6	2.37	2.40	-3.41	-3.39
7	1.16	1.20	-4.08	-4.06
8	-2.14	-2.12	-4.36	-4.26

Fig. 4. Left panel: Schematic side view of the geometric structure of the Si(111) 2×1 surface. Surface atoms 1 and 2 form π-bonded chains. The solid and open circles denote atoms in different (1 1 0) planes. Right panel: Atomic coordinates from energy minimization as compared to data from LEED [23]. (After Northrup et al. [22].)

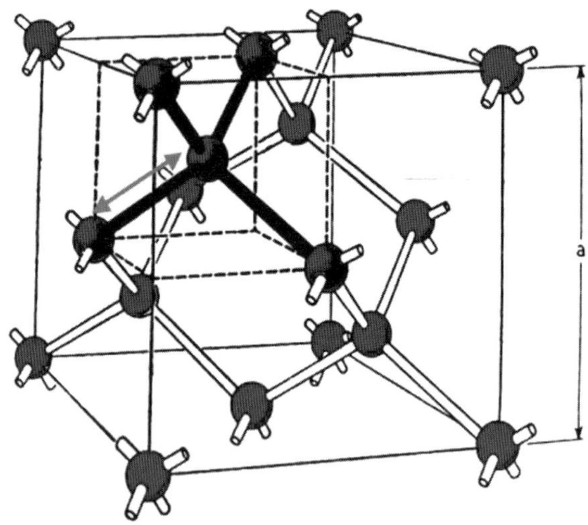

Fig. 5. Schematic picture of a frozen phonon calculation, corresponding to distortion of a zone center optical phonon mode in the diamond structure.

the ideal structure of a crystal and computes the change in the total energy of the system $\Delta E(\mathbf{u}_0)$ as a function of the distortion amplitude $\mathbf{u}_0$. Upon an expansion of the change in energy in a Taylor series,

$$\Delta E = K_2\mathbf{u}_0^2 + K_3\mathbf{u}_0^3 + K_4\mathbf{u}_0^4 + \cdots, \tag{17}$$

the harmonic contribution to the phonon frequency is given by the coefficient K_2 and the higher-order coefficients determine the anharmonic contributions. The advantage of the frozen phonon approach is that anharmonic contributions may be studied in a straightforward manner. However, since the size of the supercell in the total energy calculation is dictated by the wavelength of the frozen distortion, only a certain discrete number of phonons with wavevectors $\mathbf{k}$ that are at symmetry points or along high symmetry directions which do not lead to a very large supercell may be studied in practice.

An alternate approach to phonons is to employ density functional perturbation theory. In this scheme, the atomic force constants,

$$C_{ij} = \frac{\partial^2 E}{\partial \mathbf{R}_i \partial \mathbf{R}_j}, \tag{18}$$

are evaluated using linear response theory, and the dynamical matrix is then diagonalized to obtain the phonon frequencies. This approach allows the computation of phonon properties throughout the Brillouin zone but is limited to the harmonic approximation. Figure 6 shows the results of a linear response calculation for the phonon dispersion relation and density of states for silicon and germanium [25]. Both frozen phonon and linear response approaches have given

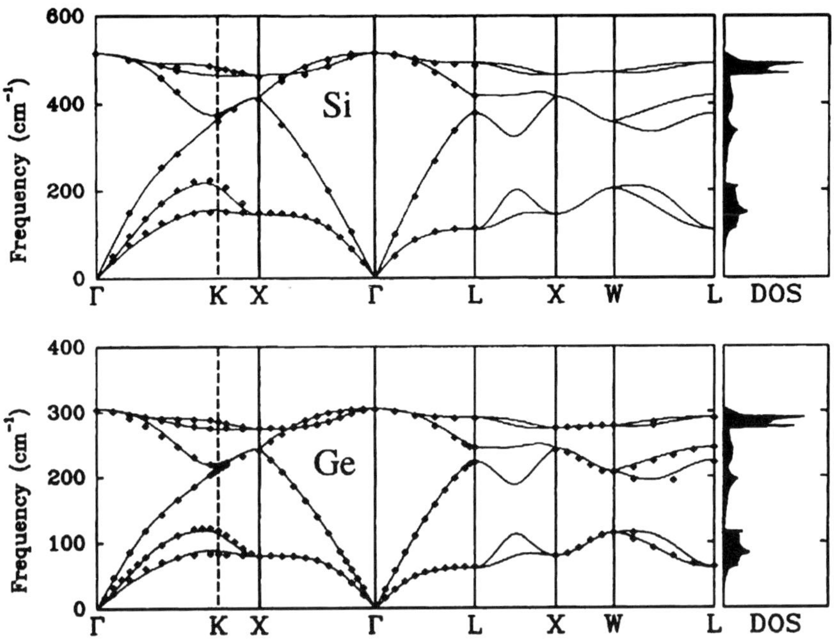

Fig. 6. Calculated phonon dispersion relations of Si and Ge (solid curves) as compared to experimental data. (After Giannozzi et al. [25].)

very accurate phonon results for a variety of materials ranging from metals to semiconductors to complex oxides.

5. ELECTRON–PHONON INTERACTION AND SUPERCONDUCTIVITY

The concepts and techniques discussed above have been used to extract the electron–phonon interaction and hence to study phenomena such as electrical resistivity and superconductivity in solids. An electron may be scattered from one electronic state $|\varphi_{n\mathbf{k}}(\mathbf{r})\rangle$ to another state $|\varphi_{n'\mathbf{k}'}(\mathbf{r})\rangle$ due to the perturbation potential created by a phonon of frequency $\omega_{s\mathbf{q}}$. Such scattering event is given by the electron–phonon coupling matrix element

$$g(n'\mathbf{k}', n\mathbf{k}; s\mathbf{q}) = \sqrt{\hbar/2M\omega_{s\mathbf{q}}}\langle\varphi_{n'\mathbf{k}'}(\mathbf{r})|H_I|\varphi_{n\mathbf{k}}(\mathbf{r})\rangle, \tag{19}$$

where H_I is the change in the potential seen by an electron in the presence of the phonon, which may be evaluated using the frozen-phonon technique or linear response theory. Since all quantities in Eq. (19) are obtainable from first principles, it has been used to provide an *ab initio* calculation of electron–phonon couplings and their contribution to various physical phenomena in materials. As an illustration,

Table 1. Comparison of calculated and measured electron–phonon coupling constant λ, electrical resistivity ρ ($\mu\Omega$ cm), and thermal resistivity w(K cm/W) at 273 K. (After Savrasov et al. [95]).

	Al	Nb	Mo
λ^{cal}	0.44	1.26	0.42
λ^{expt}	0.42	1.33	0.44
ρ^{cal}	2.35	13.67	4.31
ρ^{expt}	2.42	13.30	4.88
w^{cal}	0.42	2.17	0.73
w^{expt}	0.42	1.93	0.72

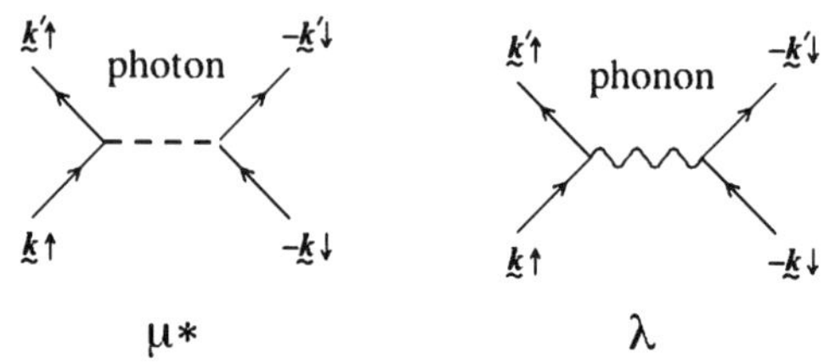

Fig. 7. Competing interactions in a conventional BCS superconductor.

Table 1 compares some calculated transport quantities with values deduced from experiment for three common metals.

Within BCS theory [26], the mechanism for superconductivity in conventional superconductors is the formation of Cooper pairs at low temperature due to an attractive interaction between electrons via the exchange of phonon. Superconductivity is possible when the electron–phonon coupling (characterized by the parameter λ) is sufficiently strong so that it overcomes the residual Coulomb repulsion (characterized by the parameter μ^*). In the strong electron–phonon coupling formalism of the BCS theory, the basic ingredient that determines the behavior of a superconductor is the momentum- and frequency-dependent Eliashberg function [27],

$$\alpha^2 F(\mathbf{k}, \mathbf{k}', \omega) = N(\varepsilon_F) \sum_j |g^j_{\mathbf{k},\mathbf{k}'}|^2 \delta(\omega - \omega_{j\mathbf{q}}), \tag{20}$$

which describes the scattering of a pair of electrons from states ($\mathbf{k}\uparrow, -\mathbf{k}\downarrow$) to states ($\mathbf{k}'\uparrow, -\mathbf{k}'\downarrow$) by phonons in the jth branch with $\mathbf{q} = \mathbf{k} - \mathbf{k}'$ (see Fig. 7). Here $N(\varepsilon_F)$ is the density of states at the Fermi level, $g^j_{\mathbf{k},\mathbf{k}'}$ is the electron–phonon matrix element, and $\mathbf{k}$ and $\mathbf{k}'$ are composite indices describing both the band and wavevector of the electronic states. The Eliashberg function in the Matsubara formalism may be re-expressed into a new set of functions (useful in finite temperature studies),

$$\lambda(\mathbf{k}, \mathbf{k}', n) = \int_0^\infty d\omega\, \alpha^2 F(\mathbf{k}, \mathbf{k}', \omega) \frac{2\omega}{\omega^2 + (2n\pi T)^2}, \tag{21}$$

where n is an integer and T the temperature. This set of functions also provides a measure of the strength of the scattering of states from $\mathbf{k}$ to $\mathbf{k}'$, and the usual electron–phonon coupling constant λ found in the literature (and in Table 1 and Fig. 7) is equal to $\lambda(\mathbf{k}, \mathbf{k}', n = 0)$ averaged over scatterings of all pairs on the Fermi surface. Given the Eliashberg function and the Coulomb parameter μ^*, one can then calculate essentially all the properties of a conventional superconductor with the BCS theory [26].

We illustrate here the *ab intio* calculation of superconducting properties with a recent investigation of multi gap superconductivity in MgB_2. This material has generated a great deal of interest since the discovery in 2001 that it is a superconductor with a rather high transition temperature of 39K [28]. Moreover, the system exhibits highly unusual behaviors in other measured quantities such as the temperature dependence of its specific heat [29–32]. The discovery was a surprise since MgB_2 is an sp-bonded material with a rather simple hexagonal layered structure of three atoms per unit cell. The boron atoms in the crystal form honeycombed layers, like carbon atoms in graphene, that stack directly on top of each other; the Mg atoms are located above the middle of each boron hexagon in between the layers.

The simplicity of the structure of MgB_2 allowed the use of *ab initio* electronic structure methods together with the full $\mathbf{k}$-dependent Eliashberg formalism to determine and predict its normal and superconducting properties, showing that it is the first known multi gap superconductor [33,34]. The band structure of MgB_2 is characterized by having both π and σ states crossing the Fermi level, leading to a Fermi surface (see Fig. 8a) with several distinct pieces of different character [35–37]. In Fig. 8a, the two cylindrical sheets wrapping the $\Gamma \rightarrow$ A line are composed of states that are 2D like and of σ character on the boron layer. The other sheets are composed of boron π states that are more 3D in character. The density of states at the Fermi level is roughly equally divided between the σ and π states. The electron–phonon coupling strength however is very different for states on the various sheets

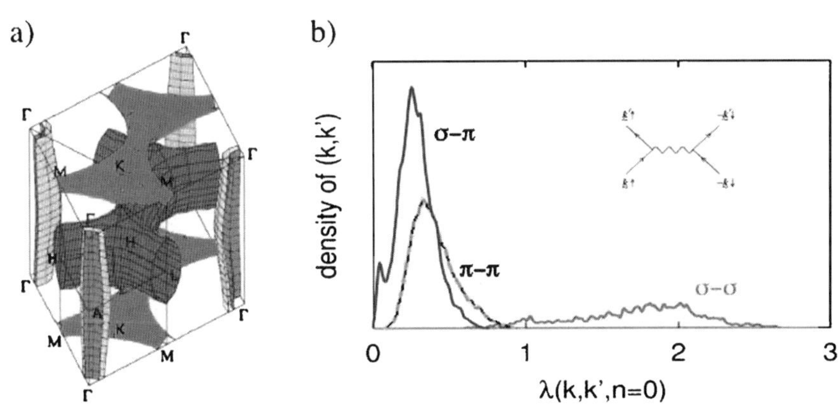

Fig. 8. (a) Calculated Fermi surface of MgB_2. (b) Density of scattering pairs $(\mathbf{k}, \mathbf{k}')$ as a function of the electron–phonon coupling strength $\lambda(\mathbf{k}, \mathbf{k}', n = 0)$ (After Choi et al. [34].)

of the Fermi surface; it is strongest on the cylindrical σ sheets. This is illustrated in Fig. 8b by the function $\lambda(\mathbf{k},\mathbf{k}',n=0)$, showing an extraordinarily large coupling strength exceeding 2 for the σ–σ scattering [34]. The averaged λ over the whole Fermi surface is only 0.61, which is in good agreement with normal state heat capacity measurements [30–32] and is a value too moderate to explain within the usual isotropic BCS model the high transition temperature and other unusual superconducting properties of this material.

Owing to the large electron–phonon coupling and its strong $\mathbf{k}$ dependency, a proper treatment of superconductivity in MgB_2 requires the full $\mathbf{k}$-dependent Eliashberg formalism. Such first-principles studies [33,34,38] show that MgB_2 is an s-wave multi gap superconductor, arising from the distinctly different electron–phonon coupling strength for the π- and σ-boron states on the Fermi surface. The predicted superconducting energy gap function $\Delta(\mathbf{k},T)$ shows a wide distribution of gap values clustering into two groups, which quantitatively explains the specific heat, tunneling and other spectroscopic data [33]. Additionally, with $\mu^* = 0.12$ (within the usual range of metallic values), the computed transition temperature T_c, mass enhancement factor λ, and isotope parameters α are in excellent accord with measurements.

Figure 9 shows the distribution of the value of the superconducting energy gap $\Delta(\mathbf{k},T)$ on the Fermi surface of MgB_2 and its temperature dependence. $\Delta(\mathbf{k},T)$ is of the same sign everywhere on the Fermi surface, but it has very distinct values on the different sheets – clustered around 2 meV on the π sheets and 7 meV on the σ sheets at low temperature. Subsequent angle-resolved photoemission (the data points on Fig. 9b) and tunneling experiments have verified the theoretical predictions in detail.

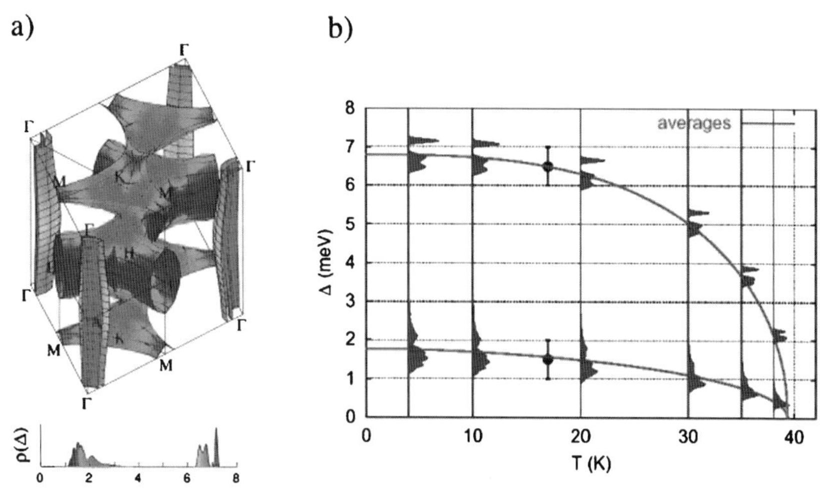

Fig. 9. (a) Value of the superconducting energy gap, $\Delta(\mathbf{k})$, on the Fermi surface of MgB_2 at $T = 4$ K. (b) Value distribution and temperature dependence of the superconducting energy gap $\Delta(\mathbf{k},T)$. (After Choi et al., [33].)

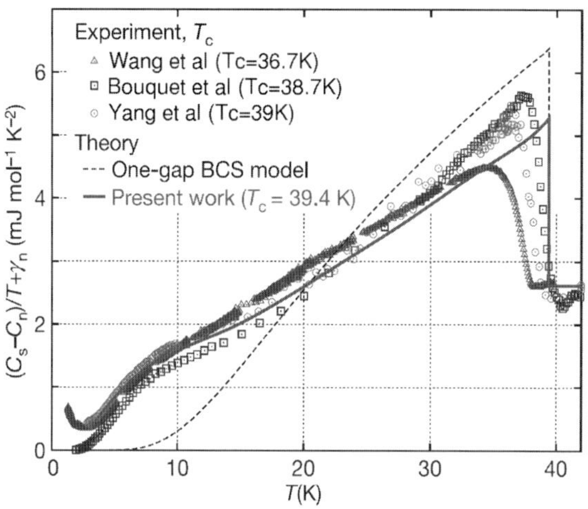

Fig. 10. Temperature dependence of the specific heat of MgB_2. (After Choi et al., [33].)

Figure 10 compares the calculated specific heat with several sets of experimental data [30–32] and with the result of a one-gap BCS model. With no adjustable parameter, the *ab initio* results reproduce the experimental data over the full temperature range from low T to above T_c. The extra amplitude in the specific heat at low temperature that cannot be reproduced by a one-gap model is a consequence of the existence of the small gaps. That is, because of the existence of the small gaps, even at low T, there are quasiparticles excited across the gap in those parts of the Fermi surface.

Methods based on *ab initio* pseudopotentials and DFT have also been developed for the study of numerous other phenomena related to the electronic ground state of condensed matter systems including bonding properties, mechanical strength, phase stability, magnetism, electric polarization, susceptibility, molecular dynamics simulations, and others. The range of interesting physical properties and systems that have been successfully studied is too large to enumerate here. However, many of these topics will be discussed in the subsequent Chapters of this volume. In the remainder of this Chapter, we will focus on phenomena related to the electronic excited states.

6. EXCITED STATES, SPECTROSCOPIC PROPERTIES, AND GREEN'S FUNCTIONS

To study electron excited states and excitation spectra, we need to go beyond ground-state theories such as the static DFT discussed above. For an interacting system, it is important to distinguish the different kinds of excitations probed in different experiments. Measurements employing photoemission, transport or tunneling techniques yield, in general, information on the individual particle-like excitations, or quasiparticles (excited electrons and holes), of a system. Due to many-electron interactions, the excited particle acquires a self-energy which changes its dispersion relation from that of an independent-particle picture. In an optical experiment, on the other hand, one creates neutral correlated electron–hole pair excitations. The electron–hole interaction effects (termed excitonic effects) become important and can in fact be dominant in many systems. A treatment of excitonic effects requires an effective two-particle approach on top of the quasiparticle picture. Self-energy, excitonic, and other many-electron interaction effects exist in all systems to various degrees. They are, however, of particular relevance in understanding the spectroscopic properties of many systems of contemporary interest including clusters, surfaces, polymers, defects, nanotubes, and other lower dimensional materials since the Coulomb interaction is enhanced in reduced dimensions.

An important parallel and complementary development in *ab initio* study of materials in the past two decades has, therefore, been the advent of first-principles methods in computing the electron excited-state properties [6]. It is now possible to calculate the quasiparticle and optical responses of many systems from first principles, in particularly those with moderately correlated electrons. The approach involves solving for the single-particle and two-particle Green's function of the interacting electron system. The advantage of the approach is that one can obtain the spectroscopic properties including the relevant self-energy and electron–hole interaction effects without any empirical parameters, and yet it is still applicable to real materials. Applications of this approach have explained and predicted the optical, photoemission, energy loss, scanning tunneling, and other spectroscopic properties of a number of systems.

Table 2 illustrates the discussion above that the energy eigenvalues from the self-consistent-field equations of ground-state theories, such as the density functional formalism and HF approximation, do not provide accurate information on electron excitation energies. Typically, HF overestimates the band gaps of semiconductors and insulators by several times the experimental value, and the LDA Kohn–Sham eigenvalues underestimate the gaps by as much as 50% or more. For Ge, even the band topology is qualitatively incorrect; LDA gives a metal. This "band gap problem" and other similar problems that exist in comparing calculated Kohn–Sham eigenvalues to experimental excitation energies do not stem from the

Table 2. Comparison of calculated LDA and HF band gaps (in eV) with experiment.

	HF	LDA	Experiment[a]
Diamond	13.6	3.9	5.48
Si	6.4	0.5	1.17
Ge	4.9	−0.26	0.74
LiCl	16.9	6.0	9.4

[a]*Source:* Kittel [98].

approximation of the exchange-correlation functional in DFT but arise from a fundamental and conceptual difference between the theoretical and measured quantities.

Rigorously, the interpretation of single-particle excitation energies such as band gaps requires the concept of quasiparticles, the long lived particle-like excitations in an interacting many-electron system [4,6,11]. It is mostly the transitions between the quasiparticle states that are probed in spectroscopic measurements. Because of electron–electron interactions, an excited particle of wavevector **k** is dressed with an electron polarization cloud resulting in a different energy E, effective mass m^*, and a finite lifetime. An accurate treatment of the exchange and dynamical correlations effects, arising from the Pauli exclusion principle and Coulomb repulsion, seen by an electron in a solid is then crucial in calculating the quasiparticle excitation energies and properties. The understanding of optical properties further involves the interaction between the excited quasi-electron and the hole that is left behind [7–10,39].

7. SINGLE-PARTICLE GREEN'S FUNCTION AND ELECTRON SELF-ENERGY

We discuss first the quasiparticle excitations since they are directly probed in a wide range of spectroscopic experiments and are the natural entities used in the theoretical understanding of optical processes. A systematic approach to quasiparticle properties is that of a Green's function formulation. The single-particle Green's function for an interacting many-electron system is defined as

$$G(\mathbf{r}, \mathbf{r}', \tau) = -i\langle 0|T\{\psi(\mathbf{r}, \tau)\psi^\dagger(\mathbf{r}', 0)\}|0\rangle \tag{22}$$

where the ψ's are the electron field operators, T the time order operator, and $|0\rangle$ the many-electron ground state. A useful alternate form is the diagonal elements of G in an orbital basis representation

$$G(p, \tau) = -i\langle 0|T\{c_p(\tau)c_p^\dagger(0)\}|0\rangle \tag{23}$$

where p denotes the quantum numbers of a single-particle state, e.g., the band index n and wavevector **k**. The power of the Green's function is that $G(p, \tau)$ gives the amplitude of finding a particle in orbital p at time $t = \tau$ if one is created into that orbital at time $t = 0$ without having to deal with the full many-electron excited-state wavefunction. For systems with long lived single-particle excitations, the quasiparticle energies and lifetime are given by the position of the poles of the interacting single-particle Green's function on the complex frequency plane [4]. In terms of the measurable spectral weight function, $A(\omega) = (1/\pi)|\mathrm{Im}\, G(\omega)|$, a dominant pole in the Green's function at a complex energy E_p corresponds to an $A(p, \omega)$ of the form

$$A(p, \omega) = \frac{(i/2\pi)Z_\mathrm{p}}{\omega - [E_\mathrm{p} - \mu]} + \text{c.c.} + \text{correction terms} \tag{24}$$

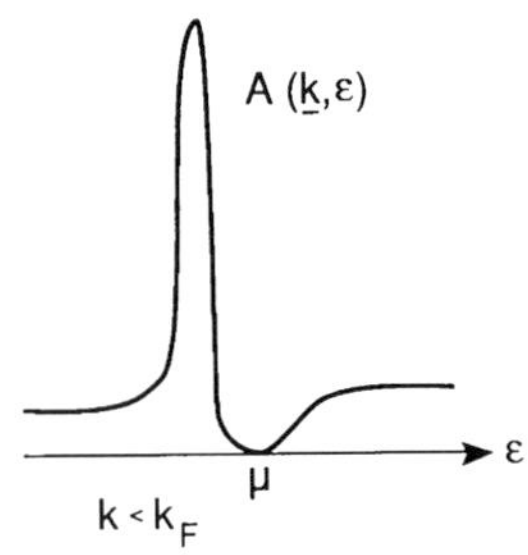

Fig. 11. Qualitative picture of the spectral weight function $A(\mathbf{k}, \varepsilon)$.

which is sharply peaked as a function of ω (see Fig. 11) and gives rise to a form for G, for positive τ,

$$G(p,\tau) = -iZ_{\rm p}{\rm e}^{-{\rm iRe}(E_{\rm p})\tau}{\rm e}^{-\Gamma_{\rm p}\tau} + \text{ correction terms} \tag{25}$$

where $\Gamma_{\rm p}$ is the imaginary part of $E_{\rm p}$. The physical content of Eq. (25) is that G, in this particular single-particle orbital basis, describes a propagation amplitude, which is oscillatory with a characteristic frequency given by the real part of $E_{\rm p}$, $\mathrm{Re}(E_{\rm p})$, and damped by $1/\Gamma_{\rm p}$. This leads to the usual interpretation that the peak position in $A(\omega)$ is the quasiparticle energy (real part of $E_{\rm p}$) and the width (imaginary part of $E_{\rm p}$) relates to the lifetime of the quasiparticle. Finding the quasiparticle properties is then equivalent to solving for the appropriate single-particle states, which give rise to sharp peaks in the diagonal matrix elements of $A(p,\omega)$, and is also equivalent to solving for the position of the poles of G in the complex energy plane.

For a many-electron system with a Hamiltonian of the form

$$H = \sum_i H_0(\mathbf{r}_i) + \sum_{i<j} V_{\rm c}(|\mathbf{r}_i - \mathbf{r}_j|), \tag{26}$$

where $H_0(\mathbf{r}) = (\mathbf{p}^2/2m) + V_{\rm ion}(\mathbf{r})$ is a one-particle term and $V_{\rm c}$ the bare Coulomb interaction, the single-particle Green's function can be shown, from the equation of motion of G, to satisfy

$$(\hbar\omega - H_0 - V_{\rm H})G(\mathbf{r},\mathbf{r}';\omega) - \int \Sigma(\mathbf{r},\mathbf{r}'',\omega)G(\mathbf{r}'',\mathbf{r}';\omega) = \delta(\mathbf{r},\mathbf{r}') \tag{27}$$

where $V_{\rm H}$ is the usual Hartree potential and Σ the electron self-energy operator, which is a functional of G. The above equation can be solved formally by expressing [4,6]

$$G(\mathbf{r},\mathbf{r}';\omega) = \sum_{n\mathbf{k}} \frac{\psi_{n\mathbf{k}}(\mathbf{r})\psi^*_{n\mathbf{k}}(\mathbf{r}')}{\omega - E_{n\mathbf{k}} - {\rm i}\delta_{n\mathbf{k}}} \tag{28}$$

where $E_{n\mathbf{k}}$ and $\psi_{n\mathbf{k}}(\mathbf{r})$ are solutions to the quasiparticle Dyson equation

$$[E_{n\mathbf{k}} - H_0(\mathbf{r}) - V_{\rm H}(\mathbf{r})]\psi_{n\mathbf{k}}(r) - \int \Sigma(\mathbf{r},\mathbf{r}';E_{n\mathbf{k}})\psi_{n\mathbf{k}}(\mathbf{r}')\,{\rm d}\mathbf{r}' = 0. \tag{29}$$

Here n and $\mathbf{k}$ are the band and wavevector quantum numbers describing electronic states in a crystal. The pole structure of $G(\omega)$ may now be identified with the solutions of Eq. (29), and, therefore, given the self-energy operator Σ, the problem of solving for the quasiparticle properties is then one of solving Eq. (29).

The quasiparticle equation, Eq. (29), is similar in form to the Schrodinger equation (such as the Kohn–Sham equation) in one-electron theories. However, because of electron–electron interactions, $\Sigma(\mathbf{r}, \mathbf{r}', \omega)$ is a non-local, non-Hermitian, and energy-dependent operator giving rise in general to a complex $E_{n\mathbf{k}}$. As seen in Eq. (25), the real part of $E_{n\mathbf{k}}$ gives the quasiparticle energy and the imaginary part corresponds to the lifetime. Often it is useful to express $E_{n\mathbf{k}}$ as a sum of a single-particle term, $E^0_{n\mathbf{k}}$, plus a self-energy $\Sigma_{n\mathbf{k}}$ containing the many-electron (exchange-correlation) effects:

$$E_{n\mathbf{k}} = E^0_{n\mathbf{k}} + \Sigma_{n\mathbf{k}}. \tag{30}$$

This physical description of the single-particle excitation spectra of an interacting electron system relies on the lifetime of the quasiparticles being sufficiently long on the time scale of the relevant experimental probes. Within Fermi liquid theory, for energies near the Fermi level of a metal or the gap region of a semiconductor or insulator, the quasiparticles are well defined, allowing us to pursue this description [40].

8. THE *GW* APPROXIMATION

The self-energy operator Σ can be systematically expanded in a series in terms of the dressed Green's function G and the screened Coulomb interaction W [3,6]

$$W(\mathbf{r}, \mathbf{r}'; \omega) = \int \varepsilon^{-1}(\mathbf{r}, \mathbf{r}''; \omega) V_c(\mathbf{r}'', \mathbf{r}')\, d\mathbf{r}'', \tag{31}$$

where $\varepsilon(\mathbf{r}, \mathbf{r}', \omega)$ is the time-ordered dielectric response function of the system. The advantage of this expansion (see Fig. 12) over a conventional one in terms of the bare Coulomb interaction V_c is that W, being a screened quantity, is in general much weaker and a series expansion in W should be more rapid in convergence.

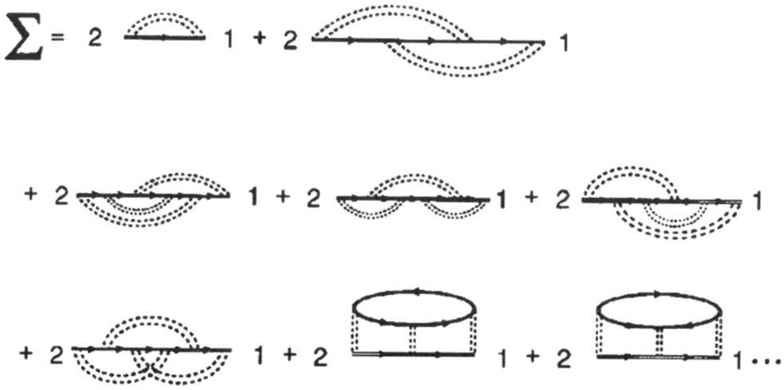

Fig. 12. Diagrammatic expansion of Σ in the screened Coulomb interaction W.

Mathematically, since W and the dressed Green's function themselves can be expressed as series expansions in the bare quantities, each diagram in Fig. 12 is a partial summation over terms in a conventional expansion, leading to more accurate physical results with only low-order terms in the expansion. Also, in perturbation theory, it is desirable in general to start from a mean-field scenario that is as close as possible to the physical system, like the Kohn–Sham DFT system, which already includes an attempt to describe exchange and correlations in an averaged way in the actual system.

The important elements in a first-principles calculation of quasiparticle energies are [6]: (1) an approximation for the self-energy operator, (2) schemes for calculating G and W, and (3) solution to the quasiparticle equation. In practical studies of real materials, inevitably, only the first term of the series in Fig. 12 for the self-energy operator is kept. This is called the GW approximation [4] since Σ is given to first order in G and W by

$$\Sigma(\mathbf{r},\mathbf{r}';E) = i\int \frac{d\omega}{2\pi} e^{-i\delta\omega} G(\mathbf{r},\mathbf{r}';E-\omega)W(\mathbf{r},\mathbf{r}';\omega) \tag{32}$$

where $\delta = 0^+$. Within this approximation, the calculation of the quasiparticle properties reduces to computing Σ using Eq. (32) and solving the quasiparticle equation (29) for $E_{n\mathbf{k}}$ and $\psi_{n\mathbf{k}}(\mathbf{r})$.

The basic ingredients in the GW theory are the dielectric response function $\varepsilon(\mathbf{r},\mathbf{r}',\omega)$ and the single-particle electron Green's function G. Both have to be treated adequately to obtain quantitative results that may be compared with experiment. If W is replaced by the bare Coulomb interaction in Eq. (32), then Σ is just the usual bare exchange operator. Equation (29) reduces to the HF equation. From this point of view, the HF eigenvalues may be considered as a lowest-order approximation to the quasiparticle energies, consistent with Koopman's theorem. The dielectric function $\varepsilon(\mathbf{r},\mathbf{r}',\omega)$ provides the dynamical and spatial screening response of the electrons that gives rise to correlation effects going beyond bare exchange. Similarly, the eigenvalues from the Kohn–Sham equations in the density functional formalism may be viewed as another set of approximate quasiparticle energies, with the exchange-correlation potential $V_{xc}(\mathbf{r})$ approximating the non-local, energy-dependent self-energy operator in Eq. (29). We must emphasize that as discussed above, in principle, both the HF and Kohn–Sham eigenvalues are only Euler–Lagrange parameters in minimizing the total energy in ground-state theories. In fact, different formulations of the DFT can lead to significantly different Kohn–Sham eigenvalues [41].

9. QUASIPARTICLE EXCITATIONS IN MATERIALS

Although the GW approximation was formulated in the 1960s, the first-principles approach to computing quasiparticle excitations in real materials was not developed until 20 years later by Hybertsen and Louie in 1985 [5,6]. This approach has since been employed with considerable success to the *ab initio* study of semiconductors, insulators, surfaces, nanostructures and other material systems. The development was made possible by the realization of the importance of local-field screening effects in the self energy. They showed that, with the crucial inclusion of the full dielectric matrix, one can obtain highly accurate quasiparticle energies. Similar to the first-principles ground-state studies, the only inputs to the calculations are the atomic numbers of the constituent elements and the geometric structure of the system which can be determined separately from a total energy calculation.

Since the screened Coulomb interaction W incorporates the dynamical many-body effects of the electrons, the dielectric response function $\varepsilon(\mathbf{r}, \mathbf{r}', \omega)$ is key in determining the electron self energy. Owing to the charge density inhomogeneity, $\varepsilon(\mathbf{r}, \mathbf{r}', \omega)$ of a solid is a two-point function of $\mathbf{r}$ and $\mathbf{r}'$ separately. In a **k**-space formulation, the crystalline dielectric function is a matrix $\varepsilon_{\mathbf{GG}'}(\mathbf{q}, \omega)$ in the reciprocal lattice vectors, **G**. The off-diagonal elements of this matrix describe the local field effects that give the variations in the electronic screening properties in different parts of the crystal [42,43]. These local fields are physically very important and are a major component in the quantitative evaluation of the self-energy operator for a real material. The frequency dependence of ε is also found to be significant for obtaining accurate results. A key factor that made possible the *ab initio* calculation of quasiparticle energies was the development of techniques for calculating the static dielectric matrices together with a scheme for extending them to finite frequencies [6].

In applications to real materials, the basic approach taken [6] has been to: (1) make the best possible approximations for G and W, (2) calculate Σ, and (3) obtain the quasiparticle energies without any adjustable parameters. In this approach, the philosophy is to consider the full Hamiltonian H in the form $H = H_{\mathrm{mf}} + (H - H_{\mathrm{mf}})$, where H_{mf} is some appropriately chosen independent particle mean-field Hamiltonian, and then to calculate, within the GW approximation, the self-energy operator due to the residual interaction $(H - H_{\mathrm{mf}})$. For most moderately correlated electron systems, the best available mean-field Hamiltonian may be taken to be the Kohn–Sham Hamiltonian. Thus, for most studies in the literature, the Green's function G is typically constructed using the Kohn–Sham wavefunctions and eigenvalues. The Kohn–Sham eigenvectors also are extremely good approximations to the quasiparticle wavefunctions with overlap between the two which is often better than 99%. The Green's function G in practice is then subsequently updated only with the quasiparticle spectrum from Eq. (29). There is very limited experience on the importance of including the detailed structure in the spectral function for the quasiparticle energies [4,44–46]. Comparison of calculated energies with experiment shows that this level of approximation is very accurate for semiconductors and insulators and for most conventional metals. However, as in the case of the effect of higher-order terms or vertex corrections to Σ, only a *aposteriori* experience truly justifies the approximation.

Figure 13 shows the calculated GW band gaps of a number of insulating materials plotted against the measured quasiparticle band gaps. A perfect agreement between theory and experiment would place the data points on the diagonal line. As we discussed before, the Kohn–Sham gaps in the LDA significantly underestimate the

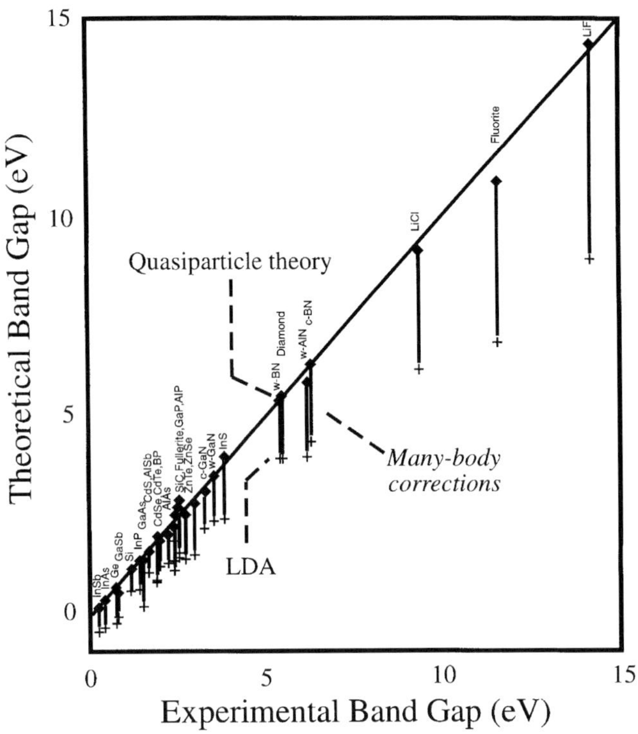

Fig. 13. Comparison of the GW band gap with experiment for a wide range of semiconductors and insulators. The Kohn–Sham eigenvalue gaps calculated within the LDA are also included for comparison.

experimental values, giving rise to the band gap problem. Some of the Kohn–Sham gaps are even negative. However, the GW quasiparticle energies (which provide an appropriate description of particle-like excitations in an interacting systems) result in band gaps that are in excellent agreement with experiments for a range of materials - from the small gap semiconductors such as InSb, to moderate gap materials, such as GaN and solid C_{60}, and to the large gap insulators such as LiF.

In addition, the *GW* quasiparticle band structures for semiconductors and conventional metals in general compare very well with data from photoemission and inverse photoemission measurements. Figure 14 depicts the calculated quasiparticle band structure of Ge [6,47] and Cu [48] as compared to photoemission data for the occupied states and inverse photoemission data for the unoccupied states. For Ge, the agreement is within the error bars of experiments. In fact, the conduction band energies of Ge were theoretically predicted [6] before the inverse photoemission measurement [47]. The results for Cu agree with photoemission data to within 0.03 eV for the highest d-band, correcting 90% of the LDA error. The energies of the other d-bands throughout the Brillouin zone are reproduced within 0.3 eV. This level of agreement for the d-bands cannot be obtained without including the self-energy

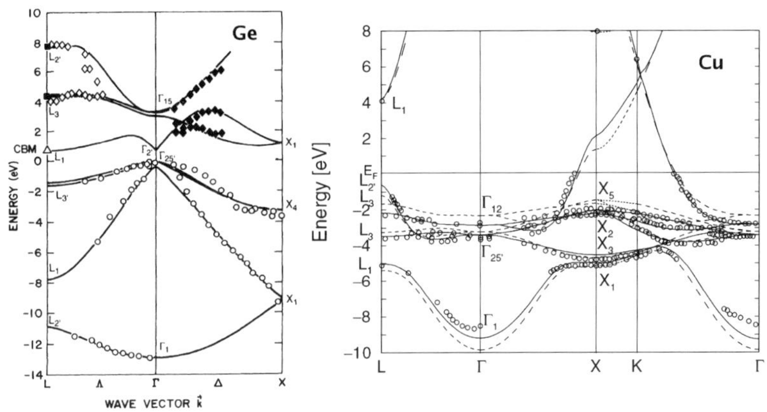

Fig. 14. Calculated GW quasiparticle band structure of Ge (left panel) and Cu (right panel) as compared with experiments (open and full symbols). In the case of Cu, we also provide the DFT–LDA band structure as dashed lines. (After Ortega and Himpsel [47]; Marini et al. [48].)

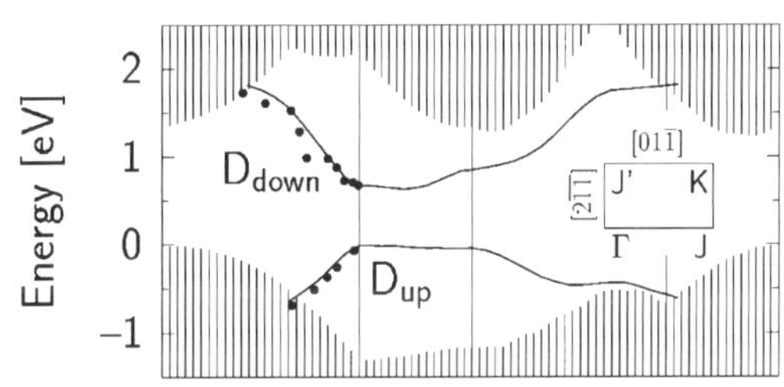

Fig. 15. Computed GW quasiparticle band structure for the Si(111) 2×1 surface compared with photoemission experimental results (dots). (After Rohlfing and Louie [49].)

contributions. Similar results have been obtained for other materials and even for some non-conventional insulating systems such as the transition metal oxides and metal hydrides.

The GW approach has also been used to investigate the quasiparticle excitation spectrum of surfaces, interfaces, clusters, defects, and other non-bulk systems. As an example of surface study, we return to the Si(111) 2×1 surface. As discussed above, this surface has a highly interesting geometric and electronic structure. The buckled π-bonded chains of the 2×1 reconstructed surface structure (see Fig. 4) give rise to an occupied and an unoccupied quasi-1D surface-state band, which are dispersive only along the π-bonded chains. These surface states lead to a quasiparticle surface-state band gap of 0.7 eV that is significantly different from the bulk Si band gap of 1.2 eV. In Fig. 15, the calculated GW quasiparticle surface-state bands are compared to photoemission and inverse photoemission data [22,49–53]. As seen from the figure, both the calculated surface-state band dispersion and surface-state band gap are in good agreement with experiment, and these results are also in accord with results from scanning tunnelling spectroscopy (STS) [54], which physically also probes quasiparticle excitations. However, a long-standing puzzle in the literature has been that the measured surface-state gap of this surface from optical experiments [55–57] is significantly smaller (by nearly 0.3 eV) than the quasiparticle gap. This discrepancy is indicative of a very strong electron–hole interaction on this surface. We shall take up this issue later when we discuss optical response.

It should be noted that the quasiparticle excitations of a material are in general not exact eigenstates of the interacting system and thus possess a finite lifetime. The lifetime of excited electrons in solids can have contributions from a variety of inelastic and elastic scattering mechanisms, such as electron–electron, electron–phonon, and electron–imperfection interactions. For many years, the theoretical treatment of the inelastic lifetime of quasiparticles due to electron–electron interaction, as manifested in the imaginary part of Σ, has been based on the electron gas model of Fermi liquids characterized by the electron-density parameter r_s. Within this simple model, for either electrons or holes with energy E very near the Fermi level E_F, the inelastic lifetime is found to be, in the high-density limit ($r_s \ll 1$), $\tau(E) = 263 r_s^{-5/2}(E - E_F)^{-2}$ fs, where E and E_F are expressed in eV [58]. As for the quasiparticle energies, a proper treatment of the electron dynamics (quasiparticle damping rates or lifetimes) needs to include band structure and dynamical screening effects in order to be in quantitative comparison with experiment. An illustrative example is given in Fig. 16 where the quasiparticle lifetimes of electrons and holes in bulk Cu and Au have been calculated within the GW method, showing an increase in the lifetime close to the Fermi level as compared to the predictions of the electron gas model. For Au, a major contribution from the occupied d-states to the screening yields lifetimes of electrons that are larger than those of electrons in a homogenous electron gas model by a factor of about 4.5 for electrons with energies 1–3 eV above the Fermi level. This finding is in accord with results from an experimental study of ultrafast electron dynamics in Au(111) films [59].

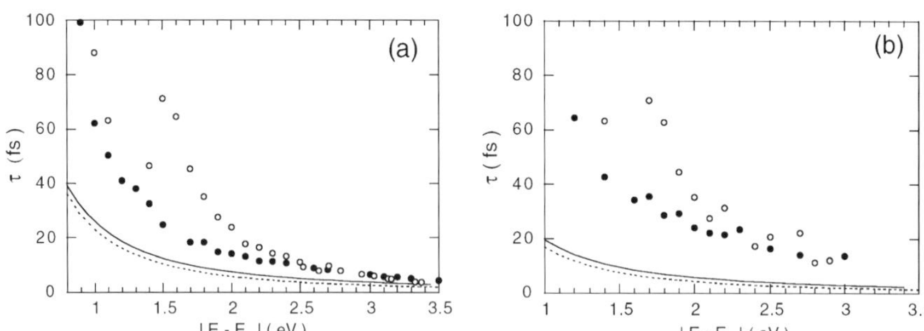

Fig. 16. Calculated GW electron and hole lifetimes for Cu (a) and Au (b). Solid and open circles represent the *ab initio* calculation of $\tau(E)$ for electrons and holes, respectively, as obtained after averaging over wavevectors and the band structure for each **k** vector. The solid and dotted lines represent the corresponding lifetime of electrons (solid line) and holes (dotted line) in an electron gas with $r_s = 2.67$ for Cu and $r_s = 3.01$ for Au. (After Campillo et al. [89]; Campillo and Rubio [90].)

Other useful applications of the first-principles quasiparticle approach include study of the spectroscopic properties of materials under pressure and of defect levels associated with point or line defects in solids. We now describe a couple of example calculations of this kind. An interesting case is that of the insulator–metal transition in xenon. Solid Xe is a large gap insulator in the fcc structure at low pressure. It is in the hcp structure at high pressure and undergoes a pressure-induced isostructural insulator–metal transition near 132 GPa [60,61]. Results from GW calculations [62,63] yield a quasiparticle band gap closure at the pressure of 128 GPa (see Fig. 17). This transition arises from the fact that, in the hcp structure, the top of the valence band of Xe is anti-bonding p-like while the bottom of the conduction band is bonding d-like. Under increasing pressure, the anti-bonding states rise in energy and the bonding states lower in energy resulting in a band gap closure. The quasiparticle results reproduce accurately the volume dependence of the observed band gap, whereas the LDA results significantly underestimate the transition volume. The theory, moreover, explains many of the salient features observed in the experimental spectra at metallization, in particular the appearance of a peak at ~ 2 eV in the optical absorption spectra which is attributed to electronic transitions to hole states at the top of the valence band made available after the band gap closed. Another important finding from the calculation is that the self-energy correction to the LDA band gap is not a constant as a function of density or

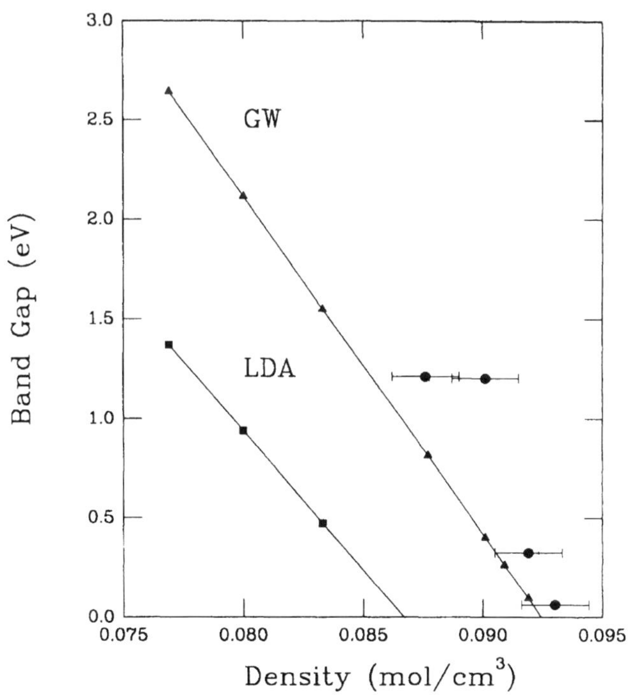

Fig. 17. LDA and GW minimum band gaps of hcp Xe as a function of density. Experimental data from [60] is shown as solid circles. (After Chacham et al. [63].)

Table 3. Bulk band gap and F-center defect excitation energies in LiCl compared to experiment (in eV). (After Surh et al. [64]).

	LDA	GW	Experiment
Band gap	6.1	9.3	9.4
$1s \rightarrow 2p$	2.4	3.4[a]	3.1–3.3
$1s \rightarrow L_c$	1.8	4.5	4.5
$1s \rightarrow \Lambda_c$	2.2	5.0	5.0
$1s \rightarrow X_c$	2.8	5.7	5.8

[a]Value included electron–hole interaction.

pressure. Near the transition, the pressure coefficient of the gap changes rapidly with pressure and is different from that of the LDA. This indicates that, in general, it is not straightforward to deduce the insulator–metal transition pressure from knowing only the pressure coefficient of the gap at low pressure and/or from LDA.

As an example of GW calculations for point defects in solids, we present results from a study of defect levels of a F-center or color center in LiCl [64]. These color centers correspond to Cl vacancies in the insulator. A neutral Cl vacancy contains a single bound electron. It is the optical transitions between the defect levels in the band gap that give rise to absorption of visible light in this otherwise colorless salt. Structurally, a neutral Cl vacancy induces only a very small lattice relaxation near the defect. Table 3 compares the calculated energies with data from experiment. The Cl vacancy produces a 1s and 2p defect level in the band gap. Excitation energies corresponding to transitions from the ground state (1s level) of the neutral F-center to the various bulk LiCl conduction band states may be compared to the calculated quasiparticle results. The 1s vacancy state is calculated to be at 4.0 eV below the conduction band minimum. Excitation energies corresponding to transitions to the conduction states at L, Λ and X are in good agreement with experimental values; but those predicted by LDA Kohn–Sham eigenvalues are off by a factor of two. For the bound $1s \rightarrow 2p$ intrasite excitation, theory also reproduces the observed transition energy if electron–hole interaction corrections are included on top of the quasiparticle transition energy [64]. In contrast, LDA predicts an unbound 2p state in the conduction band and a $1s \rightarrow 2p$ optical transition that is off by 1 eV.

10. ELECTRON–HOLE EXCITATIONS AND THE BETHE–SALPETER EQUATION

We now turn to the optical properties. For an interacting system, we must include the interaction between the quasielectron and the hole created in the optical excitation process. Thus, the study of the optical response of materials requires knowledge of the electron–hole excitations, $|N,0\rangle \rightarrow |N,S\rangle$, that do not change the total number of electrons N in a system. (Here $|N,S\rangle$ denotes the Sth excited state). Figure 18 illustrates how important such effects are for the case of crystalline SiO_2 (α-quartz). The calculated absorption spectrum, assuming interband transitions and

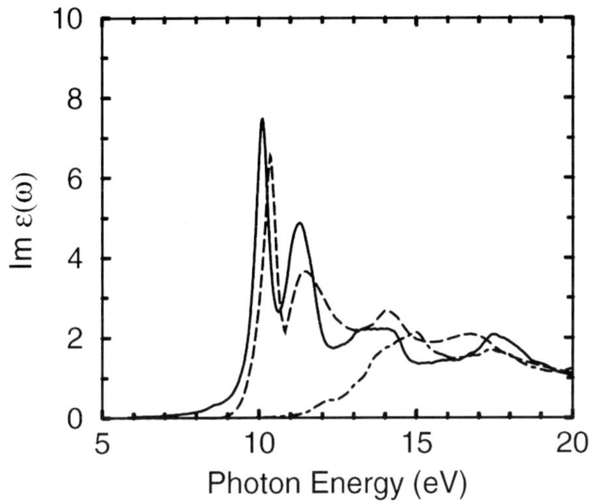

Fig. 18. Calculated absorption spectrum of α-quartz (SiO_2) with excitonic effects (solid line), compared to interband transition theory (dotted-dashed line) and experimental data [65] (dashed line). (After Chang et al. [67].)

no interaction between the excited electron and the hole, shows very little resemblance with the experimental spectrum [65]. For nearly 40 years, there has been on-going debate on the nature of the four rather sharp peaks observed in the experimental spectrum of this technologically very important material.

Just as the quasiparticle excitations are given by the one-particle Green's function G, the electron–hole excitations may be obtained by investigating the interacting two-particle Green's function G_2 (a two-particle generalization of Eq. (22)) and solving its equation of motion. In first-principles study of materials, we typically assume that the electron–hole excitations are long lived (in an analogous way to the quasiparticle approximation for the single-particle problem) and the Tamm–Dancoff approximation [11] is valid. The excited state $|N, S\rangle$ with energy Ω_S (referenced to the ground state energy) may then be written, to a very good approximation, as a linear combination of non-interacting electron–hole configurations plus correction terms that are not important in the evaluation of the optical transition strength:

$$|N, S\rangle = \sum_{v}^{\text{hole}} \sum_{c}^{\text{elec}} A_{vc}^{S} a_{v}^{\dagger} b_{c}^{\dagger} |N, 0\rangle + \cdots \tag{33}$$

Here $a_v^{\dagger}$ and $b_c^{\dagger}$ are the creation operators for holes and electrons, respectively. For simplicity of notation, we use v and c as composite indices containing both band and wavevector quantum numbers and $|vc\rangle = a_v^{\dagger} b_c^{\dagger} |N, 0\rangle$ as a configuration in which a hole with wavevector **k** is created in the valence band v and a quasielectron is created in the conduction band c with wavevector $\mathbf{k} + \mathbf{q}$, as illustrated in Fig. 19. Electron–electron interaction mixes electron–hole configurations of same center-of-mass **q** to

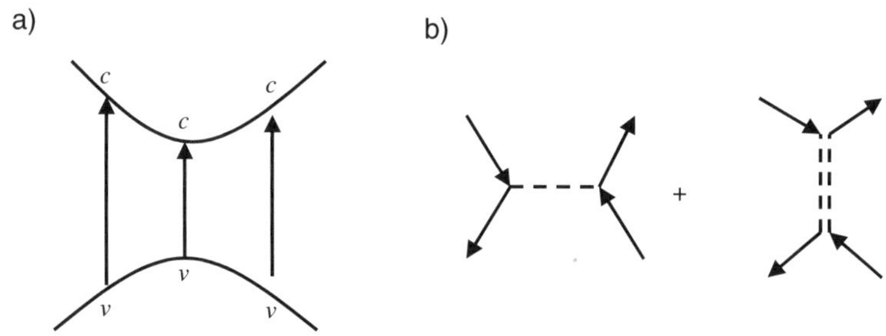

Fig. 19. Schematics of (a) optical transitions as superposition of quasielectron–hole pair configurations due to electron–electron interaction, and (b) electron–hole interaction kernel consisting of a repulsive exchange term and a screened attractive direct term.

form the excited (excitonic) state $|N, S\rangle$. The electron–hole amplitude or exciton wavefunction in real space, which describes the spatial correlation of the electron and hole, is then given by

$$\chi_S(\mathbf{r}, \mathbf{r}') = \sum_{cv} A^S_{cv} \psi_c(\mathbf{r}) \psi^*_v(\mathbf{r}'), \tag{34}$$

with $\psi(\mathbf{r})$ being the quasiparticle amplitude.

From the equation of motion of the two-particle Green's function, the excitation energy Ω_S and coefficients A^S_{cv} can be shown to satisfy a BSE of the form [39,66]

$$(E_c - E_v)A^S_{cv} + \sum_{cv,c'v'} K_{cv,c'v'}(\Omega_S) A^S_{c'v'} = \Omega_S A^S_{cv}. \tag{35}$$

Here the E's are quasiparticle energies taken from a GW calculation (e.g., using the method of Hybertsen and Louie [6]) and $K_{cv,c'v'} = \langle cv|K|c'v'\rangle$ describes the interaction between the excited electron and hole. Solving this Bethe–Salpeter equation yields the excitation energies and the excited-state wavefunctions from which one can compute the optical absorption spectrum, exciton binding energies and wavefunctions, and other related optical quantities.

The electron–hole interaction kernel K is an operator that describes the scattering of an electron–hole pair from configuration $|vc\rangle$ to $|v'c'\rangle$. Using the notation $1 = (\mathbf{r}_1, \sigma_1, t_1)$, K is given by the functional derivative

$$K(12, 34) = \frac{\delta[V_{\mathrm{H}}(1)\delta(1,3) + \Sigma(1,3)]}{\delta G(4,2)}. \tag{36}$$

To be consistent with the quasiparticle band structure calculation, the self-energy operator Σ is evaluated within the GW approximation; and with the additional assumption that the functional derivative of W with respect to G can be neglected [66], K is simplified to

$$\begin{aligned} K(12, 34) &= -i\delta(1,3)\delta(2^-,4)V(1,4) + i\delta(1,4)\delta(3,2)W(1^+,3) \\ &= K_x + K_d. \end{aligned} \tag{37}$$

Thus, the electron–hole interaction consists of two leading terms as illustrated in Fig. 19: a bare repulsive exchange interaction K_x and a screened attractive interaction K_d.

The methods for the evaluation of the matrix elements of the kernel K and the solution to the BSE have been discussed in detail in the literature (e.g., in [66]). In the absence of spin–orbit interaction, the repulsive exchange term is only effective on spin singlet states. Dynamical screening can further be neglected if the energies Ω_S are close to the energy of the non-interacting pairs. From the solution of the BSE (Eq. (35)), the optical absorption spectrum and other optical properties are obtained from the imaginary part of the macroscopic dielectric function,

$$\varepsilon_2(\omega) = \frac{16\pi^2 e^2}{\omega^2} \sum_S |\langle N, 0|\hat{\mathbf{e}} \cdot \mathbf{v}|N, S\rangle|^2 \delta(\Omega_S - \hbar\omega), \tag{38}$$

where $\hat{\mathbf{e}}$ is the normalized polarization vector of the light and $\mathbf{v} = (1/\hbar)[H, \mathbf{r}]$ the single-particle velocity operator. An important effect of the electron–hole interaction is the coupling of different electron–hole configurations $|vc\rangle$ in the excited state, $|N, S\rangle$, leading to optical transition matrix elements that are given by

$$\langle N, 0|\hat{\mathbf{e}} \cdot \mathbf{v}|N, S\rangle = \sum_{cv} A^S_{cv} \langle c|\hat{\mathbf{e}} \cdot \mathbf{v}|v\rangle, \tag{39}$$

i.e., by a coherent sum of the transition matrix elements of the contributing electron–hole pair configurations weighted by the coupling coefficients A^S_{cv}, which often leads to interesting interference effects.

The Bethe–Salpeter Equation approach to the two-particle excited states is, therefore, a natural extension of the GW approach to the one-particle excited state properties, within the same theoretical framework and set of approximations. As seen below, this GW–BSE approach has helped elucidate the optical response of a wide range of systems from nanostructures to bulk semiconductors to surfaces and defects to one-dimensional systems, such as polymers and nanotubes.

11. OPTICAL PROPERTIES OF SOLIDS, SURFACES, AND NANOSTRUCTURES

In this section, we present some results of application of the GW–BSE method to several prototypical systems to illustrate the accuracy and versatility of the approach. Both crystalline solids and reduced dimensionality systems are discussed. For many of the systems considered, especially for polymers, nanotubes, and clusters, inclusion of electron–hole interaction or excitonic effects is essential in the understanding of the observed optical spectra.

Figure 20 gives the optical spectrum $\varepsilon_2(\omega)$ of bulk semiconductors GaAs and Si. As seen from the figure, only with the inclusion of electron–hole interaction there is a good agreement between theory and experiment. Excitonic effects enhance the optical oscillator strength by nearly a factor of two in the low frequency range for both materials. Also, the electron–hole interaction shifts and heightens the second

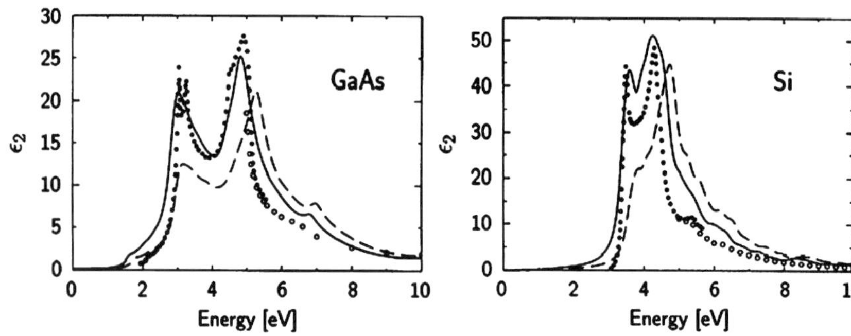

Fig. 20. Calculated optical absorption of GaAs (left panel) and Si (right panel) with (solid line) and without (dashed line) electron–hole interaction, compared to experimental data [91,92]. (After Rohlfing and Louie [8]; Rohlfing and Louie [66].)

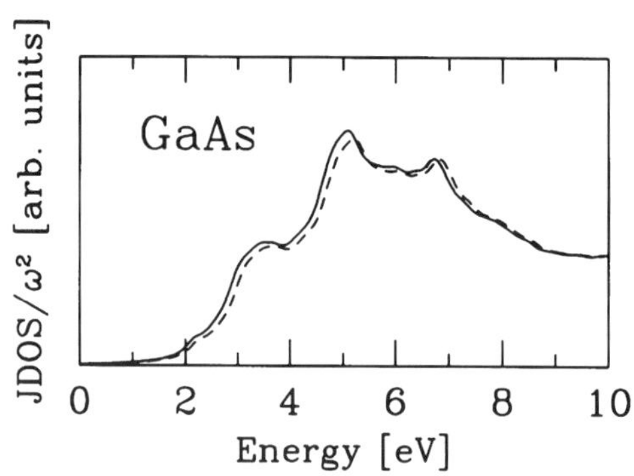

Fig. 21. Density of electron–hole excited states in GaAs with (solid line) and without (dashed line) electron–hole interaction. (After Rohlfing and Louie [66].)

main peak, yielding much closer agreement with experiment. The very large shift of nearly 0.5 eV in the second peak (near 5 eV) is not due to a negative shift of the transition energies as one might naively expect from an attractive electron–hole interaction. In fact, as is illustrated in Fig. 21 for GaAs, the density of states for the photo-excited states remains nearly unchanged by the electron–hole interaction. The changes in the position and height of the second peak in the optical spectrum originate mainly from the coupling of different electron–hole configurations in the photo-excited state, which leads to a constructive coherent superposition of oscillator strengths for states on the lower-energy side of the peak and a destructive superposition for states on the high-energy side. Spin–orbit interaction was not included in the calculated results in Fig. 20, and, hence, the fine structures in the experimental spectrum of GaAs are not reproduced in the theory.

With the GW–BSE approach, in addition to the continuum part of the spectrum, discrete bound exciton states near the absorption edge may also be obtained from first principles without making use of any effective mass or other approximations.

Table 4. Exciton binding energies in GaAs. The GW–BSE results from [8] are compared with experimental data from optical absorption [96] and from two-photon absorption measurements [97].

	Theory (meV)	Experiment (meV)
E_{1s}	4.0	4.2[a]
E_{2s}	0.9	1.05[a]
E_{2p}	0.2–0.7	~ 0.1[b]

[a]Shell [96].
[b]Michaelis et al. [97].

Table 4 shows the calculated binding energy of the lowest energy exciton states in GaAs [8]. The theory has reproduced essentially all the observed excitonic structures to a good level of accuracy. Thus, it is possible to obtain the binding energy of bound excitons very accurately from Eq. (35), even if the binding energy is only of the order of few meVs. In this case, a careful sampling of the Brillouin zone and inclusion of spin–orbit interaction are required so that the quasiparticle energy bands are described with the necessary accuracy. These results give us confidence that the approach can, therefore, be employed for situations in which simple empirical techniques do not apply.

Similarly accurate results have been obtained for the optical spectrum of other semiconductors and insulators. For larger gap materials, exciton effects are even more dramatic in their optical response as we have seen for the case of SiO_2 [67]. Figure 18 shows the calculated optical absorption of α-quartz as compared to experiment, with the polarization of light perpendicular to the hexagonal plane. The agreement between the experimental spectrum and the calculated spectrum, with excitonic effects included, is again excellent. The quasiparticle gap of α-quartz is 10 eV. From the calculation, we learn that all the prominent peaks found in experiment are due to transitions to excitonic states. The much-debated peaks in the experimental spectrum are in fact due to the strong correlations between the excited electron and hole in resonant excitonic states. These excited states have energies that are bigger than the quasiparticle band gap and are inside the energy range of the free electron–hole continuum. Theory also reveals that there is a bound spin-singlet exciton state at 8.5 eV, but it has a vanishing optical matrix element and is, therefore, not visible in the absorption spectrum.

Owing to the importance of electron–electron interaction in lower dimensional systems, the GW–BSE approach has been particularly valuable in explaining and predicting the quasiparticle excitations and optical response of reduced-dimensional systems and nanostructures. As illustrated below, self-energy and electron–hole interaction effects can be orders of magnitude larger in such systems than in bulk materials made up of the same elements.

Conjugated polymers serve as a good example of reduced dimensional materials. The optical properties of these technologically important systems are considerably less understood than conventional semiconductors. For example, there has been much argument in the literature regarding the binding energy of excitons in polymers such as poly-phenylene-vinylene (PPV). Binding energies ranging from 0.1 to 1.0 eV had been suggested. *Ab initio* calculations using the GW–BSE approach show that excitonic effects in PPV are indeed very strong and qualitatively change the optical spectrum of the material [68]. This is illustrated in Fig. 22 where each of the 1D van Hove singularity derived features in the interband absorption spectrum is replaced by a series of sharp peaks due to excitonic states. The lowest optically active exciton is a bound exciton state, but the others are strong resonant exciton states giving rise to peak structures that agree well with experiment. In particular, when compared to the quasiparticle gap of 3.3 eV, the theoretical results in Fig. 22 yield a large binding energy of nearly 1 eV for the lowest energy bound exciton in PPV.

The reduced dimensionality at a surface also often gives rise to enhanced excitonic effects. As an example of this phenomenon, we return to the case of the Si(111) 2 × 1 surface [49] discussed above. For this surface, it is found that the optical spectrum at low frequencies is dominated by a surface-state exciton (derived from the surface-state bands depicted in Fig. 15) that has a binding energy that is an order of magnitude bigger than that of the excitons in bulk Si. One cannot interpret the experimental spectrum without considering the excitonic effects [22,49]. Figure 23 compares the measured differential reflectivity [69] with the GW–BSE results. Not only is the onset of electron–hole pair creation shifted significantly to a lower energy than the quasiparticle gap value, but the experimental spectrum consists only of one asymmetric peak as opposed to a broader two-peak structure predicted by interband transition theory (dashed curves in Fig. 23).

As in PPV, due to electron–hole interaction, the optical response of the Si(111) 2 × 1 surface is considerably changed from the non-interacting case. Above the surface-state quasiparticle gap, the differential reflectivity is greatly reduced due to a destructive coupling of interband oscillator strength caused by the interaction. Below the surface-state gap, a number of discrete exciton states are found in the calculation [49]. The optical oscillator strength is, however, mostly concentrated in the lowest-energy exciton at 0.43 eV, which now dominates the spectrum and,

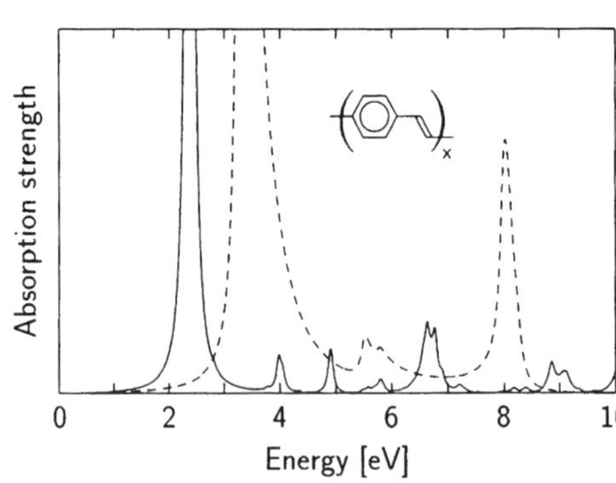

Fig. 22. Optical absorption spectrum of the polymer PPV. Theoretical results with (continuous line) and without (dashed line) excitonic effects. (After Rohlfing and Louie [68].)

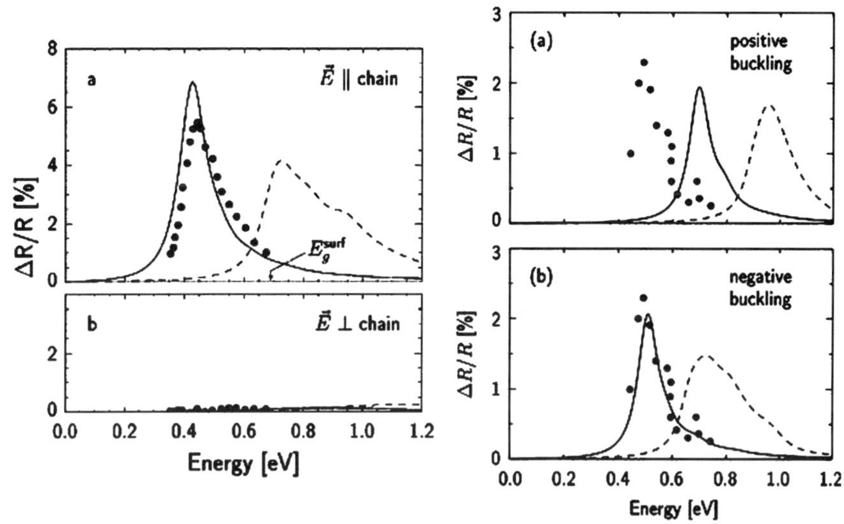

Fig. 23. Comparison between experiments and the computed differential reflectivity spectra with and without electron–hole interaction for the Si(111) 2 × 1 surface (left panel) [49] and for Ge(111) 2 × 1 (right panel) (After Rohlfing and Louie [49]; Rohlfing, et al. [93].)

together with the other excitonic states, gives rise to one slightly asymmetric main peak. This surface-state exciton has a binding energy of 0.23 eV compared to the excitonic binding energy in bulk Si which is of 15 meV. Moreover, the calculated dipole oscillator strengths of the excitations are highly anisotropic with respect to the polarization of the light due to the anisotropic nature of the π-bonded surface atomic chains, which is in agreement with experiment.

The large enhancement in the electron–hole interaction at this particular surface arises from the quasi-1D nature of the surface states, which are localized along the π-bonded atomic chains on the surface. The electron and hole are strongly confined to the surface and to the individual chains, leading to strong overlap of the quasiparticle wavefunctions and, therefore, a large electron–hole interaction. The spatial correlation of the electron and hole in the lowest surface-state exciton state is illustrated in Fig. 24, which shows a side view of a contour plot of the square of the electron–hole amplitude $\chi_S(\mathbf{r}, \mathbf{r}')$ (given by Eq. (34)) evaluated on a plane perpendicular to the Si(111) surface and to the π-bonded chains, with the position of the hole fixed near an up atom on the chain. We see that the electron amplitude is highly localized on the surface and on the chain where the hole is located. Moreover, on the surface layer, the electron–hole amplitude (or exciton wavefunction) is highly anisotropic for the surface-state exciton. The mean electron–hole distance along the chains is over 40 Å, but across the chains it is only about 8 Å. It is this quasi-1D nature of the Si(111) 2 × 1 surface that makes the electron–hole interaction so strongly manifested in the surface optical properties of this system.

Similar calculations for the Ge(111) 2 × 1 reconstructed surface also show very large excitonic effects and demonstrate how optical differential reflectivity spectra

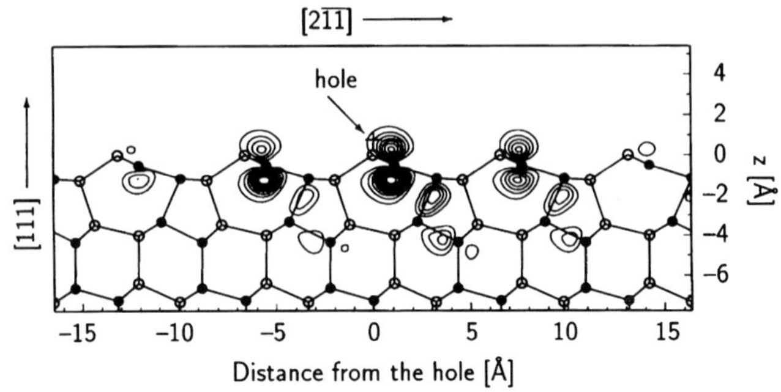

Fig. 24. Contour plots of the electron–hole amplitude squared for the lowest energy spin-singlet exciton on the Si(111) 2 × 1 surface, evaluated on a plane perpendicular to the surface. The hole is fixed at the position indicated by the cross. (After Rohlfing and Louie [49].)

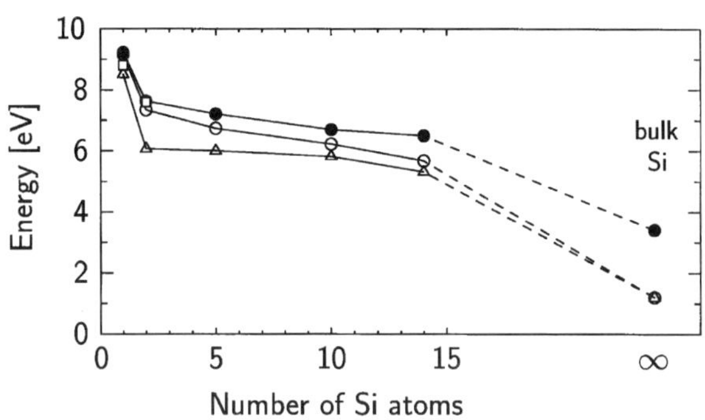

Fig. 25. Excitation energy of the lowest spin triplet (open triangles), spin singlet (open circles) and dipole allowed spin singlet (closed circles) excitations of the Si_mH_n clusters. The experimental data (open squares) are from [94]. (After Rohlfing and Louie [7].)

can be used to distinguish between two possible structural isomers (positive vs. negative buckling of the surface chains) of the reconstructed surface (see right panel of Fig. 23). This distinction has been enabled by the fact that a quantitative comparison between the calculated and experimental spectrum is possible when electron–hole effects are treated correctly [66].

The approach has also been successfully applied to the study of optical excitations in molecules and clusters. As expected, electron–hole interaction effects in these "zero"-dimensional systems are even further enhanced. In the case of the clusters, the combination of quantum confinement and excitonic effects often dramatically change the optical properties. Figure 25 depicts the calculated excitation energies of the various spin singlet and triplet excitations for a series of Si_mH_n clusters [7]. The theoretical results show that the electron–hole interaction energies are strongly size dependent and can be as large as several eVs for the very small clusters. Moreover, excitonic effects often change the shape of the spectrum as compared to the non-interacting case, owing to the mixture of different electron–hole pair configurations in the excited states.

12. SPECTROSCOPIC PROPERTIES OF NANOTUBES – A NOVEL 1D SYSTEM

In this section, we discuss another class of 1D systems of current interest – the carbon [70] and BN [71] nanotubes. Nanotubes are tubular structures with diameters in the range of a nanometer, but with lengths hundreds of microns (to centimeters) long, which possess highly unusual structural and electronic properties [72]. Moreover, extraordinarily strong excitonic effects in their optical properties have been predicted by first-principles theory [73] and subsequently confirmed by experiment [74,75].

Single-walled carbon nanotubes (SWCNTs) can be metals or semiconductors depending sensitively on their geometric structure, which is indexed by a pair of numbers (m, n) where m and n are the two integers specifying the circumferential vector in units of the two primitive translation vectors of graphene [72]. Recent experimental advances have allowed the measurement of the optical response of well-characterized individual, SWCNTs [76–78]. Owing to the reduced dimensionality of the nanotubes, many-electron (both quasiparticle and excitonic) effects [73,79,80] are shown to be extraordinarily important in these systems.

The self-energy corrections to the quasiparticle excitation energies of the carbon nanotubes, particularly for semiconducting tubes, can be very large. Figure 26 shows the GW quasiparticle corrections to the LDA Kohn–Sham energies for the metallic (3, 3) carbon nanotube and the semiconducting (8, 0) carbon nanotube. For the metallic tubes, the corrections are in general relatively straightforward. Essentially, the correction simply stretches the bands by $\sim 20\%$, as in the case of graphite. But, the self-energy corrections to the quasiparticle energies of the semiconducting tubes are more significant. The corrections cause a large opening of the minimum band gap, as well as a stretching of the bands. As seen in Fig. 26, the self-energy corrections cause the minimum quasiparticle gap of the (8, 0) carbon nanotube to open up by nearly 1 eV.

Electron–hole interaction effects play a central role in the optical response of the carbon nanotubes. The calculated optical spectrum of the metallic (3, 3) SWCNT with a diameter of 4 Å is presented in Fig. 27. The left panel shows the quasiparticle density of states. Owing to the symmetry of the states, only certain transitions between states (indicated by the arrow A) are optically allowed. The right panel shows the calculated imaginary part of the macroscopic dielectric response function for the case with and without electron–hole interactions. The optical spectrum of the (3, 3) nanotube is changed qualitatively due to the existence of a bound exciton, even though the system is metallic. This surprising result is a consequence of two factors related to dimensionality: (1) it is much easier to create a bound quantum

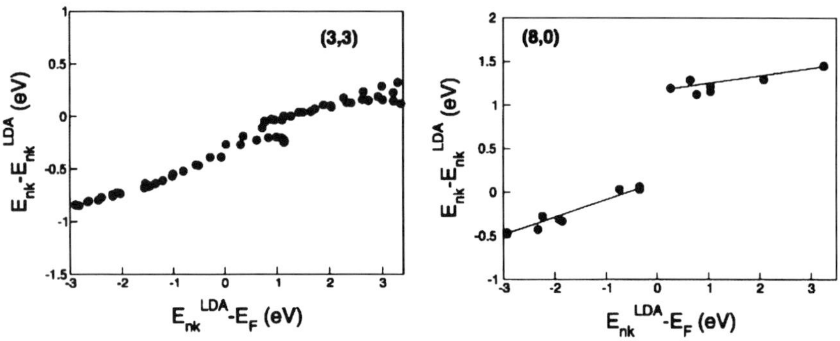

Fig. 26. Quasiparticle corrections to the LDA Kohn–Sham eigenvalues due to self-energy effects as a function of the energy of the states for the metallic (3, 3) carbon nanotube (left panel) and the semiconducting (8, 0) carbon nanotube (right panel). (After Spataru et al. [79].)

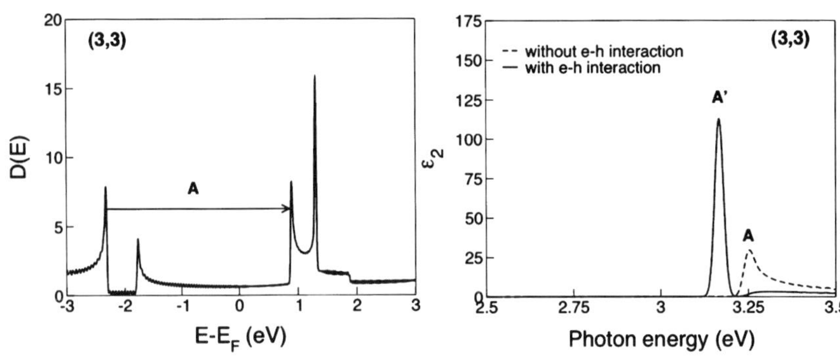

Fig. 27. Calculated quasiparticle density of states (left panel) and optical absorption spectrum (right panel) for the (3, 3) carbon nanotube. (After Spataru et al. [73].)

state in a 1D potential than in a 3D one; and (2) although the tube is a metallic system, there is a symmetry gap in the electron–hole spectrum (i.e., there are no unbound electron–hole states of the same symmetry possible in the energy range of the excitonic state). The symmetry gap is possible here because the (3, 3) tube is a 1D metal – i.e., all **k** states can have well-defined symmetry. To our knowledge, this is the first bound exciton known to exist in a truly metallic system. Calculations on other metallic SWCNT, such as (10, 10) and (12, 0), yield similar bound exciton states, showing that this is a general phenomenon.

Figure 28 depicts the results for a (5, 0) SWCNT, which is another metallic tube of 4 Å in diameter. Another surprise is found. For the range of frequencies considered, the electron–hole interaction in this tube is actually a net repulsion between the excited electron and hole. This is possible because of the two competing terms in the electron–hole interaction kernel discussed above – the attractive direct

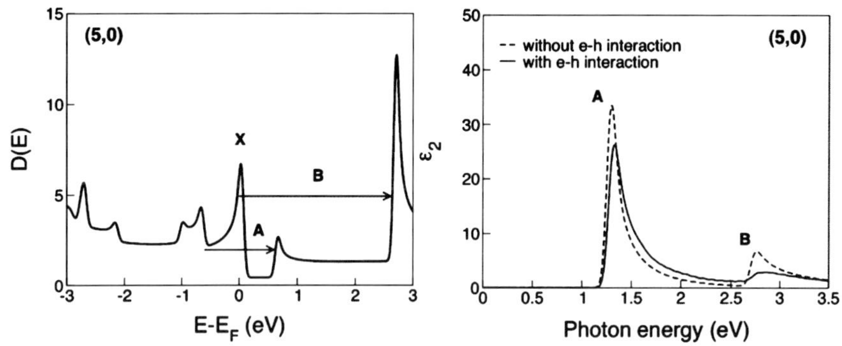

Fig. 28. Calculated quasiparticle density of states (left panel) and optical absorption spectrum (right panel) for the (5, 0) carbon nanotube. (After Spataru et al. [73]; Spataru et al. [79].)

term K_d and the repulsive exchange term K_x. Unlike bulk semiconductors, in this case, owing to the symmetry of the states involved and to the metallic screening, the repulsive exchange term dominates over the attractive direct term. As a consequence, there is no bound exciton state in Fig. 28, but, instead, there is actually a suppression of the optical strength at the van Hove singularities.

Excitonic effects are even more dramatic in the optical response of semiconducting carbon nanotubes. Figure 29 compares the calculated absorption spectrum of a (8, 0) SWCNT between the cases with and without electron–hole interactions. The two resulting spectra are radically different. When electron–hole interaction effects are included, the spectrum is dominated by bound and resonant exciton states. With electron–hole interactions included, each of the structures derived from a van Hove singularity in the non-interacting joint density of states gives rise to a series of exciton states. The bottom two panels in Fig. 29 give the spatial correlation between the excited electron and hole for two of the exciton states, one bound and one resonant state. The extent of the exciton wavefunction is about 20 Å for both of these states. For the (8, 0) tube, the lowest-energy bound exciton has a binding energy of nearly 1 eV. Note that the exciton binding energy for bulk semiconductors of similar size band gap is in general only of the order of tens of meVs, illustrating again the dominance of many-electron Coulomb effects in reduced dimensional systems. Similar results have been obtained for other semiconducting carbon nanotubes [73,79]. These predicted extraordinarily large exciton binding energies, although first met with skepticism, have been verified by recent experiments [74,75]. Owing to the unique electronic structure of the carbon nanotubes, in addition to the optically active (or bright) excitons shown in Fig. 29, there exist also a number of optically inactive (or dark) excitons associated with each of the bright ones. These dark excitons also play an important role in the optical properties of the nanotubes; for example, they strongly affect the radiative lifetime of the excitons in semiconducting carbon nanotubes [81].

These first-principles studies have given researchers a new paradigm in thinking about optical excitations in nanotubes as extraordinarily strong excitons. They have

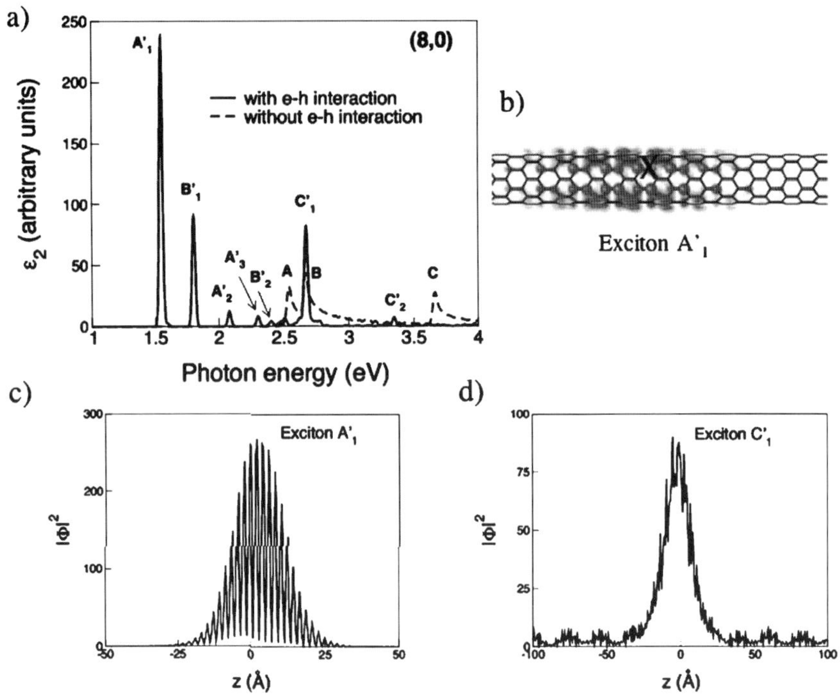

Fig. 29. (a) Optical absorption spectra for the (8, 0) carbon nanotube. (b) Spatial extent of the excitonic wavefunction (square of electron–hole amplitude) in real space with hole position fixed. (c) Square of electron–hole amplitude along the tube axis for a bound exciton state. (d) Same as (c) but for a resonant exciton state. (After Spataru et al. [79].)

provided predictions and yielded quantitative results that explain experiments in detail. For example, samples of well-aligned SWCNTs with uniform diameter of 4 Å have been obtained by growing them in the channels of zeolite ($AlPO_4$-5) crystals. Three prominent peaks were observed in the optical absorption spectrum of such samples [76]. Table 5 compares the calculated optical excitation energies to the experimental data. Given a diameter of 4 Å, there are only three possible structures for the carbon nanotubes – (5, 0), (4, 2) and (3, 3), which are all expected to be present in the samples. As seen from the table, the first-principles results quantitatively explain all three observed peaks and identify their physical origin. The first absorption peak (at 1.37 eV) is due to interband transitions from the (5, 0) tubes, with the feature arising from a 1D van Hove singularity in the joint density of states. The second peak (at 2.1 eV) and third peak (at 3.1 eV) are due to the formation of excitons in the (4, 2) and (3, 3) tubes, respectively [73,82].

The theoretical work on larger diameter semiconducting SWCNTs [73,79] has elucidated the findings of photoluminescence excitation measurements, which yield detailed information on optical transitions in well-characterized individual single-walled nanotubes. Table 6 gives a comparison between experiment and theory for

Table 5. Comparison between experimental [76] and calculated [73,67] main absorption peak energies for all possible 4 Å carbon nanotubes – (5, 0), (4, 2) and (3, 3).

CNT	Theory (eV)	Experiment (eV)	Character
(5, 0)	1.33	1.37	Interband
(4, 2)	2.0	2.1	Exciton
(3, 3)	3.17	3.1	Exciton

Table 6. Calculated lowest two-optical transition energies for the (8, 0) and (11, 0) carbon nanotubes [73] compared to experimental values [77,78].

	(8, 0)		(11, 0)	
	Experiment (eV)	Theory (eV)	Experiment (eV)	Theory (eV)
E_{11}	1.6	1.6	1.2	1.1
E_{22}	1.9	1.8	1.7	1.6
E_{22}/E_{11}	1.19	1.13	1.42	1.45

the particular cases of the (8, 0) and (11, 0) tubes. The measured transition energies are in excellent agreement with theory. There are no adjustable parameters in the calculation – the only inputs being the geometric structure and the atomic number of carbon. Calculations done on a number of other carbon nanotubes have obtained similar agreement. There are several important physical effects that go into these transition energies – namely, the band structure, the quasiparticle self energies and the excitonic effects. Each of these effects can give rise to significant changes in the excitation energies, with shifts equal to a large fraction of an eV. In fact, the many-electron effects alter the basic character of the transitions. Thus, one must include all these factors to have an understanding of the optical response of carbon nanotubes. Further, the inclusion of excitonic effects is crucial to the understanding of phenomena such as the radiative lifetime of the photo-excited states of the carbon nanotubes [81].

We find that excitonic effects are even stronger in the Boron–Nitride (BN) nanotubes [83]. Figure 30a shows the calculated optical spectra of an (8, 0) BN nanotube. Again, the electron–hole interaction leads to a series of sharp lines in the absorption spectrum due to strongly bound excitonic states, and the optical strength is virtually completely shifted to these states away from the continuum part of the spectrum. For the (8, 0) BN tube, which has the same diameter as the (8, 0) SWCNT, the exciton binding energy of the lowest energy exciton is over 2 eV. This is consistent with the fact that BN nanotubes are wide gap insulators. Unlike the carbon case, the lowest-energy exciton here is composed almost in equal weight of four sets of interband transitions, as indicated in Fig. 30b. An examination of the exciton wavefunction shows that the spatial extent of the lowest energy exciton in a (8, 0) BN

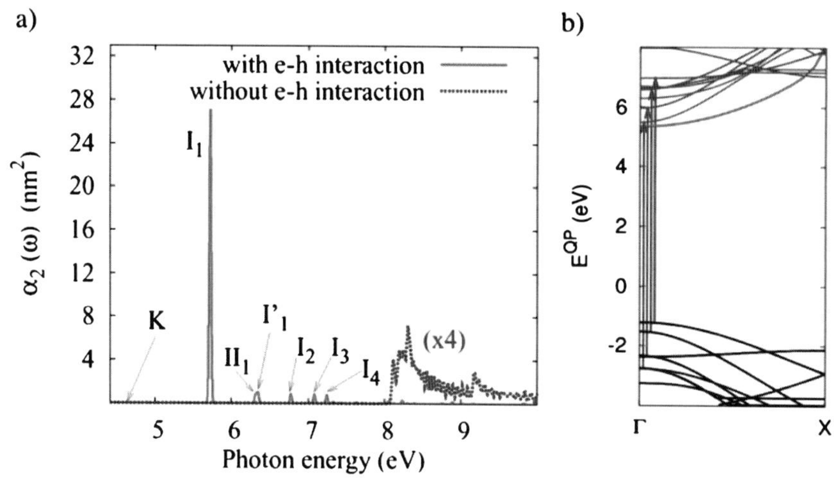

Fig. 30. (a) Absorption spectrum of the (8, 0) single-walled BN nanotube. The imaginary part of the polarizability per tube is shown, with a Gaussian broaden factor of 0.0125 eV. (b) Different sets of interband transitions contributing to the lowest-energy excitons are indicated. (After Park et al. [83].)

nanotube has a root-mean-square radius along the tube axis of only 3.7 Å. This is to be compared to a root-mean-square radius of 8.6 Å for the (8, 0) carbon nanotube. Thus, as a result of the larger-binding energy, the exciton in BN has a much smaller spatial extent. Further, we find that the exciton wavefunction for the BN case does not even extend around the circumference of the tube. This spatial localization of the electron–hole separation around the circumference is a consequence of the mixing of different subband transitions in forming the exciton state.

13. SUMMARY AND PERSPECTIVES

We have presented in this Chapter some of the basic concepts and theoretical developments in the understanding and calculation of the ground- and excited-state properties of materials. Since the properties of condensed matter are fundamentally dictated by the behavior of the electrons, advances in electronic structure theory have been of key importance. The developments of density-functional formalism and *ab initio* pseudopotentials, together with progress in computational methodologies, have allowed accurate determination of the ground-state properties of many materials from first principles. A parallel advance has been that of a first-principles Green's function approach to electron excited-state properties based on the GW approximation to the electron self-energy. The latter development has provided the proper theoretical framework and inclusion of relevant many-electron interaction effects in the description of the excited states, resulting in an accurate description of the spectroscopic properties of real materials. These are versatile approaches with predictive power. Their applications have helped in elucidating our conceptual and quantitative understanding of a wide range of material systems and are expected to continue to play a central role in future materials studies.

At present, our ability to address phenomena involving large numbers of atoms (e.g., tens of thousands or more atoms) and very long time scales from first principles is still limited. Bridging the length and time scales thus remains a major challenge (see Chapters 3 and 4 of this volume). With the available approximate exchange-correlation functionals, DFT energies have yet to reach the desired accuracy for some important applications such as chemical accuracy in the reaction path and barrier height in chemical reactions. Also, for highly correlated electron systems such as the transition metal oxides and the high T_c superconducting cuprates, local spin density approximation (LSDA) often gives incorrect ground states. The construction of better exchange-correlation functionals is needed and is currently an active and important endeavor in the field.

A related topic is the use of time-dependent density functional theory (TDDFT) for excited-stated properties [12] (see Chapter 4 of this volume). In principle, TDDFT should be able to provide exact results for certain classes of spectroscopic properties such as the optical response. However, current implementations of TDDFT (e.g., at the time-dependent LDA level) work well only for finite systems, such as molecules and small clusters, but fail for larger or extended systems owing to our lack of accurate knowledge of the time-dependent exchange-correlation functional. For ground-state properties, alternatives to DFT have also been attempted. A promising approach is that of quantum Monte Carlos simulations [84,85] which in principle can provide a systematic scheme to arrive at the exact ground state and related quantities, although one now is back to dealing with the full many-electron wavefunction.

For many systems, upon promotion to an excited state, structural change is an important and integral part of their response to excitation. Familiar examples include Stokes shifts in optical spectra, self-trapped excitons, molecular or defect conformation changes, photo-induced desorption on surfaces, and so on. Recent development in calculating forces on atoms in the excited state within the GW–BSE formalism [86] has given promise to facililating first-principles study of these phenomena. Another topic of considerable current interest is the employment of first-principles Green's function techniques such as the GW–BSE approach to study the excited-state properties of correlated electron systems from *ab initio*. Since such studies are based on perturbation theory, having a good mean-field solution to the problem is essential. Combining the GW approach with methods such as LDA + U [87] or dynamical mean-field approximations [88] would likely be a fruitful avenue to pursue.

ACKNOWLEDGMENTS

This work was supported by National Science Foundation Grant No. DMR04-39768 and by the director, Office of Science, Office of Basic Energy Sciences, Division of Materials Sciences and Engineering, U.S. Department of Energy under Contract No. DE-AC02-05CH11231.

REFERENCES

[1] P. Hohenberg and W. Kohn, Inhomogeneous electron gas, *Phys. Rev. B* **136**(3B), 864 (1964).
[2] W. Kohn and L. Sham, Self-consistent equations including exchange and correlation effects, *Phys. Rev.* **140**, A1133 (1965).
[3] L. Hedin, New method for calculating the one-particle Green's function with application to the electron-gas problem, *Phys. Rev.* **139**, A796 (1965).
[4] L. Hedin and S. Lundqvist, Effects of electron–electron and electron–phonon interactions on the one electron states of solids, *Solid State Phys.* **23**, 1 (1969).
[5] M.S. Hybertsen and S.G. Louie, First-principles theory of quasiparticles: Calculation of band gaps in semiconductors and insulators, *Phys. Rev. Lett.* **55**, 1418 (1985).
[6] M.S. Hybertsen and S.G. Louie, Electron correlation in semiconductors and insulators: Band gaps and quasiparticle energies, *Phys. Rev. B* **34**, 5390 (1986).
[7] M. Rohlfing and S.G. Louie, Excitonic effects and the optical absorption spectrum of hydrogenated Si clusters, *Phys. Rev. Lett.* **80**, 3320 (1998).
[8] M. Rohlfing and S.G. Louie, Electron–hole excitations in semiconductors and insulators, *Phys. Rev. Lett.* **81**, 2312 (1998).
[9] S. Albretch, L. Reining, R.D. Sole and G. Onida, *Ab initio* calculation of excitonic effects in the optical spectra of semiconductors, *Phys. Rev. Lett.* **80**, 4510 (1998).
[10] L.X. Benedict, E.L. Shirley and R.B. Bohm, Optical absorption of insulators and the electron–hole interaction: An *ab initio* calculation, *Phys. Rev. Lett.* **80**, 4514 (1998).
[11] L.P. Kadanoff and G. Baym, *Quantum Statistical Mechanics: Green's Function Methods in Equilibrium and Nonequilibrium Problems* (Perseus Books, New York, 1999).
[12] R. Dreizler and E. Gross, *Density Functional Theory* (Plenum Press, New York, 1995).
[13] R.G. Parr and W. Yang, *Density-Functional Theory of Atoms and Molecules* (Oxford University Press, New York, 1989).
[14] E. Fermi, On the pressure shift of the higher levels of a spectral line series, *Nuovo Cimente* **11**, 157 (1934).
[15] J.C. Phillips and L. Kleinman, New method for calculating wave functions in crystals and molecules, *Phys. Rev.* **116**, 287 (1959).
[16] W.E. Pickett, Pseudopotential methods in condensed matter applications, *Comput. Phys. Rep.* **9**, 115 (1989).
[17] D. Hamann, M. Schluter and C. Chiang, Norm-conserving pseudopotentials, *Phys. Rev. Lett.* **43**, 1494 (1979).
[18] M.L. Cohen, Pseudopotentials and total energy calculations, *Phys. Scripta* **T1**, 5 (1982).
[19] M.T. Yin and M.L. Cohen, Theory of static structural properties crystal stability, and phase transformations: Application to Si and Ge, *Phys. Rev. B* **26**, 5568 (1982).
[20] C. Kittel and H. Kroemer, *Thermal Physics*, 2nd ed. (Freeman, San Francisco, 1980).
[21] M.L. Cohen, M. Schluter, J.R. Chelikowsky and S.G. Louie, Self-consistent pseudopotential method for localized configurations: Molecules, *Phys. Rev. B* **12**, 5575 (1975).
[22] J.E. Northrup, M.S. Hybertsen and S.G. Louie, Many-body calculation of the surface-state energies for Si(111) 2×1, *Phys. Rev. Lett.* **66**, 500 (1991).
[23] F.J. Himpsel, P.M. Marcus, R.M. Tromp, I.P. Batra, M. Cook, F. Jona and H. Liu, Structure-analysis of Si(111) 2×1 with low-energy electron diffraction, *Phys. Rev. B* **30**, 2257 (1984).
[24] M.T. Yin and M.L. Cohen, Microscopic theory of the phase-transformation and lattice-dynamics of Si, *Phys. Rev. Lett.* **45**, 1004 (1980).
[25] P. Giannozzi, S. Degironcoli, P. Pavone and S. Baroni, *Ab-initio* calculations of phonon dispersions in semiconductors, *Phys. Rev. B* **43**, 7231 (1991).
[26] J.R. Schrieffer, *Theory of Superconductivity* (Perseus, 1999).
[27] G.M. Eliashberg, Interactions between electrons and lattice vibrations in a superconductor, *Sov. Phys. JETP [Engl. Transl.]* **11**, 696 (1960).
[28] J. Nagamatsu, N. Nakagawa, T. Muranaka, Y. Zenitani and J. Akimitsu, Superconductivity at 39 K in magnesium diboride, *Nature* **410**, 63 (2001).

[29] C. Buzea and T. Yamashita, Review of the superconducting properties of MgB_2, *Superconductor Sci. Technol.* **14**(11), R115–R146 (2001).
[30] Y. Wang, T. Plackowski and A. Junod, Specific heat in the superconducting and normal state (2-300K, 0-16 T), and magnetic susceptibility of the 38 K superconductor MgB_2, *Physica C* **355**, 179 (2001).
[31] F. Bouquet, R. Fisher, N.E. Phillips, D.G. Hinks and J.D. Jorgensenm, Specific heat of MgB_2: Evidence for a second energy gap, *Phys. Rev. Lett.* **87**, 47001 (2001).
[32] H.D. Yang, J.-Y. Lin, H.H. Li, F.H. Hsu, C.J. Liu, S.-C. Li, R.-C. Yu and C.-Q. Jin, Order parameter of MgB_2: A fully gapped superconductor, *Phys. Rev. Lett.* **87**, 167003 (2001).
[33] H.J. Choi, D. Roundy, H. Sun, M.L. Cohen and S.G. Louie, The origin of the anomalous superconducting properties of MgB_2, *Nature* **418**, 758 (2002).
[34] H.J. Choi, D. Roundy, H. Sun, M.L. Cohen and S.G. Louie, First-principles calculation of the superconducting transition in MgB_2 within the anisotropic Eliashberg formalism, *Phys. Rev. B* **66**, 20513 (2002).
[35] J. Kortus, I. Mazin, K. Belashchenko, V. Antropov and L. Boyer, Superconductivity of metallic boron in MgB_2, *Phys. Rev. Lett.* **86**, 4656 (2001).
[36] J.M. An and W.E. Pickett, Superconductivity of MgB_2: Covalent bonds driven metallic, *Phys. Rev. Lett.* **86**, 4366–4369 (2001).
[37] K.-P. Bohnen, R. Heid and B. Renker, Phonon dispersion and electron–phonon coupling in MgB_2 and AlB_2, *Phys. Rev. Lett.* **86**, 5771–5774 (2001).
[38] H.J. Choi, M.L. Cohen and S.G. Louie, Anisotropic Eliashberg theory of MgB_2: T_c, isotope effects, superconducting energy gaps, quasiparticles, and specific heat, *Physica C* **385**, 66 (2003).
[39] G. Strinati, Application of the Green's-function method to the study of the optical-properties of semiconductors, *Riv. Nuovo Cimento* **11**(12), 1 (1988).
[40] G.D. Mahan, *Many-Particle Physics* (Plenum, New York, 1981).
[41] A. Seidl, A. Goerling, P. Vogl, J.A. Majewski and M. Levy, Generalized Kohn–Sham schemes and the band-gap problem, *Phys. Rev. B* **53**, 3764 (1996).
[42] M.S. Hybertsen and S.G. Louie, *Ab initio* static dielectric matrices from the density-functional approach. I. Formulation and application to semiconductors and insulators, *Phys. Rev. B* **35**, 5585 (1987).
[43] M.S. Hybertsen and S.G. Louie, *Ab initio* static dielectric matrices from the density-functional approach. II. Calculation of the screening response in diamond, Si, Ge, and LiCl, *Phys. Rev. B* **35**, 5602 (1987).
[44] B. Holm and U. von Barth, Fully self-consistent GW self-energy of the electron gas, *Phys. Rev. B* **57**, 2108 (1998).
[45] F. Aryasetiawan and O. Gunnarsson, GW method, *Rep. Prog. Phys.* **61**, 3 (1998).
[46] W.G. Aulbur, L. Jnsson and J. Wilkins, Quasiparticle calculations in solids, *Solid State Phys.* **54**, 1 (2000).
[47] J.E. Ortega and F.J. Himpsel, Inverse-photoemission study of Ge(100), Si(100) and GaAs(100) – bulk bands and surface states, *Phys. Rev. B* **47**, 2130 (1993).
[48] A. Marini, G. Onida and R.D. Sole, Quasiparticle electronic structure of Copper in the GW approximation, *Phys. Rev. Lett.* **88**, 16403 (2002).
[49] M. Rohlfing and S.G. Louie, Excitons and optical spectrum of the Si(111)-(2 × 1) surface, *Phys. Rev. Lett.* **83**, 856 (1999).
[50] F.J. Himpsel, P. Heimann and D. Eastman, Surface states on Si(111)-(2 × 1), *Phys. Rev. B* **24**, 2003 (1981).
[51] F. Houzay, G.M. Guichar, R. Pinchaux and Y. Petroff, Electronic states of Si(111) surfaces, *J. Vac. Sci. Technol.* **18**, 860 (1981).
[52] R.I.G. Uhrberg, G.V. Hansson, J.M. Nicholls and S.A. Flodstrom, Experimental evidence for one highly dispersive dangling-bond band on Si(111) 2 × 1, *Phys. Rev. Lett.* **48**, 1032 (1982).
[53] P. Perfetti, J.M. Nicholls and B. Reihl, Unoccupied surface-state band on Si(111) 2 × 1, *Phys. Rev. B* **36**, 6160 (1987).

[54] R.M. Feenstra, Scanning tunneling microscopy and spectroscopy of cleaved and annealed GE(111) surfaces, *Surf. Sci.* **251**, 401 (1991).
[55] M.A. Olmstead and N.M. Amer, Direct measurement of the polarization dependence of Si(111) 2 × 1 surface-state absorption by use of photothermal displacement spectroscopy, *Phys. Rev. Lett.* **52**, 1148 (1984).
[56] J. Bokor, R. Storz, R.R. Freeman and P.H. Bucksbaum, Picosecond surface electron dynamics on photoexcited Si(111) (2 × 1) surfaces, *Phys. Rev. Lett.* **57**, 881 (1986).
[57] F. Ciccacci, S. Selci, G. Chiarotti and P. Chiaradia, Electron–phonon interaction in optical absorption at the Si(111) 2 × 1 surface, *Phys. Rev. Lett.* **56**, 2411 (1986).
[58] P.M. Echenique, J.M. Pitarke, E. Chulkov and A. Rubio, Theory of inelastic lifetimes of low-energy electrons in metals, *Chem. Phys.* **251**, 1 (2000).
[59] J. Cao, Y. Gao, H.E. Elsayed-Ali, R.D.E. Miller and D.A. Mantell, Femtosecond photoemission study of ultrafast dynamics in single-crystal Au(111) films, *Phys. Rev. B* **50**, 10948 (1998).
[60] K.A. Goettel, J.H. Eggert, I.F. Silvera and W.C. Moss, Optical evidence for the metallization of Xenon at 132(5) GPa, *Phys. Rev. Lett.* **62**, 665 (1989).
[61] R. Reichlin, K.E. Brister, A.K. McMahan, M. Ross, S. Martin, Y.K. Vohra and A.L. Ruoff, Evidence for the insulator–metal transition in Xenon from optical, X-ray, and band-structure studies to 170 GPa, *Phys. Rev. Lett.* **62**, 669 (1989).
[62] H. Chacham, X. Zhu and S.G. Louie, Metal–insulator transition in solid xenon at high-pressures, *Europhys. Lett.* **14**, 65 (1991).
[63] H. Chacham, X. Zhu and S.G. Louie, Pressure-induced insulator–metal transitions in solid xenon and hydrogen: A first-principles quasiparticle study, *Phys. Rev. B* **46**, 6697 (1992).
[64] M.P. Surh, H. Chacham and S.G. Louie, Quasiparticle excitation energies for the F-center defect in LiCl, *Phys. Rev. B* **51**, 7464 (1995).
[65] H. Philipp, Optical transitions in crystalline and fused quartz, *Solid State Commun.* **4**, 73 (1966).
[66] M. Rohlfing and S.G. Louie, Electron–hole excitations and optical spectra from first principles, *Phys. Rev. B* **62**, 4927 (2000).
[67] E.K. Chang, M. Rohlfing and S.G. Louie, Excitons and optical properties of quartz, *Phys. Rev. Lett.* **85**, 2613–2616 (2000).
[68] M. Rohlfing and S.G. Louie, Optical excitations in conjugated polymers, *Phys. Rev. Lett.* **82**, 1959 (1999).
[69] P. Chiaradia, A. Cricenti, S. Selci and G. Chiarotti, Differential reflectivity of Si(111) 2 × 1 Surface with polarized light: A test for surface structure, *Phys. Rev. Lett.* **52**, 1145 (1984).
[70] S. Iijima, Helical microtubules of graphitic carbon, *Nature* **354**, 56 (1991).
[71] N.G. Chopra, R.J. Luyken, K. Cherrey, V.H. Crespi, M.L. Cohen, S.G. Louie and A. Zettl, Boron–Nitride nanotubes, *Science* **269**, 966 (1995).
[72] M.S. Dresselhaus, G. Dresselhaus, P. Avouris (Eds), *Carbon Nanotubes* (Springer, New York, 2000).
[73] C.D. Spataru, S. Ismail-Beigi, L.X. Benedict and S.G. Louie, Excitonic effects and optical spectra of single-walled Carbon nanotubes, *Phys. Rev. Lett.* **92**, 077402 (2004).
[74] F. Wang, G. Dukovic, L.E. Brus and T.F. Heinz, The optical resonances in carbon nanotubes arise from excitons, *Science* **308**, 838 (2005).
[75] Y.Z. Ma, L. Valkunas, S.M. Bachilo and G.R. Fleming, Exciton binding energy in semiconducting single-walled carbon nanotubes, *J. Phys. Chem. B* **109**, 15671 (2005).
[76] Z. Li, Z. Tang, H. Liu, N. Wang, C. Chan, R. Saito, S. Okada, G. Li, J. Chen, N. Nagasawa and S. Tsuda, Polarized absorption spectra of single-walled 4 Å Carbon nanotubes aligned in channels of an $AlPO_4$-5 single crystal, *Phys. Rev. Lett.* **87**, 127401 (2001).
[77] M. O'Connell, S. Bachilo, C. Huffman, V. Moore, M. Strano, E. Haroz, K. Rialon, P. Boul, W. Noon, C. Kittrell, J. Ma, R. Hauge, B. Weisman and R. Smalley, Band gap fluorescence from individual single-walled Carbon nanotubes, *Science* **297**, 593 (2002).
[78] S. Bachilo, M. Strano, C. Kittrell, R. Hauge, R. Smalley and R. Weisman, Structure-assigned optical spectra of single-walled carbon nanotubes, *Science* **298**, 2361 (2002).

[79] C.D. Spataru, S. Ismail-Beigi, L.X. Benedict and S.G. Louie, Quasiparticle energies, excitonic effects and optical absorption spectra of small-diameter single-walled Carbon nanotubes, *Appl. Phys. A* **78**, 1129 (2004).
[80] T. Ando, Excitons in carbon nanotubes, *J. Phys. Soc. Jpn* **66**, 1066 (1997).
[81] C.D. Spataru, S. Ismail-Beigi, R.B. Capaz and S.G. Louie, Theory and *ab initio* calculation of radiative lifetime of excitons in semiconducting carbon nanotubes, *Phys. Rev. Lett.* **95**, 247402 (2005).
[82] E. Chang, G. Bussi, A. Ruini and E. Molinari, Excitons in carbon nanotubes: An *ab initio* symmetry-based approach, *Phys. Rev. Lett.* **92**, 196401 (2004).
[83] C.-H. Park, C.D. Spataru and S.G. Louie, Excitons and many-electron effects in the optical response of single-walled boron nitride nanotubes, *Phys. Rev. Lett.* **96**, 126105 (2006).
[84] S. Fahy, X.W. Wang and S.G. Louie, Variational quantum Monte–Carlo nonlocal pseudopotential approach to solids – formulation and application to diamond, graphite, and silicon, *Phys. Rev. B* **42**, 3503 (1990).
[85] W.M.C. Foulkes, L. Mitas, R.J. Needs and G. Rajagopal, Quantum Monte–Carlo simulations of solids, *Rev. Mod. Phys.* **73**, 33–83 (2001).
[86] S. Ismail-Beigi and S.G. Louie, Excited-state forces within a first-principles Green's function formalism, *Phys. Rev. Lett.* **90**, 076401 (2003).
[87] V.I. Anisimov, J. Zaanen and O.K. Andersen, Band theory and Mott insulators – Hubbard-U instead Stoner-I, *Phys. Rev. B* **44**, 943 (1991).
[88] A. Georges, G. Kotliar, W. Krauth and M.J. Rozenberg, Dynamical mean-field theory of strongly correlated fermion systems and the limit of infinite dimensions, *Rev. Mod. Phys.* **68**, 13–125 (1996).
[89] I. Campillo, J.M. Pitarke, A. Rubio, E. Zarate and P.M. Echenique, Inelastic lifetimes of hot electrons in real metals, *Phys. Rev. Lett.* **83**, 2230 (1999).
[90] I. Campillo, A. Rubio, J.M. Pitarke, A. Goldmann and P.M. Echenique, Hole dynamics in noble metals, *Phys. Rev. Lett.* **85**, 3241 (1999).
[91] D.E. Aspnes and A. Studna, Dielectric functions and optical parameters of Si, Ge, GaP, GaAs, GaSb, InP, InAs and InSb from 1.5 to 6.0 eV, *Phys. Rev. B* **27**, 985 (1983).
[92] P. Lautenschlager, M. Garriga, S. Logothetidis and M. Cardona, Interband critical points of GaAs and their temperature dependence, *Phys. Rev. B* **35**, 9174 (1987).
[93] M. Rohlfing, M. Palummo, G. Onida and R.D. Sole, Structural and optical properties of the Ge(111)-(2 $\times$ 1) surface, *Phys. Rev. Lett.* **85**, 5440 (2000).
[94] U. Itoh, Y. Toyoshima and H. Onuki, Vacuum ultraviolet absorption cross sections of SiH, GeH, SiH, and SiH, *J. Chem. Phys.* **85**, 4867 (1986).
[95] S.Y. Savrasov, D.Y. Savrasov and O.K. Andersen, Linear-response calculations of electron–phonon interactions, *Phys. Rev. Lett.* **72**(3), 372 (1994).
[96] D. Sell, Resolved free-exciton transitions in the optical-absorption spectrum of GaAs, *Phys. Rev. B* **6**, 3750 (1972).
[97] J.S. Michaelis, K. Unterrainer, E. Gornik and E. Bauser, Electric and magnetic dipole two-photon absorption in semiconductors, *Phys. Rev. B* **54**, 7917 (1996).
[98] C. Kittel, *Introduction to Solid state Physics*, 8th ed. (Wiley, hoboken, NJ, 2005).

Chapter 3
AB INITIO MOLECULAR DYNAMICS: DYNAMICS AND THERMODYNAMIC PROPERTIES

R. Car

1. MOLECULAR DYNAMICS

Molecular dynamics is a powerful technique to simulate classical many-body systems [1]. It amounts to solving numerically the equations of motion. Observable properties are calculated as temporal averages over the trajectories. The averages allow us to calculate correlation functions for static and dynamic properties. Molecular dynamics simulations have become possible with the advent of high-speed digital computers, and, from the early pioneering papers, molecular dynamics simulations have gained vast popularity and are now common in physics, chemistry, materials science, and biochemistry/biophysics [2].

Molecular dynamics assumes that the atoms behave like classical particles. Given the value of the atomic masses this is usually a good approximation in materials and molecules when the temperature is not too low. Typically, in solids the temperature should be higher than Debye's temperature, in liquids the thermal wavelength should be smaller than the shortest interparticle separations. The equations of motion are derived from a Lagrangian $L(\{\mathbf{R}(t)\}, \{\dot{\mathbf{R}}(t)\})$ that depends on the particle coordinates $\{\mathbf{R}\} \equiv \{\mathbf{R}_1, \mathbf{R}_2, \mathbf{R}_3, \ldots, \mathbf{R}_N\}$ and velocities $\{\dot{\mathbf{R}}\} \equiv \{\dot{\mathbf{R}}_1, \dot{\mathbf{R}}_2, \dot{\mathbf{R}}_3, \ldots, \dot{\mathbf{R}}_N\}$ at time t:

$$L(\{\mathbf{R}\}, \{\dot{\mathbf{R}}\}) = K(\{\dot{\mathbf{R}}\}) - \Phi(\{\mathbf{R}\}) \tag{1}$$

Here, $K(\{\dot{\mathbf{R}}\}) = \frac{1}{2}\sum_{I=1,N} M_I \dot{\mathbf{R}}_I^2$ is the kinetic energy, M_I are the atomic masses, $\Phi(\{R\})$ the potential energy, and N the number of particles. The equations of

Contemporary Concepts of Condensed Matter Science
Conceptual Foundations of Materials: A Standard Model for Ground- and Excited-State Properties

ISSN: 1572-0934/doi:10.1016/S1572-0934(06)02003-8

motion follow from

$$\frac{\mathrm{d}}{\mathrm{d}t}\frac{\partial L}{\partial \dot{\mathbf{R}}_I} - \frac{\partial L}{\partial \mathbf{R}_I} = 0 \,(I = 1, N) \tag{2}$$

This gives

$$M_I \ddot{\mathbf{R}}_I = -\frac{\partial \Phi}{\partial \mathbf{R}_I} \; (I = 1, N) \tag{3}$$

These are N-coupled ordinary second-order differential equations in time, the Newton's equations, which conserve the total energy $H = K + \Phi$. If the dynamics is *ergodic*, equilibrium statistical mechanics follows [1]. Statistical properties are associated to temporal averages taken along the trajectories generated by Eq. (3). For example, the temperature is related to the average kinetic energy of the particles via

$$T = \frac{2}{3Nk_\mathrm{B}} \bar{K} \tag{4}$$

Here, the bar indicates temporal average: $\bar{K} = \lim_{\tau\to\infty} 1/\tau \int_0^\tau K(\{\dot{\mathbf{R}}(t)\})\,\mathrm{d}t$. In practice, $\bar{K}$ is approximated by $1/\tau \int_0^\tau K(\{\dot{\mathbf{R}}(t)\})\,\mathrm{d}t$, where τ is a sufficiently long time interval, i.e. it is a time interval longer than the relaxation time for the physical property under consideration. More generally, correlation functions [1,3] are given by temporal averages like

$$C_{AB}(t) = \overline{A(t)B(0)} \approx \frac{1}{\tau}\int_0^\tau A(t+t')B(t')\,\mathrm{d}t' \tag{5}$$

Here $A(t) \equiv A(\{\mathbf{R}(t)\}, \{\dot{\mathbf{R}}(t)\})$ and $B(t) \equiv B(\{\mathbf{R}(t)\}, \{\dot{\mathbf{R}}(t)\})$ are physical properties (observables) at time t. Common examples of correlation functions are the velocity autocorrelation function $C_{vv}(t) = \overline{1/N\sum_{I=1,N} \dot{\mathbf{R}}_I(t)\cdot\dot{\mathbf{R}}_I(0)}$ and the 2-particle distribution function $\rho_N^{(2)}(\mathbf{R},\mathbf{R}') = \overline{\sum_{I\neq J}^{N} \delta(\mathbf{R}-\mathbf{R}_I)\delta(\mathbf{R}'-\mathbf{R}_J)}$. The velocity autocorrelation function is a two-point time correlation function, associated to a *dynamic* property of the system. Its power spectrum $\tilde{C}_{vv}(\omega) = 1/2\pi \int_{-\infty}^{\infty} C_{vv}(t)\mathrm{e}^{\mathrm{i}\omega t}\,\mathrm{d}t = 1/\pi \int_0^\infty C_{vv}(t)\cos(\omega t)\,\mathrm{d}t$ gives spectral information on the single-particle dynamics. In a fluid, the time integral of $C_{vv}(t)$ gives the diffusion coefficient: $D = \frac{1}{3}\int_0^\infty C_{vv}(t)\,\mathrm{d}t$. On the other hand, the 2-particle distribution function is an equal time correlation function, and we omit the time-dependence in its definition. The 2-particle distribution function is associated to a *static* property of the system, directly related to the pair correlation function that one extracts from diffraction experiments. So far, the particle number N, or equivalently the size of the system, has not been specified. The thermodynamic limit, $N \to \infty$, is clearly a condition that cannot be realized in numerical simulations. Fortunately, it is usually not necessary to model on a computer an exceedingly large number of particles to get equilibrium statistical information. Away from critical points, spatial correlations have finite range. It is then sufficient to consider a finite system having a size larger than the relevant spatial correlations. To avoid surface effects in bulk systems periodic boundary conditions are adopted, i.e. a periodic cell is the common choice of simulation cell.

From the above brief discussion, it should be evident that molecular dynamics is a numerical implementation of Boltzmann's formulation of equilibrium statistical mechanics [4]. A crucial condition for the validity of the whole procedure is ergodicity [1,4]. This is more easily achieved in fluid systems. In simple liquids, typical relaxation times are of the order of a picosecond and spatial correlations do not extend beyond a few shells of neighbors: these are time spans and spatial dimensions that are well within the reach of computer simulations. Not surprisingly, molecular dynamics simulations play a particularly important role in liquid-state physics. Liquids are disordered systems for which Bloch's theorem does not hold, but the finite range of interparticle correlations makes a periodic cell approach meaningful. Liquid dynamics is strongly anharmonic and approximations commonly adopted in solid-state physics, like harmonic lattice dynamics, cannot be used. On the other hand, strong anharmonicity leads to rapid establishment of thermal equilibrium. Under these circumstances, molecular dynamics is the most accurate available computational approach. Modern applications of molecular dynamics extend well beyond the physics of liquids and solids, to complex molecular systems, and chemical reaction dynamics. In situations where the anharmonicity is small, thermal equilibrium can be enforced by special thermostatting techniques. Under ergodic conditions temporal averages along the trajectories generated by Eq. (3) are equivalent to ensemble averages, i.e. to averages calculated according to Gibbs' formulation of equilibrium statistical mechanics [4]. The ensemble corresponding to the phase space points visited along the trajectories generated by Eq. (3) is the microcanonical ensemble (N, Ω, H), in which the particle number N, the volume Ω, and the energy H are fixed. The average value of an observable $A(\{\mathbf{R}\}, \{\dot{\mathbf{R}}\})$ satisfies:

$$\overline{A(\{\mathbf{R}\}, \{\dot{\mathbf{R}}\})} = \langle A(\{\mathbf{R}\}, \{\dot{\mathbf{R}}\})\rangle_{N\Omega H} \tag{6}$$

Here we use the brackets $\langle \cdots \rangle$ to indicate ensemble average and the subscript refers to the microcanonical ensemble. By suitably modifying the equations of motion it is possible to sample different ensembles, like the isobaric-isoenthalpic ensemble (N, P, H_P) where P is the pressure and $H_P = H + P\Omega$ the enthalpy, the canonical ensemble (N, Ω, T), and the isobaric-isothermal ensemble (N, P, T). In microcanonical simulations, the volume and the energy are fixed control parameters, while the pressure and the temperature fluctuate. In isobaric-isothermal simulations, pressure and temperature are fixed control parameters, while volume and energy fluctuate. Thus isobaric-isothermal simulations are closer to experimental conditions in which one typically controls pressure and temperature. Two general methodologies allow us to sample different ensembles by molecular dynamics. One is the extended Lagrangian approach [1], in which one specifies the Lagrangian of an extended dynamical system that includes few additional variables (like e.g. Ω and $\dot{\Omega}$ in constant pressure simulations) in addition to the particle coordinates and velocities. The other approach is based on dynamically coupling the system to fictitious heat and/or volume reservoirs, which act on the system via a small set of dynamical variables called thermostats, when they control the temperature, and barostats, when they control the pressure [1].

In order to simulate a specific material, we need a realistic model of its potential energy $\Phi(\{\mathbf{R}\})$ as a function of the atomic coordinates $\{\mathbf{R}\}$. A common practice in molecular dynamics applications is to represent the potential energy function, or potential energy surface, in terms of a restricted set of *empirical* few-body potentials, whose parameters are fitted to experiment and/or to the results of more accurate calculations, when these are available. A simple example is the 2-body Lennard–Jones potential $\varphi(R_{IJ}) = 4\varepsilon((\frac{\sigma}{R_{IJ}})^{12} - (\frac{\sigma}{R_{IJ}})^{6})$, where $R_{IJ} = |\mathbf{R}_I - \mathbf{R}_J|$ is the distance between particles I and J, and ε and σ are adjustable parameters. The corresponding potential energy function is $\Phi(\{\mathbf{R}\}) = \frac{1}{2}\sum_{I\neq J}\varphi(R_{IJ})$. This is a reasonable approximation for weakly bonded systems like liquid argon, but is not sufficient to capture the specificity of the interatomic interactions in the general case. In order to achieve this goal, more general forms of the potential energy function are adopted. For instance, 3-body terms can be added to 2-body terms to describe angular dependent forces. Then $\Phi(\{\mathbf{R}\}) = \frac{1}{2}\sum_{I\neq J}\varphi^{(2)}(R_{IJ}) + \frac{1}{3!}\sum_{I\neq J\neq K}\varphi^{(3)}(R_{IJ}, R_{JK}, R_{KI})$ in terms of 2-body and 3-body potentials $\varphi^{(2)}$ and $\varphi^{(3)}$. A potential of this kind, the Stillinger–Weber potential [5], is quite popular to describe covalent interactions in silicon and germanium. Different potentials are available to model different interactions, like ionic interactions, covalent bonds in bio-molecules, and hydrogen bonds in water. The need for potentials that describe specific bonds in specific materials has prompted a large research effort. There is however a basic difficulty with this approach: empirical potentials tend to have a limited transferability, i.e. a potential parameterization that works well in one situation may not work well in another. This is because interatomic interactions incorporate cooperative effects that depend on the local atomic environment. For instance, the interactions in silicon change dramatically upon melting. Crystalline silicon is a covalent semiconductor with a local atomic coordination of 4. Liquid silicon is a metal, with a local coordination larger than 6. In liquid silicon, the only remnants of the tetrahedral bond network of the crystal are short-lived local fluctuations [6]. This implies that the strength of the 3-body interactions that stabilize the tetrahedral network should change upon melting. A re-parameterization of the empirical potential may therefore be necessary whenever the thermodynamic state of a system undergoes an important change.

2. POTENTIAL ENERGY SURFACE AND ELECTRONIC STRUCTURE

2.1. Density Functional Theory

A more fundamental approach that avoids empirical fitting altogether consists in deriving the potential energy surface directly from the elementary interactions. At the atomic scale, matter is composed of electrons and nuclei. The atomic coordinates and velocities of the previous section are in fact nuclear coordinates and velocities. While it is often sufficient to treat the nuclei classically, the electrons need quantum mechanics, but, for our purpose, time-independent quantum mechanics is sufficient. This is a consequence of the large difference in mass between nuclei and electrons. To a very good approximation the light electrons *adiabatically* follow the heavy nuclei. This is the essence of the Born–Oppenheimer adiabatic approximation [7], according to which electrons that are initially in the ground state remain at each subsequent instant in the ground state corresponding to the nuclear configuration visited at that particular instant. The adiabatic principle leads to separation of electronic and nuclear dynamics. As a consequence, the nuclear coordinates $\{\mathbf{R}\}$ in the many-body electronic Hamiltonian $\hat{H}$ can be regarded as external parameters:

$$\hat{H} = \frac{-\hbar^2}{2m}\sum_i \nabla_i^2 + \sum_{I,i}\frac{-Z_I e^2}{|\mathbf{r}_i - \mathbf{R}_I|} + \frac{1}{2}\sum_{i \neq j}\frac{e^2}{|\mathbf{r}_i - \mathbf{r}_j|} \tag{7}$$

In Eq. (7), lower case indices refer to electrons, upper case indices to nuclei, Z_I are atomic numbers, e is the absolute value of the electron charge, m the electron mass, and the sums run over nuclei and electrons. The ground state energy of the electrons $E_{\mathrm{GS}}(\{\mathbf{R}\})$ is found by minimization:

$$E_{\mathrm{GS}}(\{\mathbf{R}\}) = \mathrm{Min}_{\Psi}\langle\Psi|\hat{H}|\Psi\rangle \tag{8}$$

Here $\Psi(\{\mathbf{r}\}) \equiv \Psi(\mathbf{r}_1, \mathbf{r}_2, \mathbf{r}_3, \ldots, \mathbf{r}_{N_e})$ is a normalized many-electron wavefunction (in which, for simplicity we have omitted the spin variables), and N_e is the total number of electrons. The nuclear potential energy surface is given by

$$\Phi(\{\mathbf{R}\}) = E_{\mathrm{GS}}(\{\mathbf{R}\}) + \frac{1}{2}\sum_{I \neq J}\frac{Z_I Z_J e^2}{|\mathbf{R}_I - \mathbf{R}_J|} \tag{9}$$

Inserting the potential energy function in Eq. (9) into Eq. (3), classical nuclear trajectories can be computed without empirical fitting of the potential. However, solving Eq. (8) is a formidable quantum many-body problem that requires further approximations. To simplify the problem, we adopt density functional theory [8,9], a formally exact scheme to map a system of interacting electrons into a fictitious system of non-interacting electrons with the same density [10–12]. According to density functional theory, the ground state energy of the interacting system, at nuclear configuration $\{\mathbf{R}\}$, can be obtained by minimizing a functional of the electron density $n\,(\mathbf{r})$ (with the constraint $\int n(\mathbf{r})\,\mathrm{d}\mathbf{r} = N_e$):

$$E_{\mathrm{GS}}(\{\mathbf{R}\}) = \mathrm{Min}_{n(\mathbf{r})} E_V[n] \tag{10}$$

Here $V \equiv V(\mathbf{r}) = \sum_I - Z_I e^2/|\mathbf{r} - \mathbf{R}_I|$ is the external potential of the nuclei acting on the electrons. The potential V depends parametrically on the nuclear configuration $\{\mathbf{R}\}$. Let us consider for simplicity a closed shell system. Its electron density can be represented by $n(\mathbf{r}) = 2\sum_{i=1,N_e/2}|\psi_i(\mathbf{r})|^2$ in terms of the $N_e/2$ doubly occupied single particle orbitals $\psi_i(\mathbf{r})$ of a fictitious non-interacting system. The functional $E_V[n]$ is a functional of the orbitals $\{\psi^*\}$ and $\{\psi\}$. Here $\{\psi\} \equiv \{\psi_1, \psi_2, \psi_3, \ldots, \psi_{N_e/2}\}$ indicates the full set of occupied orbitals and it is convenient to consider $\{\psi^*\}$ and $\{\psi\}$ as formally independent. The minimum problem in Eq. (10) can be cast in terms of the orbitals:

$$E_{\mathrm{GS}}(\{\mathbf{R}\}) = \mathrm{Min}_{\{\psi^*\}} E_V[\{\psi^*\}, \{\psi\}] \tag{11}$$

Going from Eq. (8) to Eq. (11) entails an enormous simplification: in Eq. (8) the basic variable is a many-body wavefunction having a numerical complexity that grows exponentially with the number of electrons, while in Eq. (11) the basic variables are $N_e/2$ independent functions of position in space. A constraint of orthonormality for the orbitals, i.e. $\langle\psi_i|\psi_j\rangle = \delta_{ij}$, is implied in Eq. (11), and it is sufficient to minimize the functional E_V with respect to $\{\psi^*\}$ since minimization with respect to $\{\psi\}$ would produce equivalent results. The energy functional in Eq. (11) can be written as

$$E_V[\{\psi^*\},\{\psi\}] = 2\sum_{i=1,N_e/2}\left\langle\psi_i\left|\frac{-\hbar^2\nabla^2}{2m}\right|\psi_i\right\rangle + \int V(\mathbf{r})n(\mathbf{r})\,d\mathbf{r} + \frac{1}{2}\int\int\frac{n(\mathbf{r})n(\mathbf{r}')e^2}{|\mathbf{r}-\mathbf{r}'|}\,d\mathbf{r}\,d\mathbf{r}' + E_{XC}[n] \tag{12}$$

Here $2\sum_{i=1,N_e/2}\langle\psi_i| - \hbar^2\nabla^2/2m|\psi_i\rangle$ is the kinetic energy associated to the single-particle orbitals, $\int V(\mathbf{r})n(\mathbf{r})\,d\mathbf{r}$ the potential energy of the electrons in the field of the nuclei, $\frac{1}{2}\iint\frac{n(\mathbf{r})n(\mathbf{r}')e^2}{|\mathbf{r}-\mathbf{r}'|}\,d\mathbf{r}\,d\mathbf{r}'$ the average Coulomb energy of the electrons, and $E_{XC}[n]$ the exchange and correlation energy, which accounts for the remaining contribution to E_{GS} in Eq. (11). $E_{XC}[n]$ is an unknown universal functional of the density [9]. The Euler–Lagrange equations for the minimum problem in Eq. (11) are

$$\frac{\delta E_V[\{\psi^*\},\{\psi\}]}{\delta\psi_i^*(\mathbf{r})} - 2\varepsilon_i\psi_i(\mathbf{r}) = 0 \tag{13}$$

Here, the orbitals ψ_i are orthogonal and the Kohn–Sham energies ε_i are Lagrange multipliers that keep the norm of the orbitals unitary. Evaluating the functional derivative in Eq. (13) gives

$$\left(\frac{-\hbar^2\nabla^2}{2m} + V(\mathbf{r}) + \int\frac{n(\mathbf{r}')e^2}{|\mathbf{r}-\mathbf{r}'|}\,d\mathbf{r}' + V_{XC}(\mathbf{r})\right)\psi_i(\mathbf{r}) = \varepsilon_i\psi_i(\mathbf{r}) \tag{14}$$

Here,

$$\hat{H}_{KS} \equiv \frac{-\hbar^2\nabla^2}{2m} + V(\mathbf{r}) + \int\frac{n(\mathbf{r}')e^2}{|\mathbf{r}-\mathbf{r}'|}\,d\mathbf{r}' + V_{XC}(\mathbf{r}) \tag{15}$$

is the Kohn–Sham Hamiltonian, and $V_{XC}(\mathbf{r}) = \frac{\delta E_{XC}[n]}{\delta n(\mathbf{r})}$ is the exchange-correlation potential. Eq. (14) are known as the Kohn–Sham equations [9]. They are self-consistent equations of the Hartree type. Formally, they map exactly a system of interacting electrons into a non-interacting system with the same energy and density. The standard way of solving Eq. (14) is by self-consistent diagonalization of the Kohn–Sham Hamiltonian. The eigenstates of the Kohn–Sham Hamiltonian are orbitals $\psi_i(\mathbf{r})$, from which one can compute the exact density $n(\mathbf{r})$ and ground state energy E_{GS} into Eqs. (10) and (9). In practice, approximations are necessary to write explicit expressions for the exchange-correlation energy and potential as functionals of the density. Commonly used approximations are the local density approximation (LDA) [9] or the generalized gradient approximation (GGA) [11,12].

The above formulation can be extended to deal with spin-dependent effects and, generally, with open shell situations in a way that closely resembles the approach followed in Hartree–Fock theory to introduce the spin-unrestricted version of the theory [13]. In the spin-unrestricted version of density functional theory the spin orbitals $\psi_{i\uparrow}(r)$ and $\psi_{i\downarrow}(\mathbf{r})$ replace the orbitals $\psi_i(\mathbf{r})$ as the basic variables. There are $N_{e\uparrow}$ spin up and $N_{e\downarrow}$ spin down orbitals, and the total number of electrons is $N_e = N_{e\uparrow} + N_{e\downarrow}$. Up and down spin densities are given by $n_\uparrow(\mathbf{r}) = \sum_{i\uparrow} |\psi_{i\uparrow}(\mathbf{r})|^2$ and by $n_\downarrow(\mathbf{r}) = \sum_{i\downarrow} |\psi_{i\downarrow}(\mathbf{r})|^2$, respectively, and the total density is $n(\mathbf{r}) = n_\uparrow(\mathbf{r}) + n_\downarrow(\mathbf{r})$. The exchange-correlation energy is a functional of the spin densities $E_{XC}[n_\uparrow(\mathbf{r}), n_\downarrow(\mathbf{r})]$, leading to different exchange-correlation potentials for up and down spin orbitals: $V_{XC\uparrow}(\mathbf{r}) = \delta E_{XC}/\delta n_\uparrow(\mathbf{r})$ and $V_{XC\downarrow}(\mathbf{r}) = \delta E_{XC}/\delta n_\downarrow(\mathbf{r})$, respectively. In what follows, we shall refer mostly to the spin-restricted version of the theory.

2.2. Pseudopotentials

When calculating the nuclear potential energy surface, one can achieve a further important simplification by eliminating the core electrons. Since the core electrons do not participate in chemical bonds, we shall assume that they follow rigidly the nuclear motion (frozen core approximation). This approximation can be implemented effectively by replacing the external potential of the nuclei acting on the electrons with a pseudopotential [14,15]:

$$V(\mathbf{r}) \rightarrow \hat{V}_{ps} = \sum_I \hat{v}_I(\mathbf{r} - \mathbf{R}_I) \tag{16}$$

Within this approach valence pseudo-orbitals replace the all electrons orbitals in Eq. (12), and the total electron density is replaced by the valence pseudocharge density. The hat in Eq. (16) indicates that the pseudopotential is in general a non-local operator. The total pseudopotential $\hat{V}_{ps}$ is a sum of atomic pseudopotentials $\hat{v}_I$ centered at the nuclear sites. The non-locality of atomic pseudopotentials is restricted to the core region of the corresponding atoms, where the pseudopotential mimics the effect of core orthogonality on the valence wavefunctions. Outside the core, atomic pseudopotentials are local and behave like $-Z_I^v e^2/|\mathbf{r} - \mathbf{R}_I|$, where $Z_I^v e$ is the total charge of the nucleus plus core electrons, i.e. Z_I^v is equal to the number of valence electrons of the (neutral) atom I. Outside the core, the pseudo-orbitals coincide with the all-electron valence orbitals for a given reference configuration, which is usually taken to be that of the free atom. Inside the core, the pseudo-orbitals differ from their all-electron counterpart. The Kohn–Sham energies of the pseudo-orbitals in the reference atomic configuration coincide with the Kohn–Sham energies of the corresponding all-electron valence orbitals. Pseudopotentials that satisfy these requirements are called *norm-conserving* because they conserve the integrated norm of the valence orbitals inside the core region [16]. Norm-conserving pseudopotentials have optimal transferability properties, i.e. outside the core the pseudo-orbitals remain close to the corresponding all-electron orbitals when an atom is placed in a molecular or a condensed environment. In the following, we shall tacitly assume a pseudopotential representation, and we shall often omit the prefix pseudo. Within pseudopotential theory Eq. (9) is replaced by:

$$\Phi(\{\mathbf{R}\}) = E_{\mathrm{GS}}(\{\mathbf{R}\}) + \frac{1}{2}\sum_{I \neq J} \frac{Z_I^v Z_J^v e^2}{|\mathbf{R}_I - \mathbf{R}_J|} \tag{17}$$

Here E_{GS} is the pseudo-ground state energy. Given the excellent transferability of the pseudopotentials, the difference in the pseudo-ground state energy of two nuclear configurations is an excellent approximation of the corresponding all-electron ground state energy difference.

Pseudo-wavefunctions are smoother than their all electron counterparts and can be efficiently expanded in plane waves [15]. This approach has been extremely successful and is often referred to as the standard model of solids. A Fourier representation of the orbitals, of the charge density and of the associated physical quantities can be effectively combined with the periodic supercell of molecular dynamics. Thus, molecular dynamics simulations extend the realm of application of the standard model to liquids and complex molecular systems. The convergence of plane wave calculations with basis set size is largely independent of the location of the atoms in the cell, a feature that is particularly important in molecular dynamics simulations. This occurs because plane waves are unaffected by the superposition errors that are typical with atom centered localized basis sets, such as Gaussian- or Slater-type orbitals.

3. *AB INITIO* MOLECULAR DYNAMICS: THE CAR–PARRINELLO APPROACH

Density functional theory in combination with pseudopotentials and plane waves provides an effective scheme to compute the nuclear potential energy surface in a non-empirical way. Potential energy surfaces from first-principles quantum mechanical theory are substantially more accurate than empirical potential energy surfaces. Breaking and formation of chemical bonds are induced by nuclear motion and changes in the atomic environment. These effects are usually well described by density functional theory. The main obstacle preventing the use of potential energy surfaces derived from density functional theory in molecular dynamics simulations is the high cost of the quantum mechanical calculation of the electronic energy and of the electronic forces on the nuclei. In a molecular dynamics simulation, the Kohn–Sham Eq. (14) need to be solved self-consistently at all the nuclear configurations visited in a trajectory. To compute meaningful statistical averages the number of nuclear configurations in a numerical trajectory must be large, in the order of several thousands and more. Until the formulation of the Car–Parrinello approach [17], it was widely believed that solving self-consistently the Kohn–Sham equations for such a huge number of nuclear configurations was simply impossible.

The Car–Parrinello approach is an extended Lagrangian formulation in which *both* nuclear and electronic degrees of freedom act as dynamical variables. The dynamics derives from the Lagrangian postulated by Car and Parrinello:

$$L_{\mathrm{CP}} = \frac{1}{2}\sum_I M_I \dot{\mathbf{R}}_I^2 + 2\mu \sum_i \langle \dot{\psi}_i | \dot{\psi}_i \rangle - \Phi_{\mathrm{CP}}[\{\mathbf{R}\}; \{\psi^*\}, \{\psi\}] + 2\sum_{i,j} \lambda_{ij}(\langle \psi_j | \psi_i \rangle - \delta_{ji}) \tag{18}$$

Here μ is a mass parameter (with dimension of a mass times a length squared) that controls the dynamical response of the electronic orbitals $\psi_i(\mathbf{r}; t)$, the fields $\dot{\psi}_i(\mathbf{r}; t) \equiv \partial\psi_i(\mathbf{r}; t)/\partial t$ describe the rate of change of the orbitals with time, and λ_{ij} are Lagrange multipliers that impose orthonormality among the orbitals. The term $2\mu\sum_i \langle \dot{\psi}_i | \dot{\psi}_i \rangle$ gives the kinetic energy associated to the time evolution of the orbitals, the factor of 2 in front of it and in front of the orthonormality constraints in Eq. (18) accounts for double occupancy of the orbitals; this factor is absent in the spin-unrestricted version of the theory. In the combined nuclear and electronic parameter space, the potential energy is

$$\Phi_{\mathrm{CP}}[\{\mathbf{R}\}; \{\psi^*\}, \{\psi\}] = E_{V(\{\mathbf{R}\})}[\{\psi^*\}, \{\psi\}] + \frac{1}{2}\sum_{I \neq J} \frac{Z_I^v Z_J^v e^2}{|\mathbf{R}_I - \mathbf{R}_J|} \tag{19}$$

Here, $E_{V(\{\mathbf{R}\})}[\{\psi^*\}, \{\psi\}]$ is the Kohn–Sham energy functional Eq. (12). *Only* at the minimum with respect to the electronic degrees of freedom, does $\Phi_{\mathrm{CP}}[\{\mathbf{R}\}; \{\psi^*\}, \{\psi\}]$ coincide with the nuclear potential energy $\Phi(\{\mathbf{R}\})$ of Eq. (17). From the Lagrangian Eq. (18) one obtains the equations of motion:

$$\begin{aligned} M_I \ddot{\mathbf{R}}_I &= -\frac{\partial \Phi_{\mathrm{CP}}[\{\mathbf{R}\}; \{\psi^*\}, \{\psi\}]}{\partial \mathbf{R}_I} \\ \mu|\ddot{\psi}_i\rangle &= -\frac{\delta \Phi_{\mathrm{CP}}[\{\mathbf{R}\}; \{\psi^*\}, \{\psi\}]}{2\delta\langle \psi_i|} + \sum_j |\psi_j\rangle \lambda_{ji} = -\hat{H}_{\mathrm{KS}}|\psi_i\rangle + \sum_j |\psi_j\rangle \lambda_{ji} \end{aligned} \tag{20}$$

Eqs. (20) are usually called Car–Parrinello equations. They generate trajectories in the extended parameter space of nuclear and electronic degrees of freedom. The trajectories consist of vectors $\mathbf{R}_I(t)$ and fields $\psi_i(\mathbf{r}; t)$. Orthonormality is preserved at all times if the orbitals are initially orthonormal and at subsequent times the Lagrange multipliers are given by

$$\lambda_{ij}(t) = \langle \psi_i(t)|\hat{H}_{\mathrm{KS}}(t)|\psi_j(t)\rangle - \mu\langle \dot{\psi}_i(t)|\dot{\psi}_j(t)\rangle \tag{21}$$

Eq. (21) follows by imposing orthonormality at time $t + \mathrm{d}t$ and taking the limit for $\mathrm{d}t \to 0$. Equations (20) conserve the total energy in extended parameter space:

$$H_{\mathrm{CP}} = K + K_{\mathrm{e}} + \Phi_{\mathrm{CP}} \tag{22}$$

Here $K(\{\mathbf{R}\}) = \frac{1}{2}\sum_I M_I \dot{\mathbf{R}}_I^2$ is the kinetic energy of the nuclei, and $K_{\mathrm{e}}[\{\psi^*\}, \{\psi\}] = 2\mu\sum_i \langle \dot{\psi}_i | \dot{\psi}_i \rangle$ the fictitious kinetic energy of the electronic orbitals under the dynamics generated by Eq. (20). The quantum mechanical kinetic energy of the Kohn–Sham electrons is included in the potential energy Φ_{CP} through the Kohn–Sham energy functional. A Car–Parrinello molecular dynamics time step requires evaluating the action of the Hamiltonian $\hat{H}_{\mathrm{KS}}$ on the electronic orbitals, an operation considerably simpler and less-time consuming than a self-consistent diagonalization of the Kohn–Sham Hamiltonian. The orbital evolution of Eqs. (20) differs from that of the time-dependent Schrödinger equation. However, the nuclear dynamics generated by Eqs. (20) is an excellent approximation of the Born–Oppenheimer time evolution when orbital dynamics is fast and follows *adiabatically* nuclear motion. A crucial point is that the fictitious electrons of Eqs. (20) have the *same* ground state of the true Kohn–Sham orbitals. Relying on adiabatic Car–Parrinello dynamics rather than on adiabatic Schrödinger dynamics has some practical advantages: the adiabaticity is controlled by the fictitious mass parameter μ, and Eq. (20) for electrons allows one to use a larger numerical integration time step than the time-dependent Schrödinger equation, because the former is second order and the latter is first order in time.

3.1. Adiabatic Behavior

Let us focus on the small oscillations of the electronic orbitals around the instantaneous ground state. To a very good approximation, these oscillations have frequencies

$$\omega_{\mathrm{cv}} = \sqrt{\frac{\varepsilon_{\mathrm{c}} - \varepsilon_{\mathrm{v}}}{\mu}} \tag{23}$$

The indices v and c refer to valence and conduction states (i.e. occupied and empty orbitals), and $\varepsilon_{\mathrm{v(c)}}$ are the ground state Kohn–Sham eigenvalues. Let $|\psi^0_{\mathrm{v(c)}}\rangle$ be the corresponding Kohn–Sham eigenstates. Small deviations around the ground state are described by $|\psi_{\mathrm{v}}(t)\rangle = |\psi^0_{\mathrm{v}}\rangle + \sum_{\mathrm{c}} \delta_{\mathrm{vc}}(t)|\psi^0_{\mathrm{c}}\rangle$, where $\delta_{\mathrm{vc}}(t)$ is small. Inserting this expression into Eqs. (20) for the electrons, linearizing with respect to δ_{vc}, and neglecting the dependence of $\hat{H}_{\mathrm{KS}}$ on δ_{vc}, one obtains expression Eq. (23) for the frequencies ω_{cv}. In insulators and closed shell molecular systems, $\omega^{\mathrm{min}}_{\mathrm{cv}}$ is finite because the minimum value of $\varepsilon_{\mathrm{c}} - \varepsilon_{\mathrm{v}}$, i.e. the Kohn–Sham gap, is finite. Thus it is possible to choose μ so that the following condition is satisfied:

$$\omega^{\mathrm{min}}_{\mathrm{cv}} \gg \omega^{\mathrm{max}}_{\mathrm{R}} \tag{24}$$

Here $\omega^{\mathrm{max}}_{\mathrm{R}}$ is the maximum frequency associated to the nuclear motion. When the condition in Eq. (24) is obeyed, the fast oscillatory motions of the electrons average out on the time scale of nuclear dynamics. Then the forces acting on the nuclei in Eqs. (20) are close to Born–Oppenheimer forces, i.e. $\frac{\partial \bar{\Phi}_{\mathrm{CP}}[\{\mathbf{R}\};\{\psi^*\},\{\psi\}]}{\partial \mathbf{R}_I} \simeq \frac{\partial \Phi(\{\mathbf{R}\})}{\partial \mathbf{R}_I}$, where the bar over Φ_{CP} indicates coarse-graining on the time scale of electron dynamics. Strict adiabatic evolution occurs in the limit $\mu \to 0$, but using a finite μ, such that Eq. (24) is satisfied, is often sufficient to ensure good adiabaticity. The typical behavior of Car–Parrinello forces for insulating systems is illustrated in Fig. 1. In panel (a), there appears to be no difference between Car–Parrinello and Born–Oppenheimer forces. To appreciate the difference we need to magnify the scale of the graph by approximately two orders of magnitude, as in panel (b). The difference has a fast oscillatory component on the time scale of the fictitious electron dynamics superimposed to a slow component on the

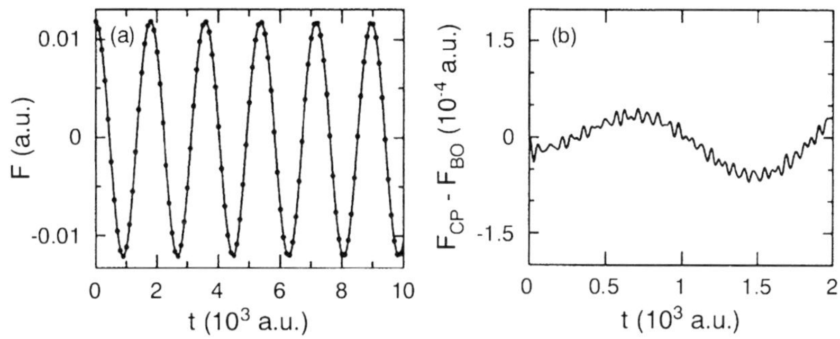

Fig. 1. (a) x-component of the force acting on one atom of a model system plotted against the simulation time. The solid line has been obtained from Car–Parrinello dynamics. The dots have been obtained from well-converged Born–Oppenheimer dynamics. (b) Magnified view of the difference between Car–Parrinello and Born–Oppenheimer forces. (Reproduced from Pastore et al. [18].)

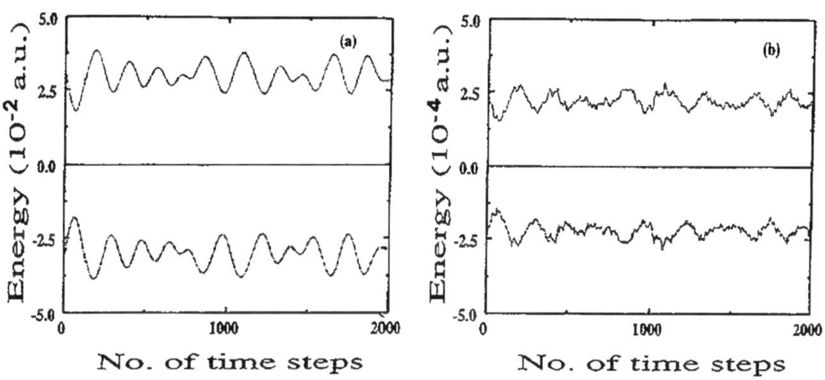

Fig. 2. Variation with time of kinetic and potential energies in Car–Parrinello molecular dynamics for a model insulating system (see text). (Reproduced from Car et al. [19].)

time scale of the nuclear motion, which is also oscillatory in this example. The fast component is due to oscillations with frequencies in Eq. (23). The slow component is due to the drag effect of the nuclei.

Under adiabatic conditions nuclear and electron dynamics are effectively decoupled, and heat transfer between nuclei and electrons is negligible over times that last up to tens of picoseconds. As a consequence, the nuclear kinetic energy K and the electron kinetic energy K_e fluctuate without appreciable drift. This is illustrated in Fig. 2. Panel (a) shows K (upper curve), Φ_{CP} (lower curve), and their sum $K + \Phi_{CP}$ (middle curve). On the energy scale of panel (a) $K + \Phi_{CP}$ is constant. This quantity approximates $H = K + \Phi$, which is an exact constant of motion of Born–Oppenheimer dynamics. Panel (b) has an energy scale magnified by a factor of 100. It shows K_e (upper curve), $K + \Phi_{CP}$ (lower curve), and their sum $H_{CP} =$

$K + K_e + \Phi_{CP}$ (middle curve). H_{CP} is an exact constant of motion of Car–Parrinello dynamics. In the time variation of K_e, one can identify fast fluctuations on the time scale of electron dynamics and slow fluctuations on the time scale of nuclear dynamics.

A temperature defined by $\bar{K} = \frac{3}{2} N k_B T$ can be associated to the nuclear dynamics, if this is ergodic. The fictitious electron dynamics is not ergodic and has two components: a fast, essentially harmonic, component consisting of small oscillations about the instantaneous ground state, and a slow component associated to electron drag by the nuclei. Both components have vanishing kinetic energy in the adiabatic limit ($\mu \to 0$). The following condition:

$$\bar{K}_e \ll \bar{K} \tag{25}$$

must be satisfied for good adiabatic behavior. The criterion in Eq. (24) considers only the fast oscillatory component of the electron dynamics, which averages to zero on the time scale of nuclear motion. The drag component has frequencies in the range of the nuclear frequencies. Therefore, even when condition (24) is satisfied, some residual coupling between electrons and nuclei is present due to the drag motion. Electron drag slows down the nuclear dynamics by softening its characteristic frequencies: the nuclei appear heavier because they have to drag electron orbitals that carry a finite mass. In order to minimize this effect, the kinetic energy associated to the electron drag must be much smaller than the kinetic energy associated to nuclear dynamics.

One can estimate the kinetic energy K_e^D of electron drag by assuming that the electrons follow rigidly the nuclei [20]. This amounts to assuming that the wavefunctions depend on time as $\psi(\mathbf{r} - \mathbf{R}(t))$ where $\mathbf{R}$ is a nuclear position vector. It follows that $\dot{\psi} = -\dot{\mathbf{R}} \cdot \nabla\psi$. The fictitious kinetic energy of the orbital ψ is $2\mu\langle\dot{\psi}|\dot{\psi}\rangle = 2\mu\dot{\mathbf{R}}_\alpha\dot{\mathbf{R}}_\beta \int dr(\nabla_\alpha\psi^*)(\nabla_\beta\psi)$, where the Greek subscripts indicate Cartesian components and summation over repeated indices is implied. Taking the average gives $2\mu\overline{\langle\dot{\psi}|\dot{\psi}\rangle} = 2\mu\overline{\dot{\mathbf{R}}^2}/3 \int dr(\nabla\psi^*) \cdot (\nabla\psi) = \frac{2\mu}{M} k_B T \int dr(\nabla\psi^*) \cdot (\nabla\psi)$. Here M is a typical nuclear mass. Finally, multiplying by the number of occupied states and taking double occupancy into account gives

$$K_e^D \approx \frac{\mu}{M} k_B T \left(\frac{2m}{\hbar^2}\right) T_{KS}[\{\psi^*\}, \{\psi\}] \tag{26}$$

Here $T_{KS}[\{\psi^*\}, \{\psi\}] = 2\sum_i \langle\psi_i| \frac{-\hbar^2\nabla^2}{2m} |\psi_i\rangle$ is the quantum kinetic energy of the Kohn–Sham orbitals. Rigid following is a reasonable assumption when the electronic structure is made of closed electronic shells tightly bound to ions or molecules: this happens in strongly ionic systems and in molecular systems like water. In all other cases, the electrons are delocalized and shared among several atoms. In this case, K_e^D should be smaller than in the estimate Eq. (26). Since μ is finite, the electronic orbitals need a fictitious kinetic energy which should be at least of the order of K_e^D to follow adiabatically the nuclear motion.

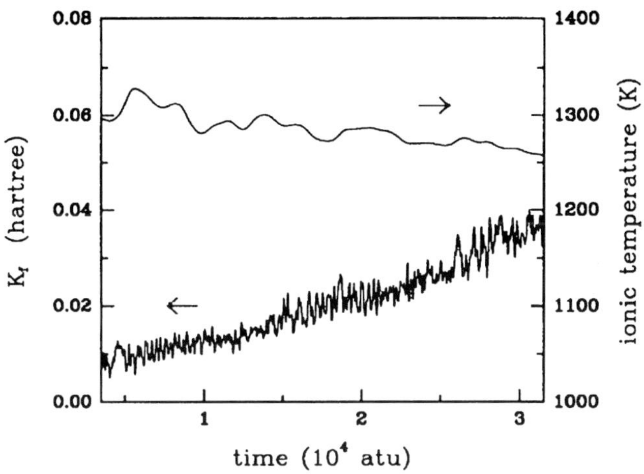

Fig. 3. Variation with time of the kinetic energy of the nuclei and of the fictitious kinetic energy of the electrons in a model metallic system. The system evolves with Car–Parrinello dynamics without thermostats. The scale on the right (Kelvin degrees) applies to the nuclear kinetic energy, the scale on the left (atomic units) applies to the fictitious kinetic energy of the electrons. (Reproduced from Pastore et al., 1991 [18].)

3.2. Equations with Thermostats

Condition (24) cannot be satisfied in metals and open shell systems for which $\omega_{\rm cv}^{\rm min} = 0$. As a consequence, when Eq. (20) are used for metals and the electrons are initially in the ground state, K and $K_{\rm e}$ drift systematically as time proceeds. The drift in kinetic energy reflects a systematic transfer of energy from nuclear to electronic degrees of freedom, i.e. K decreases and $K_{\rm e}$ increases, as shown in Fig. 3. Then, after some time, condition (25) is no longer satisfied. Eventually, when energy equipartition among all the degrees of freedom, nuclear and electronic, would be achieved, this process would end. Long before this can happen, the nuclear trajectories generated by Eq. (20) will cease to be physically meaningful. In order to generate physically meaningful Born–Oppenheimer nuclear trajectories, thermal equilibration among nuclear and electronic degrees of freedom must be counteracted by *systematically* subtracting from the electrons the energy that they acquire from the nuclei and transferring the same amount of energy back to the nuclei. This can be achieved, in a dynamical way, by separately coupling the nuclear and the electronic degrees of freedom to thermostats [20]. Equations (20) with Nose'–Hoover thermostats [1,21] become

$$M_I \ddot{\mathbf{R}}_I = -\frac{\partial \Phi_{\rm CP}[\{\mathbf{R}\};\{\psi^*\},\{\psi\}]}{\partial \mathbf{R}_I} - \xi_{\rm R} M_I \dot{\mathbf{R}}_I$$

$$\mu|\ddot{\psi}_i\rangle = -\frac{\delta\Phi_{\mathrm{CP}}[\{\mathbf{R}\};\{\psi^*\},\{\psi\}]}{2\delta\langle\psi_i|} + \sum_j |\psi_j\rangle\lambda_{ji} - \xi_{\mathrm{e}}\mu|\dot{\psi}_i\rangle = -\hat{H}_{\mathrm{KS}}|\psi_i\rangle + \sum_j |\psi_j\rangle\lambda_{ji} - \xi_{\mathrm{e}}\mu|\dot{\psi}_i\rangle \tag{27}$$

Here ξ_{R} and ξ_{e} are thermostat variables that act as dynamical friction coefficients on the nuclei and the electrons. The dynamics of the thermostats is governed by two equations:

$$\begin{aligned} Q_{\mathrm{R}}\dot{\xi}_{\mathrm{R}} &= \left(\sum_I M_I \dot{\mathbf{R}}_I^2 - gk_{\mathrm{B}}T\right) \\ Q_{\mathrm{e}}\dot{\xi}_{\mathrm{e}} &= \left(2\mu\sum_i \langle\dot{\psi}_i|\dot{\psi}_i\rangle - K_{\mathrm{e}}^{\mathrm{ref}}\right) \end{aligned} \tag{28}$$

Here Q_{R} and Q_{e} are thermostat "masses", g the number of independent internal nuclear degrees of freedom ($g = 3N - 3$ in molecular dynamic simulations of extended systems with periodic boundary conditions), and $K_{\mathrm{e}}^{\mathrm{ref}}$ a reference fictitious electronic kinetic energy. The masses Q_{R} and Q_{e} control the dynamical response of the thermostats; their values should be chosen to ensure good dynamical coupling between nuclei or electrons and the corresponding thermostats. Unlike Eqs. (20), Eqs. (27) and (28) are not equivalent to Hamiltonian equations of motion and cannot be derived from a Lagrangian like (19). There is still, however, a conserved quantity associated to the dynamics of Eqs. (27) and (28), namely:

$$H_{\mathrm{CP}}^{NH} = K_{\mathrm{e}} + K + \Phi_{\mathrm{CP}} + \frac{Q_{\mathrm{e}}}{2}\xi_{\mathrm{e}}^2 + \frac{Q_{\mathrm{R}}}{2}\xi_{\mathrm{R}}^2 + K_{\mathrm{e}}^{\mathrm{ref}}\int^t \mathrm{d}t'\xi_{\mathrm{e}}(t') + gk_{\mathrm{B}}T\int^t \mathrm{d}t'\xi_{\mathrm{R}}(t') \tag{29}$$

Equation (28) eliminates the systematic drift of K and K_{e} in simulations of metallic systems. The value adopted for $K_{\mathrm{e}}^{\mathrm{ref}}$ should be in the order of, or larger than, $K_{\mathrm{e}}^{\mathrm{D}}$ and condition (25) should be satisfied. A good choice of $K_{\mathrm{e}}^{\mathrm{ref}}$ minimizes the rate of energy transfer between nuclei and electrons, as illustrated in Fig. 4. In this figure, the arrow indicates the value of $K_{\mathrm{e}}^{\mathrm{D}}$ estimated from Eq. (26). The effect of the equations with thermostats on the fictitious kinetic energy of the electrons in a metallic system is illustrated in Fig. 5. With a good choice of parameters the nuclear dynamics generated by Eq. (27) is a good approximation of the adiabatic dynamics of the nuclei subject to a Nose'–Hoover thermostat ξ_{R}. In fact, even though condition (24) is not satisfied, dynamical coupling between nuclei and electrons is weak because the nuclear frequencies overlap only with a tiny fraction of the electron frequencies. The overlap goes to zero in the limit $\mu \to 0$. When the dynamics is ergodic Nose'–Hoover time evolution samples the canonical ensemble. This is desirable for nuclear dynamics and, if a single Nose'–Hoover thermostat is not sufficient one should resort to a chain of thermostats [22] to achieve this goal. On the other hand, ergodic evolution of the electronic orbitals should be avoided. The role of the electron thermostat is not to

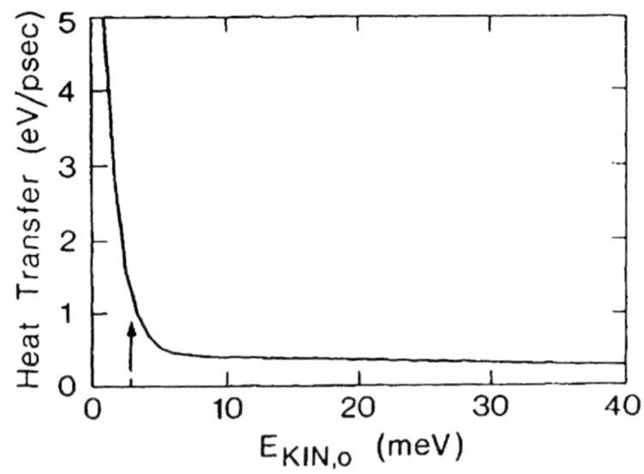

Fig. 4. Heat transfer from the nuclei to the electrons in a model metallic system which evolves according to Car–Parrinello dynamics with thermostats. The horizontal axis reports the reference electronic kinetic energy in the equations with thermostats. (Reproduced from Blöchl and Parrinello [20].)

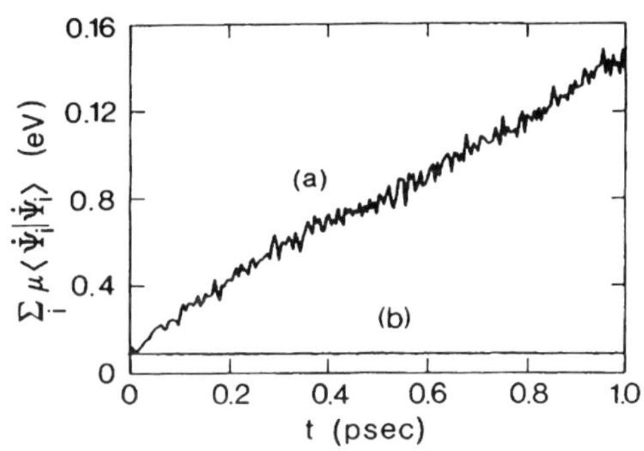

Fig. 5. Time variation of the fictitious electronic kinetic energy in a model metallic system using the Car–Parrinello equations (a) without thermostats, and (b) with thermostats. (Reproduced from Blöchl and Parrinello [20].)

control the temperature, but only the total fictitious kinetic energy, limiting fluctuations as much as possible. Iso-kinetic control schemes [21] different from a Nose'–Hoover thermostat could also be used in this context.

In conclusion, in insulators microcanonical and canonical Born–Oppenheimer nuclear trajectories can be generated by the Car–Parrinello equations. In metals, only canonical Born–Oppenheimer trajectories can be generated with this approach. Constant pressure simulations sampling the isobaric ensemble are possible in all cases by adopting techniques developed in the context of classical molecular dynamics simulations. These include schemes in which the volume of the cell is allowed to fluctuate as proposed by Andersen [23], or schemes in which both the volume and the shape of the cell fluctuate as proposed by Parrinello and Rahman [24,25]. The latter approach has been used successfully in a number of *ab initio* molecular dynamics applications to study structural phase transitions in bulk materials under pressure [26,27].

4. NUMERICAL IMPLEMENTATION

4.1. Verlet Algorithm and Lagrange Multipliers

The most common algorithm to integrate molecular dynamics equations is the Verlet algorithm, which exists in several variants [1]. By applying the standard form of this algorithm to Eq. (20) we get, for the electrons:

$$|\psi_i(t+\Delta t)\rangle = -|\psi_i(t-\Delta t)\rangle + 2|\psi_i(t)\rangle - \frac{\Delta t^2}{\mu}\left(\hat{H}_{\mathrm{KS}}(t)|\psi_i(t)\rangle - \sum_j |\psi_j(t)\rangle\lambda_{ji}\right) \quad (30)$$

Since the integration time step Δt is finite, we cannot use Eq. (21) for the Lagrange multipliers: this would lead to a systematic degradation of orthonormality due to contributions of higher order in Δt that are neglected in (21). The correct way of imposing the constraints is by a procedure called SHAKE [28], developed in the classical molecular dynamics context to impose rigidity constraints. Orthonormality constraints can be viewed as generalized rigidity constraints. Let us indicate by $|\tilde{\psi}_i(t+\Delta t)\rangle \equiv -|\psi_i(t-\Delta t)\rangle + 2|\psi_i(t)\rangle - (\Delta t^2/\mu)\hat{H}_{\mathrm{KS}}(t)|\psi_i(t)\rangle$ the predicted electron orbitals without imposing the constraints at time $t+\Delta t$. Introducing the hermitian matrix $X_{ij} \equiv (\Delta t^2/\mu)\lambda_{ij}$, Eq. (30) reads:

$$|\psi_i(t+\Delta t)\rangle = |\tilde{\psi}_i(t+\Delta t)\rangle + \sum_k |\psi_k(t)\rangle X_{ki} \quad (31)$$

The condition $\langle\psi_i(t+\Delta t)|\psi_j(t+\Delta t)\rangle = \delta_{ij}$ results in an equation for the unknown matrix X:

$$I = A + XB^\dagger + BX + X^2 \quad (32)$$

Here, I is the identity matrix, A has elements $A_{ij} = \langle\tilde{\psi}_i(t+\Delta t)|\tilde{\psi}_j(t+\Delta t)\rangle$, and B has elements $B_{ij} = \langle\tilde{\psi}_i(t+\Delta t)|\psi_j(t)\rangle$. Solving Eq. (32) for X determines the Lagrange multipliers that satisfy the constraints at each time step, consistent with the numerical integrator adopted in Eq. (30). Equation (32) can be solved iteratively by constructing X as a power series in Δt. To lowest order in Δt, X is given by

$$X^{(0)} = \frac{1}{2}(I - A) \quad (33)$$

This estimate can be improved by iterating

$$X^{(n)} = \frac{1}{2}(I - A) + X^{(n-1)}(I - B^\dagger) + (I - B)X^{(n-1)} - (X^{(n-1)})^2 \; (n = 1, n_{\mathrm{iter}}) \quad (34)$$

Usually, a few iterations ($n_{\mathrm{iter}} \sim 5$) are enough to satisfy orthonormality to an accuracy of 10^{-6}, sufficient in electronic structure calculations. The SHAKE procedure is applied systematically at each molecular dynamics time step to preserve orthonormality. When Eqs. (20) are integrated numerically, the total energy (22) is no longer an exact constant of motion. It remains, however, an approximate constant of motion. The Verlet algorithm is a symplectic integrator, which preserves the phase space density, and is remarkably stable in this respect [1].

4.2. Plane Waves and Pseudopotentials

The crucial numerical step in Eq. (30) consists in acting with the Kohn–Sham Hamiltonian on the electronic wavefunctions. These are expanded in plane waves:

$$\psi_n(\mathbf{r}) \equiv \psi_{(n,\mathbf{k})}(\mathbf{r}) = \frac{1}{\sqrt{\Omega}} e^{i\mathbf{k}\cdot\mathbf{r}} \sum_{\mathbf{G}} c_{\mathbf{G}}^{(n,\mathbf{k})} e^{i\mathbf{G}\cdot\mathbf{r}} \tag{35}$$

Here $\mathbf{k}$ is a wave vector in the Brillouin Zone, $\mathbf{G}$ are reciprocal lattice vectors, and $n \equiv (n, \mathbf{k})$ a composite band and wave vector index. Typically, *ab initio* molecular dynamics simulations are performed on relatively large supercells, for which the k dependence can often be neglected. Thus we take $k = 0$ and drop the wave vector index. Equation (35) becomes

$$\psi_n(\mathbf{r}) = \frac{1}{\sqrt{\Omega}} \sum_{\mathbf{G}} c_{\mathbf{G}}^{n} e^{i\mathbf{G}\cdot\mathbf{r}} \tag{36}$$

Care must be taken in applications to check that the supercell is large enough to justify the $\mathbf{k} = 0$ approximation. In metals, band dispersion effects may be not negligible even for supercells with 100 atoms and more. In this case, several $\mathbf{k}$ points must be included in Eq. (35) for a good sampling of the Brillouin Zone.

When using the plane wave representation (36) (or (35)), we take advantage of the fast Fourier transform algorithm to switch rapidly from a wavefunction evaluated on a regular grid in real space to its Fourier components on the corresponding regular grid in reciprocal space, and vice versa. Let M be the number of grid points. The numerical cost of the fast Fourier transform algorithm scales like $M \ln M$, i.e. it is essentially linear in the number of grid points. The Kohn–Sham Hamiltonian (15) has kinetic and potential energy contributions. The latter are both local and non-local in space. The local potential contributions include the Hartree potential, the exchange-correlation potential, and the local reference component of the pseudopotential. The action on wavefunctions of the kinetic energy operator is calculated most efficiently in reciprocal space, where the kinetic energy operator is diagonal. The action on wavefunctions of the local potential operator is calculated most efficiently in real space, where the local potential operator is diagonal. These calculations require M operations per electron orbital, giving a total numerical cost that scales like $N_e M$, if one ignores the logarithmic dependence on M of the fast Fourier transforms.

The non-local component of a norm-conserving atomic pseudopotential has the form:

$$\Delta\hat{v} = \sum_{lm} \Delta v_l \hat{P}_{lm} \tag{37}$$

Here $\hat{P}_{lm}$ are projectors on the angular momentum eigenstates $|lm\rangle$, and $\Delta v_l = v_l - v_{\rm loc}$ is the difference between the pseudopotential in the l channel and the local reference pseudopotential. Δv_l is zero outside the core. Numerically the semi-local form (37) is not convenient. A fully non-local form introduced by Kleinman and Bylander [29] is superior in this respect. Following Kleinman and Bylander:

$$\Delta\hat{v} = \sum_{lm} |\Delta v_l \varphi^0_{lm}\rangle \frac{1}{\langle\varphi^0_{lm}|\Delta v_l|\varphi^0_{lm}\rangle} \langle\varphi^0_{lm}\Delta v_l| \tag{38}$$

Here $|\varphi^0_{lm}\rangle$ are the atomic reference states used to construct the pseudopotential. Since a wavefunction $|\psi_i\rangle$ is well represented in the core region by a superposition of $|\varphi^0_{lm}\rangle$ states, the operator (38) is physically equivalent to the operator (37). Acting with (38) on a wavefunction requires $n_{\rm ref}M$ operations. The number of atomic reference states $n_{\rm ref}$ is size-independent. Multiplying by the number of atoms and by the number of wavefunctions gives $(Nn_{\rm ref})N_{\rm e}M$. Thus, the total number of operations scales cubically with system size. Notice that $M \gg N_{\rm e} \sim (Nn_{\rm ref})$. With the form (37) the numerical cost is also cubic, but the number of operations is $N_{\rm e}M^2$, significantly larger than $(Nn_{\rm ref})N_{\rm e}M$. Calculating the action of the non-local pseudopotential on the wavefunctions can have only quadratic cost if one exploits the fact that the states $|\Delta v_l \varphi_{nlm}\rangle$ vanish outside the core. This requires a calculation in real space, and should become advantageous for relatively large system sizes.

In addition to acting with the Hamiltonian on the wavefunctions, one needs to calculate the Lagrange multipliers. This requires $N_{\rm e}^2M$ operations to construct the matrices A and B, and $N_{\rm e}^3$ operations for the iterative procedure (33) and (34).

4.3. Beyond Norm-Conserving Pseudopotentials

Norm-conserving pseudopotentials need large plane wave cutoffs to describe tightly bound electrons, like the valence electrons of second row atoms or the *d*-electrons of the first group of transition elements. In these cases, a substantial saving of computational effort can be achieved by replacing norm-conserving pseudopotentials with Vanderbilt ultrasoft pseudopotentials [30]. Within ultrasoft pseudopotentials only the subset of Fourier components that describe the smooth part of the valence wavefunctions are treated as dynamical variables. The remaining part of the electronic charge density is treated as augmented core charges. Thus, the plane wave cutoff can be substantially reduced at the price of a more complicated electronic structure scheme with augmented charges and a generalized orthonormality condition. Ultrasoft pseudopotentials can be efficiently combined with Car–Parrinello molecular dynamics [31,32]. In this context, ultrasoft pseudopotentials have the additional advantage of mitigating the requirements for adiabatic evolution by significantly reducing the number of dynamical electronic variables. Augmentation concepts are used elegantly in Blöchl's Projector Augmented Wave (PAW) method [33]. PAW can be combined with Car–Parrinello molecular dynamics in a way that parallels very closely the approach adopted with ultrasoft pseudopotentials. PAW, however, is an all electron scheme rather than a valence only pseudopotential method.

4.4. Verlet Algorithm with Thermostats

The Verlet step (30) can be easily adapted to equations with thermostats like (27). The time derivative of an orbital should be calculated as $|\dot{\psi}_i(t)\rangle = \frac{|\psi_i(t+\Delta t)\rangle - |\psi_i(t-\Delta t)\rangle}{2\Delta t}$. Inserting this expression in the Verlet integrator gives:

$$|\psi_i(t+\Delta t)\rangle = -\frac{1-f_{\mathrm{e}}(t)}{1+f_{\mathrm{e}}(t)}|\psi_i(t-\Delta t)\rangle + \frac{2}{1+f_{\mathrm{e}}(t)}|\psi_i(t)\rangle - \frac{\Delta t^2}{\mu(1+f_{\mathrm{e}}(t))}\left(\hat{H}_{\mathrm{KS}}(t)|\psi_i(t)\rangle - \sum_j |\psi_j(t)\rangle\lambda_{ji}\right) \tag{39}$$

Here, $f_{\mathrm{e}}(t) \equiv \frac{\xi_{\mathrm{e}}(t)\Delta t}{2}$. Similarly, the integrator for the other equation in (27) is

$$\mathbf{R}_I(t+\Delta t) = -\frac{1-f_{\mathrm{R}}(t)}{1+f_{\mathrm{R}}(t)}\mathbf{R}_I(t-\Delta t) + \frac{2}{1+f_{\mathrm{R}}(t)}\mathbf{R}_I(t) - \frac{\Delta t^2}{\mu(1+f_{\mathrm{R}}(t))}\frac{\partial \Phi_{\mathrm{CP}}(t)}{\partial \mathbf{R}_I} \tag{40}$$

Here, $f_{\mathrm{R}}(t) \equiv \frac{\xi_{\mathrm{R}}(t)\Delta t}{2}$. When $f_{\mathrm{e}} = f_{\mathrm{R}} = 0$, Eqs. (39) and (40) reduce to the standard Verlet step.

The initial condition for the Verlet integrator requires specification of the nuclear coordinates and of the corresponding ground state Kohn–Sham orbitals at two subsequent time steps, i.e. at $t = -\Delta t$ and at $t = 0$. The ground state orbitals at an arbitrary nuclear configuration $\{\mathbf{R}\}$ can be found by damped molecular dynamics:

$$\mu|\ddot{\psi}_i\rangle = -\hat{H}_{\mathrm{KS}}|\psi_i\rangle + \sum_j |\psi_j\rangle\lambda_{ji} - \gamma_{\mathrm{e}}\mu|\dot{\psi}_i\rangle \tag{41}$$

The friction coefficient γ_{e} in Eq. (41) is time-independent and positive. Starting from an initial configuration, usually a random orthonormal orbital configuration, Eq. (41) is integrated numerically using Eq. (39), in which we set $f_{\mathrm{e}} = \frac{\gamma\Delta t}{2}$. At steady state, $|\ddot{\psi}_i\rangle = |\dot{\psi}_i\rangle = 0$, and the solution of Eq. (41) coincides, within a unitary transformation in the occupied orbital subspace, with the solution of the Kohn–Sham Eq. (14). When $f_{\mathrm{e}} = 1$, the Verlet integrator coincides with the Euler integrator of steepest descent dynamics. When $f_{\mathrm{e}} = 0$, one recovers the Verlet integrator of free (Newtonian) dynamics. Thus we adopt $0 < f_{\mathrm{e}} \leq 1$ for minimization dynamics. The criteria for an optimal choice of f_{e} are discussed in Ref. [34].

Combined damped molecular dynamics for electrons and nuclei is an effective approach for the local optimization of molecular structures. An efficient alternative for the electrons is minimization by conjugate gradients [35–38].

4.5. Numerical Integration Step

The Verlet time step Δt must be much smaller than the shortest period of a dynamical fluctuation. In Car–Parrinello dynamics, nuclear and electronic fluctuations are present, but the electrons are faster. Thus the electrons determine Δt. The maximum electronic frequency $\omega_{cv}^{max} = (\sqrt{\frac{\varepsilon_c - \varepsilon_v}{\mu}})^{max}$ is well approximated by $\omega_{cv}^{max} \sim \sqrt{\frac{E_{cut}^{PW}}{\mu}}$, where E_{cut}^{PW} is the kinetic energy cutoff for the plane wave expansion ((36) or (35)). The maximum wave vector in expansion (36) satisfies the condition $\frac{\hbar^2 \mathbf{G}_{max}^2}{2m} \leq E_{cut}^{PW}$. Therefore, Δt must satisfy:

$$\Delta t \ll \frac{2\pi}{\omega_{cv}^{max}} \tag{42}$$

Choosing for Δt the maximum value compatible with (42) reduces the number of time steps needed to span a given time interval. As usual in molecular dynamics simulations, we can assess the accuracy of the integration by monitoring drift and fluctuation of quantities like (22) and (29), which are constants of motion of the exact dynamics. A typical time step in Car–Parrinello simulations is $\Delta t \sim 10^{-1}$ fs. Larger values, such as $\Delta t \sim 1$ fs, may be possible to integrate the nuclear dynamics alone. Thus, a time step of Car–Parrinello molecular dynamics is more expensive than a time step of classical molecular dynamics for the same system not only because of the costly operations necessary to update the electronic orbitals, but also because about 10 times more steps are necessary to span the same time interval.

The time step can be optimized, by exploiting the arbitrariness of the fictitious electronic mass. This approach is called mass preconditioning or Fourier acceleration [34,36]. It amounts to replacing the parameter μ with a wave vector-dependent mass $\mu(G)$ that reduces the maximum electronic frequency ω_{cv}^{max} but leaves the minimum electronic frequency, ω_{cv}^{min}, essentially unchanged. Mass preconditioning tends to increase the average mass of the electronic orbitals. As a consequence, it also tends to increase the drag component of the electronic kinetic energy K_e^D.

4.6. Car–Parrinello and Born–Oppenheimer Schemes

The heart of the matter is the "on the fly" calculation of the potential energy surface for nuclear motion, without performing a self-consistent diagonalization of the Kohn–Sham Hamiltonian at each time step. In the Car–Parrinello scheme, occupied electronic orbitals and nuclear coordinates are updated *simultaneously* with an efficient iterative scheme. Variants of this basic strategy are possible. For instance, one could adopt Eq. (3) for nuclear dynamics and use an iterative scheme to extrapolate and re-optimize the occupied electronic wavefunctions at each nuclear time step. This approach is often called Born–Oppenheimer molecular dynamics because it enforces explicitly the minimum condition for the electrons at each update of the nuclear coordinates [38]. Born–Oppenheimer molecular dynamics could be achieved, for instance, by using Eq. (30) (without the term corresponding to the action of the Hamiltonian on the wavefunctions) to *extrapolate* the electrons to time $t + \Delta t$ from their *ground state* configuration at the times t and $t - \Delta t$. Extrapolation should be followed by damped molecular dynamics (Eq. (41)) or conjugate gradient iterations to re-minimize the electrons at time $t + \Delta t$. Then, the new nuclear coordinates at $t + 2\Delta t$ are calculated, and the entire procedure for the electrons is repeated. This procedure requires several electronic steps per step of nuclear dynamics, but the additional cost is partially offset by using a Δt larger than in standard Car–Parrinello simulations. This is possible because in Born–Oppenheimer dynamics the time step is dictated by the nuclear dynamics and not by the fictitious electron dynamics. More elaborate strategies based on fast electron minimization algorithms in combination with optimal wavefunction extrapolation techniques can be used to improve the efficiency of Born–Oppenheimer schemes. The Car–Parrinello approach has remarkable time stability, which derives from energy conservation in the extended parameter space of electrons and nuclei. In the Car–Parrinello scheme, the electrons oscillate rapidly around the instantaneous minimum. Very little drift in the temperature of the nuclei is observed after tens of picoseconds in microcanonical Car–Parrinello simulations of insulating systems. Considerably larger drifts occur in Born–Oppenheimer simulations due to accumulation of the systematic errors in the electron minimization. As a consequence, Born–Oppenheimer simulations may need the use of a thermostat to keep the average nuclear temperature constant even for insulating systems. *Ab initio* molecular dynamics simulations require good adiabatic nuclear trajectories. In Born–Oppenheimer schemes, the quality of the nuclear trajectories is controlled by the accuracy of the electronic minimization at each nuclear time step. In the Car–Parrinello approach, the quality of the trajectories is controlled by the mass parameter μ, which sets the relative time scales of nuclear and fictitious electron dynamics.

5. AN ILLUSTRATIVE APPLICATION: LIQUID WATER

Since its inception in 1985, *ab initio* molecular dynamics has been used extensively in a large number of applications in various areas of physics, chemistry, materials science, and biochemistry/biophysics. It is well outside our scope to review this vast literature. Here we limit ourselves to discuss the application of the approach to a representative liquid system: water, which is arguably the most important liquid on earth.

Water molecules bind via hydrogen-mediated, or hydrogen bond (H-bond), interactions. To visualize H-bond interactions, let us consider the densities of the four valence orbitals of an isolated water molecule shown in Fig. 6. Panel (a) shows isodensity plots of the four valence orbitals in the maximally localized Wannier function (MLWF) representation [39]. This is obtained by a unitary transformation in the occupied orbital subspace that minimizes the spread of the orbitals in real space [39,40]. In most cases, maximally localized Wannier functions conform to simple chemical intuition. For instance, in Fig. 6(a) we see that two orbitals correspond to bond pairs of electrons, while the other two correspond to lone pairs of electrons. The four orbital pairs point into approximately tetrahedral directions. Water molecules are polar molecules (with a dipole moment $|\mathbf{p}| \sim 1.8$ D in the gas phase): excess positive charge is present near the protons, whereas excess negative

Fig. 6. Binding in water. Panel (a) shows isodensity plots of the four MLWF of an isolated water molecule: two MLWF are bond pairs, the other two MLWF are lone pairs. Panel (b) shows the characteristic hydrogen bond geometry of a water dimer, in which a bond pair of the left molecule (donor) points toward a lone pair of the right molecule (acceptor). Panel (c) shows schematically the local tetrahedral environment found in condensed water phases: the central molecule forms two donor and two acceptor bonds with the neighboring molecules (Pauling ice rules).

charge is present in the region of the lone pairs. When two water molecules are brought together, electrostatic forces favor the characteristic H-bond configuration shown in Fig. 6(b). In the condensed phase, H-bonds are arranged in a local tetrahedral network, shown schematically in Fig. 6(c). In ice phases, up to very high pressure, this pattern is intact and periodically replicated. In water, at normal pressure and temperature, the network is still largely intact, although H-bonds keep breaking and reforming to allow molecular diffusion to take place. Water at normal pressure and temperature is a network liquid. It is precisely the H-bond network that confers to water many of its unusual properties, like its thermodynamic anomalies, its large dielectric constant, and its peculiar character as a solvent.

Under supercritical conditions, the network collapses and water becomes a more standard fluid. The changes occurring in the H-bond network when regular water becomes supercritical are well described by *ab initio* molecular dynamics. This is illustrated in the following figures. Figure 7 reports partial pair correlation functions of standard water from Refs. [41,42] and compares the *ab initio* molecular dynamics results with neutron and x-ray scattering data. The simulations used a box volume corresponding to the experimental density at standard pressure and temperature. The run time was about 10 ps, during which the average temperature of the simulation was 303 K. The simulations were performed on heavy water, in which deuterion (D) replaces hydrogen (H). Within classical statistical mechanics there is no difference in the static properties of light (hydrogenated) and heavy (deuterated) water. The only difference is in the dynamical properties and originates from the mass difference between protons and deuterons. Experimentally small differences in the static properties of light and heavy water are detected, reflecting quantum mechanical effects on the nuclear dynamics. These effects are outside the realm of classical molecular dynamics simulations. Partial pair correlation functions for supercritical water (density 0.73 g/cm^3 and temperature 653 K) are reported in Fig. 8, where they are compared to neutron scattering data [43] that were taken at very similar thermodynamic conditions. In water, there are three distinct 2-body

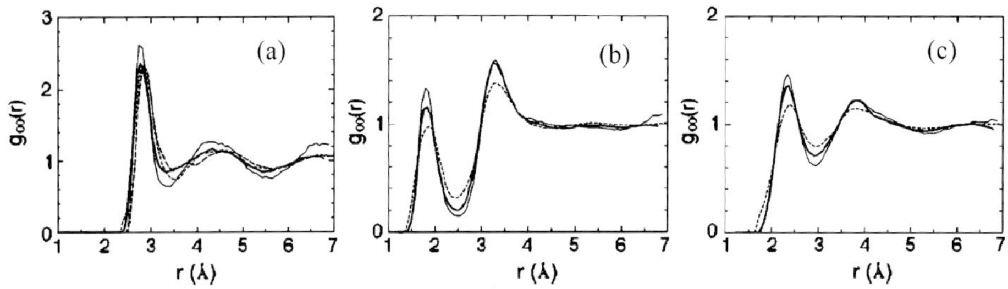

Fig. 7. Calculated partial pair correlation functions of liquid water (thick solid line, corresponding to a cell with 64 molecules; thin solid line, corresponding to a cell with 32 molecules). Panel (a) shows OO correlations; the calculations are compared to experiment (dashed line, neutron scattering data; long-dashed line, x-ray scattering data). Panels (b) and (c) show OH and HH correlations, respectively; the calculations are compared to experiment (dashed-line, neutron scattering data). (Reproduced from Silvestrelli and Parrinello [41,42], where references to x-ray and neutron scattering data can be found.)

correlations: the oxygen–oxygen (OO) correlation, the hydrogen–hydrogen correlation (HH) and the oxygen–hydrogen (OH) correlation. These correlations are described by 2-particle distribution functions. In a macroscopically isotropic and homogenous liquid, the 2-particle distribution functions depend only on the interparticle separation $r \equiv |\mathbf{R} - \mathbf{R}'|$, where $\mathbf{R}$ and $\mathbf{R}'$ are particle positions. For instance, in the case of OO correlations we have $\rho^{(2)}_{OO}(r) = \overline{\sum_{I \neq J}^{N_O} \delta(\mathbf{R} - \mathbf{R}_I)\delta(\mathbf{R}' - \mathbf{R}_J)}$, and the sum is restricted to the oxygen atoms. The OO pair correlation function is given by $g_{OO}(r) = \rho^{-2}\rho^{(2)}_{OO}(r)$, where ρ is the molecular density of the fluid. Similar definitions hold for the two other pair correlation functions, i.e. $g_{HH}(r)$ and $g_{OH}(r)$, that are reported in Figs. 7 and 8. In the same simulations [41,44], the diffusion coefficient D was calculated from the mean square displacement of the molecules using the Einstein relation: $\overline{\frac{1}{N}\sum_I |\mathbf{R}_I(t) - \mathbf{R}_I(0)|^2} = 6Dt + \text{const}$, which holds for large time t. In this way, one obtained a diffusion coefficient $D = 2.8 \pm 0.5 \times 10^{-5}\,\text{cm}^2/\text{s}$ in the simulation with 64 molecules per cell for standard water. This should be compared with the experimental value of $D = 2.4 \times 10^{-5}\,\text{cm}^2/\text{s}$ for water at standard thermodynamic conditions. In the case of supercritical water at the thermodynamic conditions reported above, the simulation gave $D = 46.2 \pm 0.2 \times 10^{-5}\,\text{cm}^2/\text{s}$ to be compared with an experimental value of $D = 47.4 \times 10^{-5}\,\text{cm}^2/\text{s}$ under similar thermodynamic conditions. The agreement between theory and experiment is excellent. It indicates that *ab initio* molecular dynamics and density functional theory (using currently available generalized gradient approximations for exchange and correlations) do not only describe well the structure and dynamics of water under normal thermodynamic conditions, but also capture the collapse of the H-bond network induced by temperature and pressure. Notice that H-bonds do not completely disappear in supercritical water, but more labile and short-lived local H-bond structures involving few molecular units are still present (Fig. 9). This is quite interesting from a physical point of view and resembles findings of *ab initio* molecular dynamics simulations of liquid silicon [6], which show the collapse of the

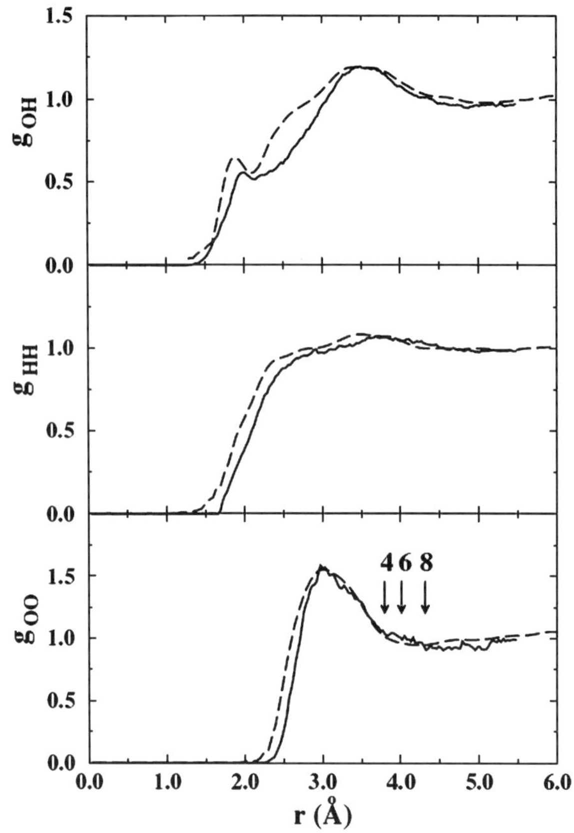

Fig. 8. Partial pair correlation functions for supercritical water (see text): the computed pair correlation functions (solid line) are compared to experiment (dashed line) at the same density. Intramolecular OH and HH contributions have been subtracted out. The arrows indicate (integer) coordination numbers obtained by integrating g_{OO}.

strong tetrahedral covalent bond network that characterizes the crystalline state. Although a covalent bond network is absent in liquid silicon, short-lived local covalent fluctuations are still present and explain the special properties of this liquid. In general, these effects are outside the reach of empirical model potentials, which do not carry information on the electronic structure and which, as a consequence, need a re-parameterization to deal with mutated thermodynamic conditions.

Detailed information on water dynamics can be gathered experimentally from the infrared spectrum (IR). The IR spectrum depends on both nuclear dynamics and electronic structure. The latter mediates the coupling of the system to a macroscopic electric field.

To extract the IR spectrum of water from classical molecular dynamics simulations, one needs to parameterize not only the interatomic interactions, but also the dynamic polarizability of the molecules in the liquid. No empirical parameters are

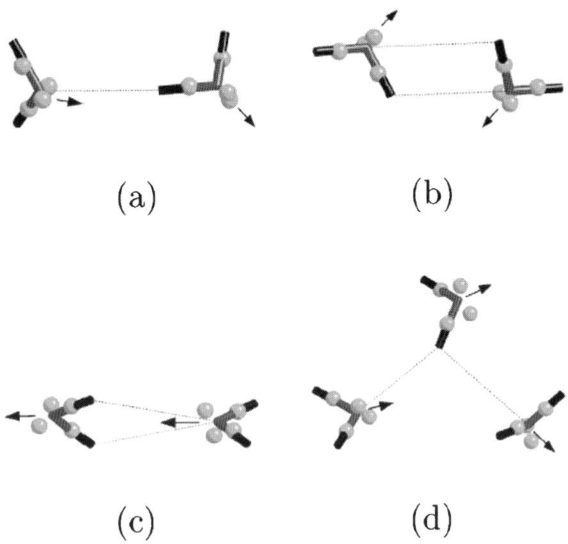

Fig. 9. H-bond configurations in supercritical water (see text). (a) linear H-bond, (b) cyclic H-bond, (c) bifurcated H-bond, (d) twofold H-bond. Water molecules are represented by V-shaped sticks with the H evidenced in black. The light gray balls show MLWF centers. H-bonds are dashed lines and the arrows indicate the direction of the dipole moment of each molecule.

needed instead to extract the IR spectrum from *ab initio* molecular dynamics simulations, which include simultaneously and consistently nuclear and electronic structure information. We consider here heavy water. Each molecule has two deuterons (D_1 and D_2), one oxygen atom (O), and four doubly occupied MLWF (W_s with $s = 1, 4$). To each molecule in the liquid, we associate a dipole moment according to the definition [42] $\mathbf{p} = e(\mathbf{R}^{D_1} + \mathbf{R}^{D_2} + 6\mathbf{R}^{\mathrm{O}} - 2\sum_{s=1,4}\mathbf{R}^{\mathrm{W}_s})$. The dipole moment $\mathbf{p}$ of a molecule in condensed phase is not a physical observable, because MLWF belonging to adjacent molecules overlap. A change of the total dipole moment per unit volume, however, is measurable and defines a change in the macroscopic polarization $\mathbf{P} = \frac{1}{\Omega}(\sum_I \mathbf{p}_I)$. The time autocorrelation function $\overline{P(t) \cdot P(0)}$ can be extracted from *ab initio* molecular dynamics trajectories. Its power spectrum is directly related to the product of the cross section $\alpha(\omega)$ for IR absorption with the refractive index $n(\omega)$ according to the formula:

$$\alpha(\omega)n(\omega) \propto \frac{\omega \tan h\left(\frac{\hbar\omega}{2k_{\mathrm{B}}T}\right)}{\Omega} \int_{-\infty}^{\infty} \mathrm{d}t \exp(-\mathrm{i}\omega t)\overline{\mathbf{P}(t) \cdot \mathbf{P}(\mathbf{0})} \tag{43}$$

An IR spectrum calculated from *ab initio* molecular dynamics using Eq. (43) is reported [45] in Fig. 10, together with the available experimental information for light and heavy water. Not only the calculated spectral features, but also their relative intensity, reflecting the IR couplings, agree well with experiment. In order of decreasing frequency, the peaks in the figure correspond to DO stretching, DO

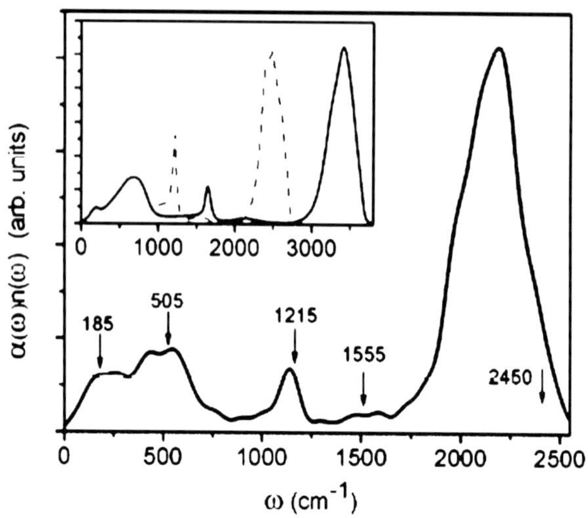

Fig. 10. Calculated IR spectrum of heavy water. The arrows indicate experimental features for the same system. The inset reports the experimental spectrum (continuous line) of standard (non-deuterated) water over the entire frequency range, and the experimental spectrum (dashed line) of heavy water over a partial frequency range.

bending, D_2O librational, and H-bond hindered translational modes, respectively. The first three features have *intramolecular* character and would be present also in the gas phase. In the condensed liquid phase, they are broadened by the mutual molecular interactions. The feature corresponding to hindered translational modes has *intermolecular* character and would not be present in the gas phase. This feature is due to the H-bond network and is particularly interesting. Experimentally, it disappears from the spectrum in supercritical water signaling network collapse. Given the high frequency of the DO stretching and bending modes, compared to the H-bond network modes, it is reasonable to analyze the hindered translations in terms of rigid molecules mutually interacting via H-bonds. This is, however, a good approximation only for the nuclear frames. The electronic cloud, which follows adiabatically the nuclear motion, does not follow *rigidly* the nuclear frames. If it did there would be no spectral feature at $\sim 200/\text{cm}$, because translations of rigid dipoles do not couple to a macroscopic electric field. The electron cloud deforms in a way that reflects the changes of the local H-bond network (see Fig. 11). Environmental effects of this nature are very difficult to capture with classical empirical models. In the *ab initio* molecular dynamics spectrum of Fig. 10, the feature at $\sim 200/\text{cm}$ is due to short-range correlations in the H-bond dynamics schematically depicted in Fig. 11. In classical simulations, in which the molecules are represented by dynamically polarizable point dipoles, the effect of the environment enters only via the long-range Coulomb interactions originating the local field at each molecular position. Interestingly, such simulations give raise to different selection rules for IR absorption than in the *ab initio* molecular dynamics simulation [45]. It is a

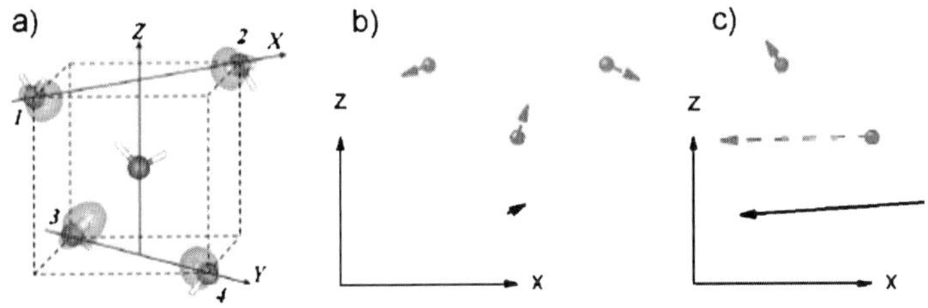

Fig. 11. (a) Instantaneous tetrahedral environment of a water molecule with 4 H-bonds in a simulation snapshot. The O neighbors are labeled 1 to 4; the x-axis connects 1 and 2; the y-axis connects 3 and 4; the z-axis connects the center of 1–2 with the center of 3–4. This local frame is not orthogonal, but the deviation is small. 1, 2, 3, and 4 identify a "cubic" cage shown by the dashed lines. The iso-density plots show MLFW of the corner molecules participating in H-bonds with the central molecule. In (b) and (c), the dashed gray arrows show molecular dipole changes induced by a displacement of the central molecule along z (b) and x (c); the full black arrows show the net effect of the coordinated dipolar change.

remarkable success of *ab initio* molecular dynamics and of the standard model that such subtle features of the H-bond network in liquid water are so well reproduced.

Recently, it has been pointed out that the good agreement between theory and experiment in the pair correlation functions shown in Fig. 7 was due in part to imperfect adiabaticity of the corresponding Car–Parrinello simulations, which for reason of numerical efficiency used large values for the mass parameter μ and the numerical integration time step Δt. More accurate calculations [46,47] using smaller values of μ and Δt which satisfy well the conditions (24) and (25) for adiabatic behavior, give water pair correlation functions that appear over structured compared to experiment. In these more accurate simulations, the experimental diffusion coefficient is no longer reproduced: *ab initio* water at room temperature exhibits sluggish diffusion indicative of glassy behavior. Only by increasing the temperature of the simulation by $\sim 100\,\mathrm{K}$ does one recover good liquid-like diffusion. These results suggest that the intermolecular bonds in *ab initio* water are too strong. It is not clear at present why *ab initio* water is over structured. It may be due to the limited accuracy of the GGA approximation of density functional theory, or it may be due to neglect of quantum effects in the nuclear motion [47]. We notice that a temperature of 100 K corresponds to an energy $k_{\mathrm{B}}T \sim 0.01\,\mathrm{eV}$. This is very small on the scale of the H-bond binding energies, which are of the order of 0.2 eV in water. Yet, a change of temperature of 100 K is enough to bring water from the freezing to the boiling point.

Finally, we remark that the imperfect adiabaticity of some of the simulations discussed above has negligible effect on electronic properties like the electron charge density. This is not surprising in view of the small magnitude of the effect on the scale of electronic energies. Thus, adjusting μ to achieve good liquid-like diffusion at room temperature can be considered as an empirical way of fixing some limitations of the current theory.

6. PHASE DIAGRAMS FROM FIRST PRINCIPLES

In many situations, the standard model of solids accurately predicts the relative stability of crystalline structures [15]. Density functional calculations have been successful in predicting the phase diagrams of solids under varying pressure conditions and/or in assessing the stability of hypothetical crystalline structures. This has been useful to guide the design of new materials. Knowing the stable thermodynamic phases of elemental and composite materials under extreme pressure and temperature conditions is of utmost importance in geophysical and planetary sciences because extremely high pressures and temperatures are present in the Earth's mantle and in the interiors of the planets. Experiments at such extreme conditions are difficult and, in some cases, impossible. In these situations, access to reliable simulation data is the best substitute to experiment. The stability of competing phases is controlled by their relative free energy. At zero temperature, the free energy coincides with the internal energy and is given by the total potential energy in Eq. (9). In crystalline phases, finite temperature corrections to the potential energy can be evaluated in terms of harmonic or quasi-harmonic expansions. These expansions fail close to the melting point and cannot be applied to liquid phases. To predict a phase diagram that includes solid and liquid phases, one need to compute the free energy exactly, avoiding perturbative expansions. This is possible with molecular dynamics simulations [1]. In this context, *ab initio* molecular dynamics extends to finite temperature and to liquid phases the capability of the standard model to predict phase diagrams from parameter free fundamental quantum theory.

Let us focus on the canonical ensemble (N, Ω, T). In terms of the canonical partition function Z the Helmholtz free energy F is given by

$$F = -k_{\mathrm{B}}T \ln Z \tag{44}$$

For an interacting classical system of N identical particles of mass M the partition function is given by $Z = Z_{\mathrm{id}}Z_{\mathrm{con}}$, where $Z_{\mathrm{id}} = \frac{\Lambda^{-3N}}{N!}\Omega^{\mathrm{N}}$ is the ideal gas partition function, $\Lambda = (\frac{2\pi\hbar^2}{Mk_{\mathrm{B}}T})^{1/2}$ the de Broglie thermal wavelength, and $Z_{\mathrm{con}} = \frac{1}{\Omega^{\mathrm{N}}}\int \mathrm{d}\{\mathbf{R}\}\exp[-\frac{\Phi(\{\mathbf{R}\})}{k_{\mathrm{B}}T}]$ the configurational partition function. The partition function Z_{con} depends on all the configurations $\{\mathbf{R}\}$ of the N particle system and cannot be directly calculated even with ultra-powerful computers. However, a derivative of the free energy with respect to a parameter λ controlling the potential energy $\Phi(\{\mathbf{R}\})$ is accessible to computer simulations. This derivative is given by

$$\frac{\partial F}{\partial \lambda} = -k_{\mathrm{B}}T\left(Z^{-1}\frac{\partial Z}{\partial \lambda}\right) = \frac{\int \mathrm{d}\{\mathbf{R}\}(\frac{\partial \Phi}{\partial \lambda})\exp[-\frac{\Phi(\{\mathbf{R}\})}{k_{\mathrm{B}}T}]}{\int \mathrm{d}\{\mathbf{R}\}\exp[-\frac{\Phi(\{\mathbf{R}\})}{k_{\mathrm{B}}T}]} = \left\langle \frac{\partial \Phi}{\partial \lambda} \right\rangle \tag{45}$$

Equation (45) is a configurational ensemble average. It can be calculated as a temporal average $\overline{\partial \Phi / \partial \lambda}$ from molecular dynamics simulations. Thus, the free energy difference between two thermodynamic states of a system, A and B, characterized by values λ_A and λ_B of a parameter λ can be calculated by integration:

$$\Delta F = F_B - F_A = \int_A^B \mathrm{d}\lambda \left\langle \frac{\partial \Phi}{\partial \lambda} \right\rangle_\lambda \tag{46}$$

In Eq. (46), the integrand is evaluated for the ensemble characterized by the value λ of the parameter, as indicated by the subscript. If A is a reference state of known free energy, Eq. (46) allows us to calculate the absolute free energy of state B as $F_B = F_A + \Delta F$. Thermodynamic integration is also used to extract free energy differences from experimental data, e.g. by integrating the measured specific heat over a given temperature interval. In numerical simulations, there is more flexibility than in experiments in the choice of the integration parameter. A common choice uses a λ-dependent potential defined by $\Phi_\lambda = \lambda\Phi_B + (1-\lambda)\Phi_A$. The path connecting A and B must be a reversible thermodynamic path. A necessary condition for this to happen is to avoid the crossing of first-order phase boundaries. Then the absolute free energy of a solid can be calculated by converting it into a harmonic solid, typically an Einstein crystal, the free energy of which is known analytically. Similarly, the absolute free energy of a liquid can be calculated by converting it into an ideal gas, the free energy of which is known analytically. In molecular dynamics simulations, the integral (46) can be obtained from a single molecular dynamics trajectory [48], by introducing a time-dependent switching function $\lambda(t)$, which varies smoothly from 0 to 1 on a time long compared to system dynamics. Equation (46) is replaced by

$$\Delta F = \int_0^{\tau_s} \mathrm{d}t \dot{\lambda}(t) \frac{\partial \Phi_\lambda}{\partial \lambda} \tag{47}$$

Here τ_s is the switching time. If the switching is *adiabatic*, the integral (47) does not depend upon the switching time. This approach has been pursued successfully in the context of *ab initio* molecular dynamics to compute absolute free energies of solid and liquid systems [49,50]. Given the large computational cost of *ab initio* simulations, it is convenient to adopt a two-stage protocol. In the first stage, the *ab initio* system is converted into an intermediate reference system described by empirical interatomic interactions. The potential parameters of the intermediate reference system should be optimized to make this system as close as possible to the *ab initio* system. Then, the free energy change in the first stage of the protocol is relatively small and the number of molecular dynamics steps necessary to switch adiabatically from the *ab initio* to the intermediate reference system is not too large. In the second stage, the empirical system is converted into a reference system of known free energy. No electronic structure calculation is involved in this stage, which has negligible numerical cost in comparison to the first stage.

In its first application, this approach has been used to study the melting transition of silicon [49]. This work adopted the LDA for exchange and correlation. The results of this study are summarized in Table 1, which shows that several properties of silicon at melting are well described by *ab initio* density functional theory, the most notable exception being the melting temperature itself, which is underestimated by more than 20%. This error has been attributed to the LDA, which stabilizes the metallic liquid more than the insulating solid [49]. Subsequent

Table 1. Thermodynamic properties of silicon at the theoretical and at the experimental melting point. The LDA calculations are from Sugino and Car [49].

	This work (LDA)	Experiment
T_m (K)	$1.35(10) \times 10^3$	$1.685(2) \times 10^{3a}$
S_s	$6.9(1)k_B$	$7.4k_B^b$
S_l	$9.9(2)k_B$	$11.0k_B^b, 10.7k_B^c$
H_s (eV/atom)	$H_s(0\,K) + 0.33(2)$	$H_s(0\,K) + 0.41^b$
ΔH_{sl} (eV/atom)	0.35(2)	$0.52^b, 0.47^c$
C_s (eV/K atom)	$3.0(4) \times 10^{-4}$	3.03×10^{-4d}
C_l (eV/K atom)	$2.7(4) \times 10^{-4}$	3.03×10^{-4d}
V_s [(a.u.)^3/atom]	$1.350(5) \times 10^2$	1.380×10^{2a}
$\Delta V/V_s$	10(1)%	$11.9\%^c, 9.5\%^c$
α_s (K^{-1})	$0.3(1) \times 10^{-5}$	0.44×10^{-5a}
α_l (K^{-1})	$4.8(5) \times 10^{-5}$	5.2×10^{-5d}
dT_m/dp (K/GPa)	$-50(5)$	-38^a

Footnotes (a–d) refer to references quoted in [49] which report the experimental data given in the table.

calculations found that the error in the melting temperature could be systematically reduced, by improving the approximation adopted for exchange and correlation. In particular, the GGA gives a melting temperature close to 1500 K [50], while the TPSS metaGGA [51,52] gives a melting temperature close to 1700 K [53]: this temperature coincides with experiment within the numerical uncertainty of the calculation. Very recently this approach has been used to predict the diamond melting line in an extended range of pressures and temperatures [54]. The results are reported in Fig. 12. So far, it has not been possible to extract the diamond melting line from experiment, given the extreme pressure and temperatures that are necessary to melt diamond. Only the graphite–diamond coexistence line and the graphite melting line, which are located at lower pressure and temperature than the diamond melting line, are known with reasonable accuracy from experiment. The diamond melting line in Fig. 12 represents therefore the best prediction available to date. Notice that the extrapolation to low temperature and pressure of the calculated melting line is in excellent agreement with the experimental location of the diamond–graphite–liquid triple point. It is also worth noticing that the diamond melting line exhibits a re-entrant behavior at high pressure. The re-entrant behavior can be understood as follows. At relatively low temperature and pressure, molten diamond is less dense than the crystal because of the presence in the liquid of low-coordinated local graphitic configurations. At sufficiently high pressure, however, molten diamond acquires a larger coordination than the solid. As a consequence, it becomes more dense than the latter, causing the melting slope to become negative. This is similar to molten silicon and germanium, which both have higher coordination than the corresponding crystal at standard pressure. Pressures around 600 GPa and temperatures around 7000 K are expected to occur in the interior of the outer planets Uranus and Neptune. The phase diagram in Fig. 12 indicates that

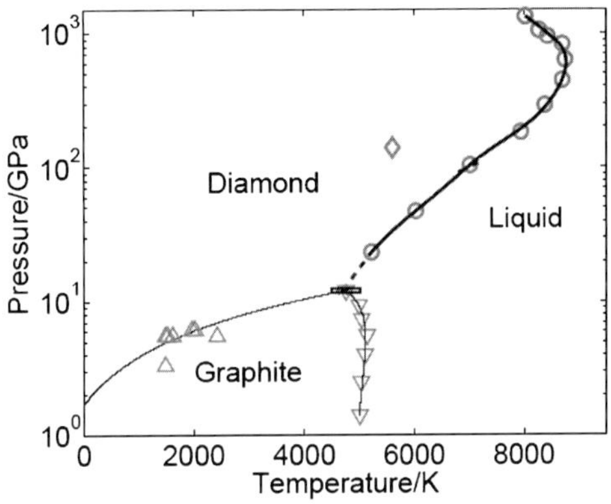

Fig. 12. Phase diagram of carbon. The graphite–diamond boundary (thin solid line and up triangles) is from experiment. The graphite–liquid boundary (thin solid line and down triangles) is from experiment. The rectangle gives the uncertainty on the experimental location of the triple point. The open diamond gives the thermodynamic condition of a shock wave experiment, in which carbon was found to be in the crystalline diamond phase. The diamond–liquid boundary (thick solid line and open circles) is from *ab initio* molecular dynamics simulations using a GGA exchange-correlation functional. The thick dashed line is an extrapolation.

any carbon present in the interior of these planets should be in the crystalline diamond form.

7. RARE EVENTS

With currently available computational resources, it is possible to generate *ab initio* molecular dynamics trajectories spanning tens of picoseconds. These times are longer than molecular vibration periods and are also longer than typical relaxation times in a liquid. Significantly longer times, the order of hundreds of nanoseconds, are ordinarily accessible in simulations based on empirical potentials. Yet, all these times are very short on a macroscopic scale. Many physical phenomena involve collective molecular motions that occur on time scales inaccessible to direct molecular dynamics simulation. These dynamical processes are called rare events because they have a low probability of occurrence on the time scale of molecular motion. Phenomena controlled by rare events include relaxation processes in glasses, conformational changes in macromolecules, nucleation processes, and chemical reactions. Understanding these phenomena at the molecular level poses a great theoretical challenge and is the focus of intense current research.

Here we limit ourselves to consider the special case of chemical reactions when we know the reactant (A) and the product (B), and there is a well-defined path of minimal energy in configuration space connecting A to B. For an activated reaction this path has to overcome an energy barrier ΔE. The probability $\wp$ to find the system at the top of the barrier obeys the Arrhenius law $\wp \propto \exp(-\frac{\Delta E}{k_B T})$ and is exponentially small for large ΔE. Under these circumstances direct molecular dynamics simulation is useless, because it would take a time much longer than is practically accessible to sample a reactive dynamical path. While a dynamical path is defined in phase space, a minimum energy path (MEP) is defined in configuration space. A MEP identifies the energetically most likely sequence of configurations visited by a system going from A to B. Finding a MEP is considerably simpler than finding a reactive dynamical path. Then the rate $R_{A\to B}$, i.e. the reaction probability per unit time, can be estimated from Transition State Theory (TST) [1,55]. Under the simplest approximations [55] the TST rate takes the form:

$$R_{A\to B} \cong \nu_A \exp(-\Delta E/k_{\mathrm{B}}T) \tag{48}$$

Here, ΔE is the activation energy barrier and ν_A is the attempt frequency, i.e. the frequency of the vibration mode of the system along the direction of the MEP at the reactant site A. Both ΔE and ν_A are easily calculated once we know the MEP. The estimate in Eq. (48) can be further refined, by using a more accurate TST expression and by including dynamical corrections to TST [1,55]. In what follows, we briefly illustrate techniques that can be used to find a MEP in the context of *ab initio* molecular dynamics.

We consider in particular the so-called string method [56], which can be seen as a variant of the nudged elastic band (NEB) method [57]. In this context, a MEP is a string (or path) φ connecting A to B, which satisfies the equation:

$$(\nabla_{\{\mathbf{R}\}}\Phi[\varphi])^{\perp} = 0 \tag{49}$$

Here $\nabla_{\{\mathbf{R}\}}\Phi \equiv (\frac{\partial\Phi}{\partial\mathbf{R}_1}, \frac{\partial\Phi}{\partial\mathbf{R}_2}, \ldots, \frac{\partial\Phi}{\partial\mathbf{R}_N})$ is the configuration space gradient of the nuclear potential energy, and $(\nabla_{\{\mathbf{R}\}}\Phi)^{\perp}$ is the component of this gradient orthogonal to the string φ. In order to write Eq. (49) in explicit form, we introduce a parametric representation $\varphi(\alpha)$ of the string φ in terms of a scalar variable α defined along the string. A convenient choice is to take the normalized arc length of a path connecting A to B. Then $0\leqslant\alpha\leqslant 1$ and the two end states, A and B, correspond to $\alpha = 0$ and $\alpha = 1$, respectively. The parameter α labels a configuration $\{\mathbf{R}\}^{\alpha}$, i.e. $\varphi(\alpha) \equiv \{\mathbf{R}\}^{\alpha}$. The tangent vector to the string is $\varphi_{\alpha} \equiv \frac{\mathrm{d}\varphi}{\mathrm{d}\alpha}$, and its norm is $|\varphi_{\alpha}| = (\varphi_{\alpha}\cdot\varphi_{\alpha})^{1/2}$. The requirement that the parameterization is preserved when the string deforms is expressed by the condition that the local arc length is constant along the string, i.e.:

$$\frac{\mathrm{d}}{\mathrm{d}\alpha}|\varphi_{\alpha}| = 0 \tag{50}$$

Eq. (50) is equivalent to

$$\varphi_{\alpha}\cdot\varphi_{\alpha} = c \tag{51}$$

Here c is a constant (independent of α) determined by the condition $\int_0^1 |\varphi_{\alpha}|^2\,\mathrm{d}\alpha = c$. With the adopted parameterization the only allowed deformations of a string are those in which the local elastic stretching energy, proportional to $|\varphi_{\alpha}|^2$, is distributed uniformly along the string. Equation (49) becomes

$$(\nabla_{\{\mathbf{R}\}}\Phi[\varphi(\alpha)])^{\perp} - \lambda(\alpha)\hat{\tau}(\alpha) = 0 \tag{52}$$

Here $\hat{\tau}(\alpha) \equiv \frac{\varphi_\alpha(\alpha)}{|\varphi_\alpha(\alpha)|}$ is the unit vector tangent to the string, $\lambda(\alpha)$ is a Lagrange multiplier that imposes the constraint of Eq. (50) (or (51)), and $(\nabla_{\{\mathbf{R}\}}\Phi)^\perp = \nabla_{\{\mathbf{R}\}}\Phi - (\nabla_{\{\mathbf{R}\}}\Phi \cdot \hat{\tau})\hat{\tau}$. Starting from an initial guess, a string that satisfies Eq. (52) can be found by damped molecular dynamics:

$$\ddot{\varphi}(\alpha; t) = -(\nabla_{\{\mathbf{R}\}}\Phi[\varphi])^\perp - \gamma_s \dot{\varphi}(\alpha; t) + \lambda(\alpha; t)\hat{\tau}(\alpha; t) \tag{53}$$

Here the "time" t labels a string following damped molecular dynamics evolution, and γ_s is a friction coefficient that can be adjusted to control convergence. In numerical implementations, the continuous parameter α is replaced by an integer $l = 0, 1, 2, \ldots, P$, so that a string becomes a chain of $P + 1$ replicae of the system (including the end points) distributed uniformly (because of the constraint) on a path connecting reactant and product in configuration space.

The string method can be extended to *ab initio* molecular dynamics within the Car–Parrinello scheme in a straightforward way [58]. In this case, the lth replica of a system has N nuclei and $N_e/2$ doubly occupied electronic orbitals (within spin restricted DFT). Damped molecular dynamics in the extended space of the electronic and nuclear degrees of freedom belonging to all the replicae allows us to simultaneously optimize the electronic state and the spatial configuration of all the replicae. At the end of this procedure, the replicae are distributed uniformly along a MEP connecting reactant and product states. The damped molecular dynamics equations of *ab initio* string dynamics are:

$$\begin{aligned} M\ddot{\mathbf{R}}_I^{(l)} &= -\left(\frac{\partial\Phi_{\mathrm{CP}}[\{\mathbf{R}\}^{(l)}, \{\psi\}^{(l)}, \{\psi^*\}^{(l)}]}{\partial\mathbf{R}_I^{(l)}}\right)^\perp - \gamma_s M\dot{\mathbf{R}}_I^{(l)} + \Lambda^{(l)}\hat{\tau}_I^{(l)} \\ \mu|\ddot{\psi}_i^{(l)}\rangle &= -\frac{\delta\Phi_{\mathrm{CP}}[\{\mathbf{R}\}^{(l)}, \{\psi\}^{(l)}, \{\psi^*\}^{(l)}]}{2\delta\langle\psi_i^{(l)}|} - \gamma_e\mu|\dot{\psi}_i^{(l)}\rangle + \sum_j |\psi_j^{(l)}\rangle\lambda_{ji}^{(l)} \end{aligned} \tag{54}$$

The first of these equations is just Eq. (53) for a discretized string and we have added a mass parameter M that controls the dynamic response of the nuclei. Since the purpose of string molecular dynamics is to find a MEP, the mass M is just an adjustable parameter to speed up convergence. The Lagrange multiplier $\Lambda^{(l)}$ ensures that the replicae stay equally spaced when the string deforms: it can be calculated with a SHAKE-like algorithm as detailed in Ref. [58]. The second equation in (54) is the usual damped Car–Parrinello equation for the electrons, which is applied here independently to each replica. The number of replicae depends on the application but a number of about 10 replicae is usually sufficient to model simple chemical reaction pathways. Distinct replicae are largely independent from each other since they are coupled only via the unit tangent vector to the string and via the Lagrange multipliers $\Lambda^{(l)}$. Thus, *ab initio* string dynamics can be efficiently implemented on parallel computers.

To illustrate the method, we recall a recent application in which *ab initio* string dynamics was used to study the reaction between an organic molecule, acetylene (C_2H_2), and a partially hydrogenated silicon (111) surface [59]. Understanding the processes by which organic molecules interact with semiconductor surfaces is

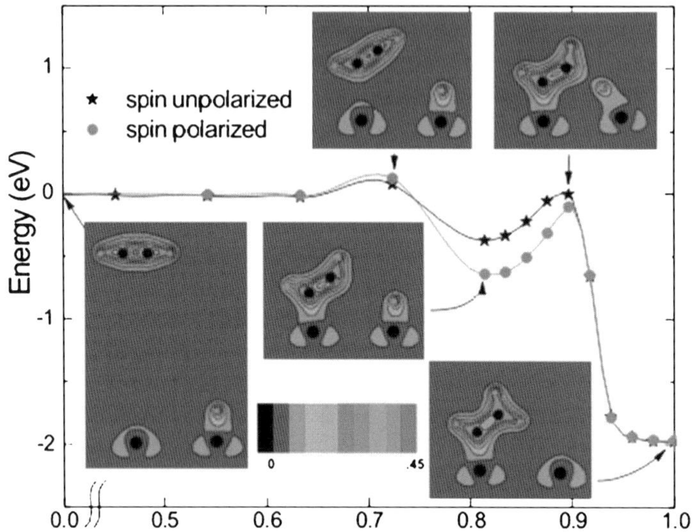

Fig. 13. Potential energy profile along the MEP for the reaction of acetylene with a H–Si(111) surface. Black and gray lines correspond to spin unpolarized and spin polarized calculations, respectively. The zero of energy corresponds to the non-interacting surface + molecule system. The panels show charge density plots at selected locations along the MEP. The total electronic charge density is represented on a plane perpendicular to the surface and passing through the C–C molecular bond.

important to functionalize semiconductors surfaces. The reaction between acetylene and a silicon surface is a particularly simple example in this class of reactions. In Ref. [59], a Si(111) surface was considered in which a single dangling bond was present in an otherwise fully hydrogenated surface. The presence of a dangling bond makes the surface reactive when acetylene approaches. The MEP for the reaction between acetylene and the silicon surface is shown in Fig. 13: it is characterized by the presence of two barriers and an intermediate metastable state. The first barrier is rather low and leads to the formation of a chemical bond between the molecule and the surface. As a result, the dangling bond initially present on the surface has disappeared. There is an energy barrier because one of the π bonds in the triple C–C bond has to break to allow one of the two carbons to bind to the silicon surface. The resulting configuration is metastable because the carbon atom that does not bind to the surface is left with one bond less, i.e. C_2H_2 has become a radical in the intermediate state. A more stable configuration is found by allowing the radical to further interact with the surface. The second (larger) barrier in Fig. 13 corresponds to the process in which a hydrogen atom is abstracted from the surface to form a more stable bond with the second carbon of the acetylene molecule. As a result, a dangling bond reappears on the surface at a location adjacent to the C_2H_2 absorption site. This is a prototype chain reaction, which has been found experimentally to play an important role in the chemical functionalization of semiconductor surfaces with organic molecules [60]. It is a chain reaction because it leaves the surface with a dangling bond, so that the entire reaction can be repeated when a second acetylene molecule approaches the surface. As a result of chain reactions of this type, clusters of organic molecules tend to form on semiconductor surfaces exposed to these molecules [60]. Figure 13 reports the results of both spin-restricted and spin-unrestricted *ab initio* string dynamics calculations. It is worth noticing that spin polarization stabilizes the radical state of the absorbed C_2H_2 molecule.

The string method works nicely in the absorption process that we have just described because this process is characterized by a well-defined MEP connecting reactant and product states. In this process, the distance between adjacent replicae of the system in configuration space is a good reaction coordinate.

8. OMISSIONS, PERSPECTIVES AND OPEN ISSUES

In reviewing *ab initio* molecular dynamics, we have focused on key concepts and a few illustrative applications. Complementary perspectives can be found in other reviews [26,27]. Among our most notable omissions are applications to biological systems, a field in which there has been a growing number of *ab initio* molecular dynamics studies in recent years [61]. *Ab initio* simulations, which describe accurately chemical reactions and solution effects, may help to elucidate the molecular underpinnings of important biological processes, such as e.g. enzymatic reactions. Biological systems pose severe challenges in terms of size and time limits of the simulations. Biological molecules contain several thousands of atoms and are typically in aqueous solution. Such large and complex systems are currently outside the reach of *ab initio* simulations. This has prompted the development of hybrid schemes in which only a fragment of the system of interest, e.g. the enzymatic active site and its immediate environment, is treated at the quantum mechanical level, while the more distant environment is treated with classical force fields [62]. These hybrid approaches are called QM/MM methods because they combine quantum mechanics (QM) with classical molecular mechanics (MM). The treatment of the boundary between quantum and classical subsystems poses difficulties, for which no general solution has been found yet. However, practical schemes that work satisfactorily under special circumstances are quite common.

The size limit would be less severe if the cost of the quantum mechanical calculations had a better scaling with size. Existing pseudopotential plane wave implementations scale as the cube of the number N of atoms. In principle, linear scaling with N should be possible, for insulating systems at least, as a consequence of the local character of the chemical bonds [63] and of the principle of nearsightedness of electronic matter [12]. Linear scaling methods are reviewed in Ref. [64]. So far, these methods have been implemented mainly in the context of simplified electronic structure calculations, but, in principle, it should be possible to develop linearly scaling *ab initio* molecular dynamics schemes without sacrificing the accuracy currently achievable within plane wave pseudopotential approaches. For instance, it may be possible to achieve this goal by exploiting the localization properties of MLWF [39]. *Ab initio* molecular dynamics schemes in the MLWF representation have been formulated (see e.g. [65]), but current implementations do not exploit the localized character of the MLWF to achieve linear scaling.

Rare events are an important challenge for atomistic simulations. In rugged potential energy landscapes, there may not be a single, or a few, isolated MEP connecting reactant and product states, but the reactive flux may proceed via an ensemble of minimum energy pathways. Proposals have been made to extend the string method to finite temperature in order to sample ensembles of pathways. More generally, the whole concept of a string consisting of equally spaced replicae in configuration space may loose its meaning. This would happen whenever the distance in configuration space between different realizations of a system is not a good reaction coordinate. Consider for instance the process in which a crystal nucleates from a liquid. Because of atomic diffusion, two equivalent microscopic realizations of a liquid can have arbitrary separation in configuration space. Thus, the distance between microscopic realizations in configuration space is not a good reaction coordinate for the nucleation process. Good reaction coordinates must be collective coordinates, or order parameters, that are able to differentiate between a liquid and a solid. Recently, Laio and Parrinello formulated a new molecular dynamics approach to explore free energy surfaces [66]. In the Laio–Parrinello scheme, called metadynamics, an ad-hoc dynamics in the space of a few order parameters forces the microscopic dynamics of a system to explore macroscopically different configurations, i.e. configurations that are characterized by different values of the collective coordinates. The metadynamics approach has been also implemented in the *ab initio* molecular dynamics context [67]. In general, we expect that finding ways to overcome the time bottleneck of rare events will remain an important area of theoretical research in the years to come. The whole field of molecular dynamics simulations will greatly benefit from progress in this domain.

Quantum effects in the nuclear coordinates are another issue. These effects, neglected in molecular dynamics simulations, are particularly important for light nuclei such as protons. In crystalline systems, quantum effects in the nuclear motion are usually treated within harmonic or quasi-harmonic expansions. There is however a general methodology to include "exactly" quantum nuclear effects in simulations of equilibrium statistical properties. This is based on the Feynman path integral representation of the partition function. At finite temperature, the path integral representation establishes an exact correspondence, or isomorphism, between the partition function of a system of N quantum particles and that of a system of NP classical particles subject to a suitable interaction potential [68]. Here P is an integer that stands for the number of replicae of a system that are needed in the discrete representation of the action in the path integral. By exploiting the isomorphism, we can use classical molecular dynamics or Monte Carlo simulations on a suitably defined fictitious system of NP particles to compute "exactly" the equilibrium statistical properties of a quantum system of N particles. This approach has been successfully combined with *ab initio* molecular dynamics in the Car–Parrinello scheme [69] to study highly fluxional molecular species [70] and proton tunneling effects in charged water complexes [71]. *Ab initio* path integral simulations have a high computational cost because a system with N atoms and N_e

electrons needs to be replicated P times. The number P is inversely proportional to temperature: the lower the temperature the higher is the number of replicae that need to be included in a simulation. Since only adjacent replicae on a Feynman path are coupled by simple harmonic forces, path integral simulations can be effectively implemented on parallel computers.

Path integrals are convenient to compute static equilibrium properties, but there is still no general way to handle satisfactorily the quantum dynamics of a system of many coupled electrons and nuclei. An important quantum dynamical effect that we need to consider in condensed phase simulations is branching of classical nuclear trajectories near conical intersections of the Born–Oppenheimer surfaces. These non-adiabatic effects can be handled with approximations such as surface hopping and related models [72]. In these models, a system evolves classically on a Born–Oppenheimer surface, either ground- or excited-state. Quantum transitions (hops) between different Born–Oppenheimer surfaces occur only near conical intersections leading to branching of the classical trajectories. Surface hopping models have been mostly used in simulations of the dynamics of a single quantum particle, typically an electron, solvated in a classical fluid environment. A difficulty arises in the many electron case because density functional theory is a ground-state theory and does not describe excited Born–Oppenheimer surfaces. The single particle approach of density functional theory can be extended to excitation phenomena in the context of time-dependent density functional theory [73], which appears therefore very promising to formulate a non-adiabatic *ab initio* molecular dynamics approach. A viable *ab initio* surface hopping scheme would be very useful to study phenomena in which structural changes are induced by electron excitation, such as, for instance, in photo-catalytic reactions.

Finally, the success of *ab initio* molecular dynamics applications owes much to the success of currently available approximations in density functional theory. These approximations work well in many situations, but there are cases in which they fail. Most notably this happens with weak physical interactions between closed shell systems (Van der Waals interactions), but there are cases in which even strong chemical bonds are not handled properly. The accuracy of density functional approximations may affect the energy ordering of isomers, and/or the value of the barriers that separate reactants from chemical products. Finding better approximate functionals for exchange and correlation has been an extremely active area of research in the recent past, and the pace of progress does not seem to have slowed down yet.

Looking back at the past two decades, the whole area of *ab initio* molecular dynamics has undergone major progress and many exciting new developments have been made since its inception in 1985. Originally conceived in the context of solid and liquid state physics, this approach has rapidly reached out to chemical physics, earth and planetary sciences, chemistry, and biochemistry/biophysics. This progress has been boosted by the phenomenal increase in the available computational resources that has occurred at the same time. Judging from the past, we should expect that progress and excitement should continue in the years to come. Theoretical concepts and computational approaches devised in the context of *ab initio* molecular dynamics have not only been useful on their own but have also been a source of inspiration in other fields.

ACKNOWLEDGEMENTS

I am indebted to all the collaborators and colleagues who have greatly contributed to develop and apply *ab initio* molecular dynamics methodologies over the past two decades. Owing to the limits of this review, it has been impossible to properly acknowledge most of these contributions. NSF support under grant CHE-0121432 is gratefully acknowledged.

REFERENCES

[1] D. Frenkel and B. Smit, *Understanding Molecular Simulation, from Algorithms to Applications* (*Computational Science, from Theory to Applications*) (Academic Press, San Diego, 2002).
[2] G. Ciccotti, D. Frenkel, and I.R. McDonald *(eds) Simulation of Liquids and Solids: Molecular Dynamics and Monte Carlo Methods in Statistical Mechanics* (North Holland, Amsterdam, 1986).
[3] J.P. Hansen and I.R. McDonald, *Theory of Simple Liquids* (Academic Press, San Diego, 1986).
[4] K. Huang, *Statistical Mechanics* (Wiley, New York, 1963).
[5] F.H. Stillinger and T.A. Weber, Computer simulation of local order in condensed phases of silicon, *Phys. Rev. B* **31**, 5262 (1985).
[6] I. Stich, R. Car and M. Parrinello, Structural, bonding, dynamical, and electronic properties of liquid silicon: An *ab-initio* molecular dynamics study, *Phys. Rev. B* **44**, 4262 (1991).
[7] M. Born and J.R. Oppenheimer, Zur quantentheorie der molekeln, *Ann. Phys.* **84**, 457 (1927).
[8] P. Hohenberg and W. Kohn, Inhomogeneous electron gas, *Phys. Rev.* **136**, B864 (1964).
[9] W. Kohn and L.J. Sham, Self-consistent equations including exchange and correlation effects, *Phys. Rev.* **140**, A1133 (1965).
[10] R.G. Parr and W. Yang, *Density Functional Theory of Atoms and Molecules* (Oxford University Press, New York, 1989).
[11] R.M. Dreizler and E.K.U. Gross, *Density Functional Theory, an Approach to the Quantum Many-Body Problem* (Springer, Berlin, 1990).
[12] W. Kohn, Nobel lecture: Electronic structure of matter-wave functions and density functionals, *Rev. Mod. Phys.* **71**, 1253 (1999).
[13] R. McWeeny, *Methods of Molecular Quantum Mechanics* (Academic Press, London, 1992).
[14] J.C. Phillips and L. Kleinman, New methods for calculating wave functions in crystals and molecules, *Phys. Rev.* **116**, 287 (1959).
[15] W.E. Pickett, Pseudopotential methods in condensed matter applications, *Comput. Phys. Rep.* **9**, 115 (1989).
[16] D.R. Hamann, M. Schlüter and C. Chiang, Norm-conserving pseudopotential, *Phys. Rev. Lett.* **43**, 1494 (1979).
[17] R. Car and M. Parrinello, Unified approach for molecular dynamics and density functional theory, *Phys. Rev. Lett.* **55**, 2471 (1985).
[18] G. Pastore, E. Smargiassi and F. Buda, Theory of *ab-initio* molecular dynamics calculations, *Phys. Rev. A* **44**, 6334 (1991).
[19] R. Car, M. Parrinello and M. Payne, Comment on error cancellation on the molecular dynamics method for total energy calculation, *J. Phys.: Condens. Matter* **3**, 9539 (1991).

[20] P. Blöchl and M. Parrinello, Adiabaticity in first-principles molecular dynamics, *Phys. Rev. B* **45**, 9413 (1992).
[21] S. Nose, Constant temperature molecular dynamics methods, *Prog. Theor. Phys. Suppl.* **103**, 1 (1991).
[22] G.J. Martyna, M.L. Klein and M.E. Tuckerman, Nose'–Hoover chains: The canonical ensemble via continuous dynamics, *J. Chem. Phys.* **97**, 2635 (1992).
[23] H.C. Andersen, Molecular dynamics simulations at constant pressure and/or temperature, *J. Chem. Phys.* **72**, 2384 (1980).
[24] M. Parrinello and A. Rahman, Crystal structure and pair potentials: A molecular-dynamics study, *Phys. Rev. Lett.* **45**, 1196 (1980).
[25] M. Parrinello and A. Rahman, Polymorphic transitions in single crystals: A new molecular dynamics method, *J. Appl. Phys.* **52**, 7182 (1981).
[26] M. Parrinello, From silicon to RNA: the coming of age of *ab-initio* molecular dynamics, *Solid State Comm.* **102**, 53 (1997).
[27] J.S. Tse, *Ab-initio* molecular dynamics with density functional theory, *Annu. Rev. Phys. Chem.* **52**, 249 (2002).
[28] J.P. Rickaert, G. Ciccotti and J.C. Berendsen, Numerical integration of the cartesian equations of motion of a system with constraints: Molecular dynamics of *n*-alkanes, *J. Comput. Phys.* **23**, 327 (1977).
[29] L. Kleinman and D.M. Bylander, Efficacious form for model pseudopotentials, *Phys. Rev. Lett.* **48**, 1425 (1982).
[30] D. Vanderbilt, Soft self-consistent pseudopotentials in a generalized eigenvalue formalism, *Phys. Rev. B* **41**, 7892 (1991).
[31] A. Pasquarello, K. Laasonen, R. Car, C-Y. Lee and D. Vanderbilt, *Ab-initio* molecular dynamics for d-electron systems: Liquid copper at 1500 K, *Phys. Rev. Lett.* **69**, 1982 (1992).
[32] K. Laasonen, A. Pasquarello, R. Car, C-Y. Lee and D. Vanderbilt, Car–Parrinello molecular dynamics with Vanderbilt ultrasoft pseudopotentials, *Phys. Rev. B* **47**, 10142 (1993).
[33] P. Blöchl, Projector augmented-wave method, *Phys. Rev. B* **39**, 4997 (1994).
[34] F. Tassone, F. Mauri and R. Car, Acceleration schemes for *ab-initio* molecular dynamics simulations and electronic structure calculations, *Phys. Rev. B* **50**, 10561 (1994).
[35] I. Stich, R. Car, M. Parrinello and S. Baroni, Conjugate gradient minimization of the energy functional: A new method for electronic structure calculation, *Phys. Rev. B* **39**, 4997 (1989).
[36] M.P. Teter, M.C. Payne and D.C. Allan, Solution of Schroedinger's equation for large systems, *Phys. Rev. B* **40**, 12255 (1989).
[37] T.A. Arias, M.C. Payne and J.D. Joannopoulos, Ab initio molecular dynamics: Analytically continued energy functionals and insights into iterative solutions, *Phys. Rev. Lett.* **69**, 1077 (1992).
[38] M.C. Payne, M.P. Teter, D.C. Allan, T.A. Arias and J.D. Joannopoulos, Iterative minimization techniques for *ab-initio* total energy calculations – molecular dynamics and conjugate gradients, *Rev. Mod. Phys.* **64**, 1045 (1992).
[39] N. Marzari and D. Vanderbilt, Maximally localized generalized Wannier functions for composite energy bands, *Phys. Rev. B* **56**, 12847 (1997).
[40] S.F. Boys, Construction of some molecular orbitals to be approximately invariant for changes from one molecule to another, *Rev. Mod. Phys.* **32**, 296 (1960).
[41] P.L. Silvestrelli and M. Parrinello, Structural, electronic, and bonding properties of *Li* from first-principles, *J. Chem. Phys.* **111**, 3572 (1999).
[42] P.L. Silvestrelli and M. Parrinello, Water dipole moment in the gas and the liquid phase, *Phys. Rev. Lett.* **82**, 3308 (1999).
[43] T. Tassaing, M-C. Bellissant-Funel, B. Guillot and Y. Guissani, The partial pair correlation functions of dense supercritical water, *Europhys. Lett.* **42**, 265 (1998).
[44] M. Boero, K. Terakura, T. Ikeshoji, C.C. Liew and M. Parrinello, Hydrogen bonding and dipole moment of water at supercritical conditions: A first-principle molecular dynamics study, *Phys. Rev. Lett.* **85**, 3245 (2000).

[45] M. Sharma, R. Resta and R. Car, Intermolecular dynamical charge fluctuations in water: A signature of the H-bond network, *Phys. Rev. Lett.* **95**, 187401 (2005).
[46] J.C. Grossman, E. Schwegler, E.K. Draeger, F. Gygi and G. Galli, Towards an assessment of the accuracy of density functional theory for first-principles simulations of water, *J. Chem. Phys.* **120**, 300 (2004).
[47] E. Schwegler, J.C. Grossman. F. Gygi and G. Galli. Towards an assessment of the accuracy of density functional theory for first-principles simulations of water. II, *J. Chem. Phys.* **121**, 5400 (2004).
[48] M. Watanabe and W.P. Reinhardt, Direct dynamical calculation of entropy and free energy by adiabatic switching, *Phys. Rev. Lett.* **65**, 3301 (1990).
[49] O. Sugino and R. Car, Ab initio molecular dynamics study of first-order phase transitions: Melting of Silicon, *Phys. Rev. Lett.* **74**, 1823 (1995).
[50] D. Alfe and M.J. Gillan, Exchange-correlation energy and the phase diagram of silicon, *Phys. Rev. B* **68**, 205212 (2003).
[51] J. Tao, J.P. Perdew, V.N. Staroverov and G.E. Scuseria. Climbing the density functional ladder: Nonempirical meta-generalized gradient approximation designed for molecules and solids, *Phys. Rev. Lett.* **91**, 146401 (1991).
[52] V.N. Staroverov, G.E. Scuseria, J. Tao and J.P. Perdew. Tests of a ladder of density functionals for bulk solids and surfaces, *Phys. Rev. B* **69**, 075102 (2004).
[53] X.F. Wang, S. Scandolo and R. Car, Melting of silicon and germanium from density functional theory, private communication, to be published (2006).
[54] X.F. Wang, R. Car and S. Scandolo, Carbon phase diagram from ab-initio molecular dynamics, *Phys. Rev. Lett.* **95**, 185701 (2005).
[55] R. Zwanzig, *Nonequilibrium Statistical Mechanics* (Oxford University Press, New York, 2001).
[56] W. E, W. Ren and E. Vanden-Ejinden, String method for the study of rare events, *Phys. Rev. B* **66**, 052301 (2002).
[57] H. Jonsson, G. Mills and K.W. Jacobsen, Nudged elastic band method for finding minimum energy paths of transitions, in *Classical Quantum Dynamics in Condensed Phase Simulations* edited by B.J. Berne, G. Ciccotti, D.F. Coker, (World Scientific, Singapore, 1998), pp. 385–404.
[58] Y.Kanai, A. Tilocca, A. Selloni and R. Car, First-principles string molecular dynamics: An efficient approach for finding chemical reaction pathways, *J. Chem. Phys.* **121**, 3359 (2004).
[59] N. Takeuchi, Y. Kanai and A. Selloni, Surface reaction of alkynes and alkenes with H-Si(111): A density functional study, *J. Am. Chem. Soc.* **126**, 15890 (2004).
[60] R.L. Cicero, C.E.D. Chidley, G.P. Lopinsky, D.D.M. Wayner and R.A. Wolkow. Olefin addition on H-Si(111): Evidence for a surface chain reaction initiated at isolated dangling bonds, *Langmuir* **18**, 305 (2002).
[61] P. Carloni, U. Rothlisberger and M. Parrinello, The role and perspective of ab initio molecular dynamics in the study of biological systems, *Acc. Chem. Res.* **25**, 455 (2002).
[62] J. Aqvist and A. Warshel, Simulation of enzyme reactions using valence bond force fields and other hybrid quantum classical approaches, *Chem. Rev.* **93**, 2523 (1993).
[63] P.W. Anderson, Chemical pseudopotentials, *Phys. Rep.* **110**, 311 (1984).
[64] S. Godecker, Linear scaling electronic structure methods, *Rev. Mod. Phys.* **71**, 1085 (1999).
[65] M. Sharma, Y. Wu and R. Car, Ab initio molecular dynamics with maximally localized wannier functions, *Int. J. Quant. Chem.* **95**, 821 (2003).
[66] A. Laio and M. Parrinello, Escaping free energy minima, *Proc. Natl. Acad. Sci. USA* **99**, 12562 (2002).
[67] M. Iannuzzi, A. Laio and M. Parrinello, Efficient exploration of reactive potential energy surfaces using Car–Parrinello molecular dynamics, *Phys. Rev. Lett.* **90**, 238302 (2003).
[68] D. Chandler and P. Wolynes, Exploiting the isomorphism between quantum theory and classical statistical mechanics of polyatomic fluids, *J. Chem. Phys.* **74**, 4078 (1981).
[69] D. Marx and M. Parrinello, Ab initio path integral molecular dynamics: Basic idea, *J. Chem. Phys.* **104**, 4077 (1996).

[70] D. Marx and M. Parrinello, Structural quantum effects and three-centre two-electron bonding in CH^{+5}, *Nature* **375**, 216 (1995).

[71] M.E. Tuckerman, D. Marx, M.L. Klein and M. Parrinello, On the quantum nature of the shared proton in hydrogen bonds, *Science* **275**, 817 (1997).

[72] J.C. Tully, *Mixed Quantum-Classical Dynamics: Mean-Field and Surface-Hopping*, edited by B.J. Berne, G. Ciccotti, D.F. Coker, (World Scientific, Singapore, 1998), pp. 489–514.

[73] E. Runge and E.K.U. Gross, Density-functional theory for time-dependent systems, *Phys. Rev. Lett.* **52**, 997 (1984).

Chapter 4
STRUCTURE AND ELECTRONIC PROPERTIES OF COMPLEX MATERIALS: CLUSTERS, LIQUIDS AND NANOCRYSTALS

J. R. Chelikowsky

1. INTRODUCTION

As discussed in previous chapters, the pseudopotential model of condensed matter is one of the most promising concepts within computational materials science. It has led the way in providing a workable science framework for describing the properties of materials [1,2], while modern computers have provided the computational resources to implement the pseudopotential method. In this chapter, we will review some of the numerical methods that have allowed the implementation of the pseudopotential method for complex systems such as clusters, nanocrystals and liquids. These are important and complex systems that were essentially impossible to quantify before the advent of the *standard model*.

We note that most progress within this area has been made by "software" advances, i.e., algorithm developments, as opposed to hardware advances, i.e., faster computers. In fact, most computational work today could be done on computers developed some 20 years ago. The computational loads would be quite demanding on these old platforms, but given enough time most of the work could be done. In contrast, even with the fastest computers of today, most of the current work could not be done without the notable advances in algorithms, such as fast diagonalization methods and new pseudopotential constructs [1].

Contemporary Concepts of Condensed Matter Science
Conceptual Foundations of Materials: A Standard Model for Ground- and Excited-State Properties

ISSN: 1572-0934/doi:10.1016/S1572-0934(06)02004-X

We will focus on such advances in this chapter for computationally demanding systems. For example, liquid systems which are important in understanding crystal growth involve systems that are temporal in nature. Unlike crystalline matter where the structure is periodic and static, liquid structures can only be characterized by ensemble averages, which require averaging over time evolved structures. Moreover, liquid structures are dependent on temperature and to a lesser extent pressure. Molecular simulations of such systems were possible before pseudopotentials through the use of empirical or classical interatomic forces. Unfortunately, these forces often did capture the quantum nature of interatomic interactions, e.g., the role of charge transfer and hybridization are often outside the scope of interatomic forces derived by classical potentials. Although pseudopotentials constructed within density functional theory (DFT) can be used to compute interatomic forces (see Section 3), it is not easy to implement such approaches.

This situation is similar for clusters and nanocrystals. These systems are very important from both a fundamental science and a technological perspective. In the case of nanocrystals, quantum confinement can radically alter the nature of the electronic properties of these systems. A notable example comes from silicon quantum dots. Such dots, typically a few nanometers in diameter, are optically active in contrast to bulk silicon. A seamless coupling of dots to bulk materials would produce an ideal optoelectric material. These dots are also of keen interest for understanding fundamental electronic interactions and dielectric properties at small length scales. However, these systems are characterized by many degrees of freedom, both nuclear and electronic, and these systems often assume structures with little symmetry. In fact, the structure for clusters may not be known and not easily measured. Pseudopotentials can be used to compute the total electronic energy of a structure, but there are myriads of possible structures. Also, one would like to understand the optical and dielectric properties of these systems using methods that are not limited to ground state formalisms. Only recently have numerical techniques been developed for such systems.

We will illustrate some contemporary numerical methods for describing these complex systems in this chapter. We will briefly review the electronic structure problem before outlining methods appropriate for liquids, clusters and nanocrystals.

2. THE ELECTRONIC STRUCTURE PROBLEM

According to the Hohenberg–Kohn density-functional formalism [3], the total energy E_{tot} of a system comprising electrons and ions (the latter in positions $\{\mathbf{R}_a\}$) can be written as a unique functional of the electron density ρ:

$$E_{\text{tot}}[\rho] = T[\rho] + E_{\text{ion}}(\{\mathbf{R}_a\}, [\rho]) + E_H[\rho] + \xi_{\text{xc}}[\rho] + E_{\text{ion-ion}}(\{\mathbf{R}_a\}) \tag{1}$$

where $T[\rho]$ is the kinetic energy (KE) of the electrons, $E_{\text{ion}}(\mathbf{R}_a, [\rho])$ the electron–ion energy, $E_H[\rho]$ the electron–electron Coulomb energy or Hartree potential energy, $\xi_{\text{xc}}[\rho]$ (the "exchange-correlation" energy) includes all the other contributions of the electrons to the total energy, and $E_{\text{ion-ion}}(\mathbf{R}_a)$ the classical electrostatic energy among the ions. Minimization of this expression by variation of the function ρ affords the correct electron density. In principle, if accurate energy functionals for T and ξ_{xc} are provided, DFT allows calculation of the ground-state density and energy of the system.

In practice, the problem is simplified further by using the Kohn–Sham reformulation [4]. The heart of the Kohn–Sham scheme is to replace the electrons of the original problem with a set of noninteracting electrons with the same total density as the original system. The energy functional can then be rewritten as

$$E_{\text{tot}}[\rho] = \tilde{T}[\rho] + E_{\text{ion}}(\{\mathbf{R}_a\}, [\rho]) + E_H[\rho] + E_{\text{xc}}[\rho] + E_{\text{ion-ion}}(\{\mathbf{R}_a\}) \tag{2}$$

where $\tilde{T}$, the KE of these noninteracting electrons, can now be treated exactly and $E_{\text{xc}} = \xi_{\text{xc}} + (T - \tilde{T})$. Finding the electron density that minimizes the energy functional is now equivalent to solving the set of one-particle Schrödinger (Kohn–Sham) equations

$$\left[-\frac{\nabla^2}{2} + V_{\text{ion}}(\mathbf{r}) + V_H(\mathbf{r}) + V_{\text{xc}}(\mathbf{r})\right]\psi_n(\mathbf{r}) = \varepsilon_n \psi_n(\mathbf{r}) \tag{3}$$

and setting

$$\rho(\mathbf{r}) = \sum_n |\psi_n(\mathbf{r})|^2 \tag{4}$$

where the sum runs over the occupied states. V_{ion} and V_H are the ionic and Hartree potentials, respectively; $V_{\text{xc}} = \delta E_{\text{xc}}/\delta\rho$. Here and in the rest of the text we use atomic units (a.u.), where the fundamental constants are set to unity: $e = m = \hbar = 1$, unless otherwise stated. Solving Eqs (3) and (4) requires finding a self-consistent solution for the charge density, and constitutes the most computationally intensive part of the electronic structure calculation.

3. SOLVING THE KOHN–SHAM PROBLEM

There are a number of methods and approaches to solving the Kohn–Sham equation (Eq. (3)) [1,5]. For example, one could use a simple basis such as a plane wave expansion [6,7] or Gaussians [8,9]. Each method has its advantages and disadvantages. For example, a Gaussian basis generally results in a smaller Hamiltonian matrix, but the matrix elements are more complex than a plane wave basis. In contrast, a plane wave basis is simple, complete and has only one convergence parameter. However, for some systems, many plane waves are required and the Hamiltonian matrix is large, and dense.

In this section, we will focus on *real-space methods* [10–17]. These methods have a number of advantages for both extended and confined systems, which possess little symmetry. As for a plane wave basis, real-space approaches are easy to implement, and convergence can be tested using a single parameter. In the case of plane waves, this parameter corresponds to the largest wave vector allowed in the expansion, or conversely the smallest wave length present. In the case of a real-space description, the grid spacing is the parameter of merit for convergence, assuming a uniform cubic grid. With either a plane wave basis or with real-space grids the basis is complete, if one chooses to call a grid a "basis." Besides, real-space methods have several advantages over plane waves. Real space methods are more efficient on parallel platforms as they require few global communications. Also, real-space methods do not require supercell geometries for confined systems, this is especially important for charged systems [18,19].

Real-space methods represent wave functions, the electron density and potentials on a grid. For simplicity, the grid is usually taken as an orthogonal three-dimensional uniform grid, but the extension to a general orthorhombic grid is straightforward. In order to construct the grid, only two parameters need to be specified: the grid spacing h (the distance between adjacent points in each of the three Cartesian directions) and the domain of the grid in which the wave functions are assumed finite.

Here we will consider two situations: confined or finite systems such as a cluster and extended or periodic systems as for a solid or liquid. For finite systems, we will assume a spherical domain with a radius of R. For extended systems, we can use supercells as for the plane wave method [1]. The system is made periodic by replicating the unit cell and the atoms it contains (the basis) throughout space. An example of an extended system in this context could be a liquid or an amorphous solids. By making the system periodic, one can consider a finite number of atoms without invoking a surface. In this case, the domain is taken to be of size L for a cubic unit cell. For either a confined or extended system, the grid is generated by the points

$$\mathbf{r}(i,j,k) \equiv (x_i, y_j, z_k) = (ih, jh, zh) \tag{5}$$

with the integers i, j and k running from 1 to $N_{\text{grid}} = L/h$ for a supercell. For a finite spherical domain, $i^2 + j^2 + k^2 \leqslant (R/h)^2$. An example of a grid for a *finite system* is presented in Fig. 1.

In order to model Eq. (3) on the real-space grid we use a higher-order finite difference expansion [20] for the Laplacian operator. We approximate the partial derivatives of the wave function at a given point on the grid by a weighted sum over its values at that point and at neighboring points. The second partial derivative in the x-direction, has the form

$$\left.\frac{\partial^2 \psi}{\partial x^2}\right|_{\mathbf{r}(i,j,k)} = \sum_{n=-N}^{N} C_n \psi(x_i + nh, y_j, z_k) \tag{6}$$

where N is the order of the expansion (typically $N = 6$–8). Under the assumption that the wave function can be approximated accurately by a power series in h, this approximation is accurate to $O(h^{2N+2})$. Algorithms are available that compute the coefficients C_n for arbitrary order in h [21].

In each iteration of the algorithm for self-consistent solution of the Kohn–Sham equations, the Hartree and exchange-correlation potentials are set up directly on the real-space grid using the electron density from the previous iteration. For V_{xc}

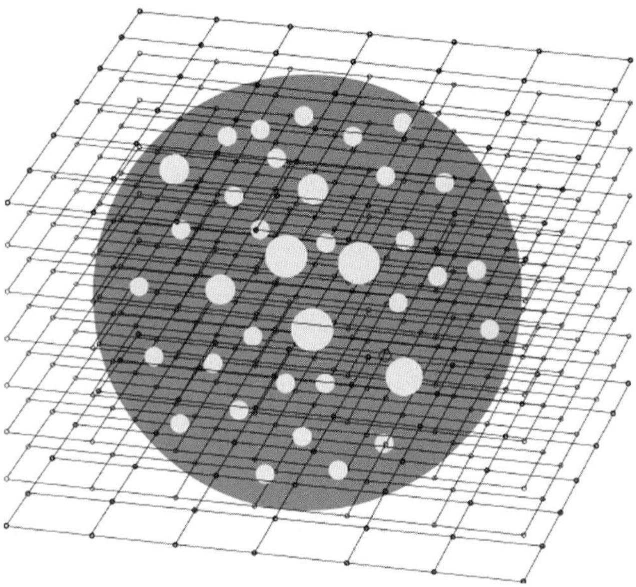

Fig. 1. Uniform grid illustrating a typical configuration for examining the electronic structure of a localized system. The dark gray sphere represents the domain where the wave functions are allowed to be nonzero. The light spheres within the domain are atoms.

we use the local density approximation (LDA): the value of V_{xc} at a given point is a function of the electron density at that point. To construct V_H, we solve the Poisson equation $[\nabla^2 V_H(\mathbf{r}) = -4\pi\rho(\mathbf{r})]$ using the matrix formalism corresponding to the higher-order finite difference method [22,23]. This procedure must be modified for extended systems. In this case, we first set the total charge in the supercell to zero in order to prevent the system from becoming infinitely charged due to the required periodicity.

The remaining potential term in Eq. (3), the ionic term, is determined using *pseudopotential* theory. If one were to use the full-electronic potential, the use of a simple grid would not be efficient as length scale of the problem would be set by the most tightly bound core state. In addition, the core states would possess cusps, which are very difficult to express in a cubic grid. Since the pseudopotential binds on valence states, both the wave functions and potentials are smoothly varying and easily handled by finite difference methods.

We employ nonlocal, norm-conserving ionic pseudopotentials [24] cast in the Kleinman–Bylander form [25]. The ionic contribution V^a_{ion}, is obtained as the sum of a local term and a nonlocal term, the latter corresponding to an angular-momentum-dependent projection [25,22,23]. This can be expressed in Eq. (3) as

$$V^a_{\text{ion}}(\mathbf{r})\psi_n(\mathbf{r}) = V_{\text{loc}}(r_a)\psi_n(\mathbf{r}) + \sum_{lm} G^a_{n,lm} u_{lm}(\mathbf{r}_a)\Delta V_l(r_a) \tag{7}$$

where $\mathbf{r}_a = \mathbf{r} - \mathbf{R}_a$; u_{lm} is the atomic pseudopotential wave function corresponding to the angular momentum quantum numbers l and m; $\Delta V_l = V_l - V_{\mathrm{loc}}$ is the difference between V_l (the l component of the ionic pseudopotential) and the local potential V_{loc}; and the projection coefficients $G^a_{n,lm}$ given by

$$G^a_{n,lm} = \frac{1}{\langle \Delta V^a_{lm} \rangle} \int u_{lm}(\mathbf{r}_a) \Delta V_l(r_a) \psi_n(\mathbf{r})\, \mathrm{d}^3 r \tag{8}$$

include the normalization factor

$$\langle \Delta V^a_{lm} \rangle = \int u_{lm}(\mathbf{r}_a) \Delta V_l(r_a) u_{lm}(\mathbf{r}_a)\, \mathrm{d}^3 r \tag{9}$$

For an *extended* system, the local and nonlocal terms in Eq. (7) must in principle be evaluated and accumulated for all the atoms in the system, i.e., for both the atoms in the basis and their periodic images. However, the summation of nonlocal terms is actually limited to the basis, because at distances greater than the pseudopotential core radius (a fraction of a bond length) V_l is $-Z/r$ for all l, where Z is the number of electrons acting as valence electrons in the pseudopotential (see the left-hand panel of Fig. 2); this makes ΔV_l short ranged, so that the nonlocal terms need only be evaluated for atoms belonging to the basis. Furthermore, the integrals in Eqs (8) and (9) can be efficiently calculated in real space by direct summation over the grid points surrounding each atom.

The situation is different for the local contribution to the ionic potential. In extended systems, this contribution involves a divergent summation of the

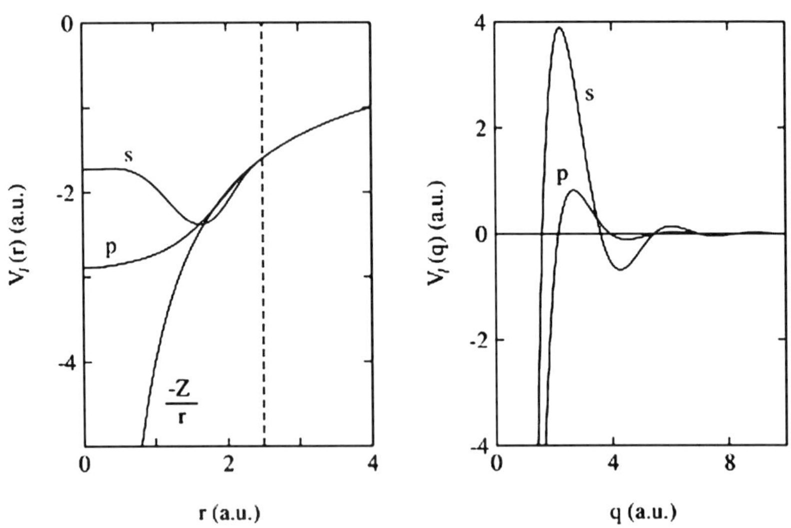

Fig. 2. Real-space (left panel) and reciprocal-space (right panel) representations of the pseudopotentials employed in our real-space MD simulation of liquid silicon. The pseudopotentials were constructed using the Troullier–Martins prescription [24]; the dashed line in the left panel corresponds to the core radius used in their generation, 2.5 a.u.

long-range Coulomb term $-Z/r$. However, this divergence can be avoided by making use of the fact that the pseudopotentials are short-ranged functions in reciprocal space as indicated in the right-hand panel of Fig. 2. The local ionic potential, $V_{\text{ion,loc}}$, can be calculated efficiently in reciprocal space and transferred to the real-space grid by fast Fourier transform (FFT). We obtain the local ionic potential in reciprocal space as in a plane wave calculation with an energy cutoff of $\pi^2/2h^2$, the cutoff for which FFTs of the wave functions and potentials require a grid of size N_{grid}^3 [26]. We first calculate the structure factor $S_{\text{ion}}(\mathbf{q})$ at wave vector $\mathbf{q} = (2\pi/L)(n_x, n_y, n_z)$ (where n_x, n_y and n_z are integers):

$$S_{\text{ion}}(\mathbf{q}) = \sum_a \exp(i\mathbf{q} \cdot \mathbf{R}_a) \tag{10}$$

where the sum is taken over the positions of all the atoms in a single unit cell. The inclusion of a periodic image of an atom in the summation on the right-hand side of Eq. (10) would give the same contribution as that of the atom since their exponential terms would differ by $2\pi n$, with n an integer.

$V_{\text{ion,loc}}$ is then calculated as

$$V_{\text{ion,loc}}(\mathbf{q}) = \frac{1}{L^3} S_{\text{ion}}(\mathbf{q}) V_{\text{loc}}(q) \tag{11}$$

and transferred to the real-space grid by FFT. We need to perform this transformation once, just before we enter the loop for self-consistent solution of the Kohn–Sham equations. Since the local ionic potential is determined by the positions of the ions, it does not change during the process of finding a self-consistent solution for ρ.

When discretized as above, Eq. (3) adopts the form

$$\begin{aligned} -\frac{1}{2} \sum_{n_1,n_2,n_3=-N}^{N} & C_{n_1,n_2,n_3} \psi_n(x_i + n_1 h, y_j + n_2 h, z_k + n_3 h) \\ & + [V_{\text{ion}}(x_i, y_j, z_k) + V_H(x_i, y_j, z_k) + V_{\text{xc}}(x_i, y_j, z_k)] \psi_n(x_i, y_j, z_k) \\ & = \varepsilon_n \psi_n(x_i, y_j, z_k) \end{aligned} \tag{12}$$

Since ΔV_l differs from zero only inside the pseudopotential core radius and the Laplacian operator extends only to a few neighbors around each grid point, the matrix representation of Eq. (13) is *very sparse*. Consequently, highly efficient diagonalization procedures can be employed to extract the required eigenvalue/eigenfunction pairs [27,28].

The total ground-state energy (Eq. (2)) is given by

$$\begin{aligned} E_{\text{tot}}[\rho] = \tilde{T}[\rho] + \int \rho(\mathbf{r}) V_{\text{ion,loc}}(\mathbf{r})\, \mathrm{d}^3 r + \sum_{a,n,lm} \langle \Delta V_{lm}^a \rangle [G_{n,lm}^a]^2 \\ + E_H[\rho] + E_{\text{xc}}[\rho] + E_{\text{ion-ion}}(\{\mathbf{R}_a\}) + \alpha \end{aligned} \tag{13}$$

where the sum on n is over the occupied states. The last term, α, appears only for *extended* systems. α is the contribution of the non-Coulomb part of the pseudopotential at $\mathbf{q} = \mathbf{0}$:

$$\alpha = \frac{ZN_a^2}{L^3} \int \left(V_{\text{loc}}(r) + \frac{Z}{r} \right) 4\pi r^2 \, \mathrm{d}r \tag{14}$$

The force on ion a is given by

$$\mathbf{F}_a = - \int \rho(r) \frac{\partial V_{\text{loc}}(r_a)}{\partial \mathbf{R}_a} \, \mathrm{d}^3 r - 2 \sum_{n,lm} \langle \Delta V_{lm}^a \rangle G_{n,lm}^a \frac{\partial G_{n,lm}^a}{\mathbf{R}_a} - \frac{\partial E_{\text{ion-ion}}}{\mathbf{R}_a} \tag{15}$$

For confined systems, this term is straightforward. For *extended* systems or supercells, the forces can be evaluated in reciprocal space. The first term on the right-hand side of Eq. (15) is the contribution from the local ionic potential, $\mathbf{F}_{a,\text{loc}}$. It involves the integral of a long-range function (Z/r^2), but in reciprocal space there is no long-range tail [29]:

$$\mathbf{F}_{a,\text{loc}} = -\mathrm{i} L^3 \sum_{\mathbf{q}} \mathbf{q} \exp(\mathrm{i}\mathbf{q} \cdot \mathbf{R}_a) V_{\text{loc}}(q) \rho(\mathbf{q}) \tag{16}$$

where $\rho(\mathbf{q})$ is obtained by an FFT from the solution of the Kohn–Sham equations on the real-space grid. The other electronic contribution to the force arises from the nonlocal components of the pseudopotential. Owing to their short range, we calculate this term in real space, in which its computation scales as the square of the system size. In reciprocal space, this term scales as N_a^3 [30]. The remaining term in Eq. (15) is the force exerted on the ion by other ions. For confined systems, this results in a simple summation. For extended or artificial-periodic systems [5,6], we evaluate this term by performing two convergent summations, one over lattice vectors and the other over reciprocal lattice vectors, using Ewald's method. As for the case of plane waves, Eq. (15) contains no term representing the derivative of the basis set with respect to the position of the ion, i.e., no "Pulay forces" are present [31].

4. SIMULATING LIQUID SILICON

In this section, we illustrate how pseudopotentials can be used to simulate liquids. We will focus on silicon, which is frequently used to test new theoretical methods [32]. Silicon constitutes a severe test as its liquid form is likely the most complex elemental liquids. Upon melting, silicon undergoes a transition from a semiconducting covalent structure to a rather unusual metallic phase. Unlike simple liquid metals, which have a coordination of $\sim$ 12 in the melt, liquid silicon corresponds to a loosely packed structure with a coordination of $\sim$ 6 [33]. The existence of covalent bonds in the metallic phase is indicative of the "many-body" nature of the interactions in liquid silicon. As such, a realistic description of the melt requires a quantum mechanical treatment. One of the first such simulations for liquid silicon was executed by Car and Parrinello using method (See Chapter 3) similar to the one outlined here, save they used a plane wave basis [34].

The most demanding aspect of modeling any liquid is to construct an accurate ensemble. Previous simulations of silicon using classical potentials have suggested that a minimum of about 50–100 atoms are required to extract reasonable statistical averages [35]. A common configuration is to consider a cubic supercell constructed by considering the diamond structure and doubling the conventional cell in all three directions. This results in a 64-atom cell. The use of a supercell allows one to consider a finite sample of silicon without considering any surface effects. If too few atoms are considered, the atom motions become highly correlated and the statistics are inadequate for an accurate description of the melt.

A truly first principles simulation method would determine the supercell size, or the density of the liquid, by calculating the equilibrium pressure for a given temperature. Given the difficulty of *ab initio* simulations, this is rarely done. Instead, the size of the supercell is chosen so that the density of the simulated liquid agrees with experiment. In the case of a 64-atom cubic supercell, the size of the cell is chosen to be $L = 19.80$ a.u. (1 a.u. $= 0.529$ Å). This cell size corresponds to a liquid density close to the experimental value of $2.59\,\mathrm{g/cm^3}$.

The grid spacing is fixed to give a converged solution of the Kohn–Sham equation. A grid spacing is $h = 0.7$ a.u., which corresponds to about a 20 Ry cutoff if plane waves are used. In the simulation illustrated here, norm-conserving pseudopotentials were generated for the reference configuration [Ne]$3s^2 3p^2$ using the Troullier–Martins prescription [24]. A radial cutoff of 2.5 a.u. was used for both s and p. This potential is illustrated in Fig. 2. The potential was made separable by the procedure of Kleinman and Bylander [25] with the s potential chosen to be the local component. The local-density functional of Ceperley and Alder [36] was used, as parameterized by Perdew and Zunger [37].

The periodic nature of the supercell is artificial. Although one can invoke Bloch's theorem and define a wave vector, such vectors are not physically meaningful. As such, provided the supercell size is sufficiently large, the zero phase ($\vec{k} = 0$ or the Γ point) wave function can be used to construct the charge density.

A key issue in constructing the ensemble is controlling the temperature of the supercell. In order to avoid issues with supercooling, the temperature of the liquid is achieved by heating the system well above the melting point and then quenching the system to the desired temperature [35]. For the example illustrated here, a temperature of 1800 K was used. The temperature is chosen to be near the experimental melting point, $T_{\mathrm{m}} = 1680$ K.

A number of methods exist in the literature for controlling the temperature of the ensemble [35,38–41]. These thermostats range from a simple rescaling of velocities to sophisticated alterations of the kinetic processes. Here we consider controlling the temperature of the ensemble using Langevin dynamics, which effectively couples the ensemble to a heat bath [35,42,40,41].

In Langevin dynamics, the ionic positions, $\mathbf{R}_j$, evolve according to

$$M_j \ddot{\mathbf{R}}_j = \mathbf{F}(\{\mathbf{R}_j\}) - \gamma M_j \dot{\mathbf{R}}_j + \mathbf{G}_j \tag{17}$$

where $\mathbf{F}(\{\mathbf{R}_j\})$ is the interatomic force on the jth particle, and $\{\mathrm{M}_j\}$ the ionic masses. The last two terms on the right-hand side of Eq. (17) are the dissipation and fluctuation forces, respectively. The dissipative forces are defined by the friction coefficient, γ. The fluctuation forces are defined by random Gaussian variables, $\{\mathbf{G}_i\}$, with a white-noise spectrum:

$$\langle G_i^\alpha(t)\rangle = 0 \quad \text{and} \quad \langle G_i^\alpha(t) G_j^\alpha(t')\rangle = 2\gamma M_i k_B T \delta_{ij}\delta(t-t') \tag{18}$$

The angular brackets denote ensemble or time averages, and α stands for the Cartesian component. The coefficient of T on the right-hand side of Eq. (18) insures that the fluctuation–dissipation theorem is obeyed, i.e., the work done on the system is dissipated by the viscous medium [43,44]. The interatomic forces can be obtained from the Hellmann–Feynman theorem using the pseudopotential wave functions.

Langevin simulations using quantum forces can be contrasted with other techniques as outlined in Chapter 3, i.e., the method by Car and Parrinello [34]. Langevin simulations as outlined above do not employ fictitious electron dynamics; at each time step the system is quenched to the Born–Oppenheimer surface and the quantum forces are determined. This approach, called "Born–Oppenheimer molecular dynamics (BOMD)," requires a fully self-consistent treatment of the electronic structure problem; however, because the interatomic forces are true quantum forces, the resulting molecular dynamics simulation can be performed with much larger time steps. Typically it is possible to use time steps, an order of magnitude larger than for methods based on fictitious dynamics [45,46].

Using a heat bath, as defined by Langevin dynamics, the temperature of the liquid is established by heating the system far above the melting point. If one starts with a crystalline environment, the solid is melted only by greatly exceeding the melting point temperature. Owing to the small size of the ensemble and perfect environment, the melting process is controlled by homogeneous melting, i.e., no surfaces or intrinsic defects are involved in initiating the melting transition. After liquifying the ensemble, the temperature is slowly lowered to the target temperature. Once the target temperature is achieved, the heat bath is removed and the microcanonical ensemble follows Newtonian dynamics. In this regime, the trajectories of the atoms in the melt are followed and the physical properties of the liquid can be calculated [35]. The time step used to integrate the equations of motion is typically chosen to be a few fs. The simulation over which statistics are collected might correspond to a thousand time steps, or several ps. A common annealing schedule is illustrated in Fig. 3.

A microcanonical molecular dynamics simulation offers a stringent test of the accuracy of calculated ionic forces because the trajectory of the system through configuration space is deterministic. Any systematic error in the force calculations will prevent conservation of the total energy of the system. In Fig. 4, we illustrate how well the energy is conserved. This figure shows that there is no significant violation of energy conservation. Highly accurate forces imply more than the validity of the scheme used in their calculation. Since errors in the Hellmann–Feynman forces are first order with respect to errors in the wave functions, accurate forces can only be obtained when the wave functions are very nearly exact

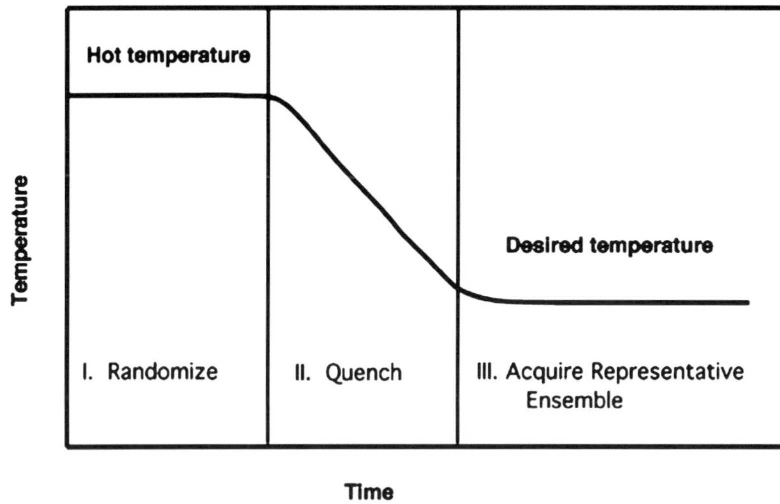

Fig. 3. A schematic temperature schedule for creating a liquid ensemble.

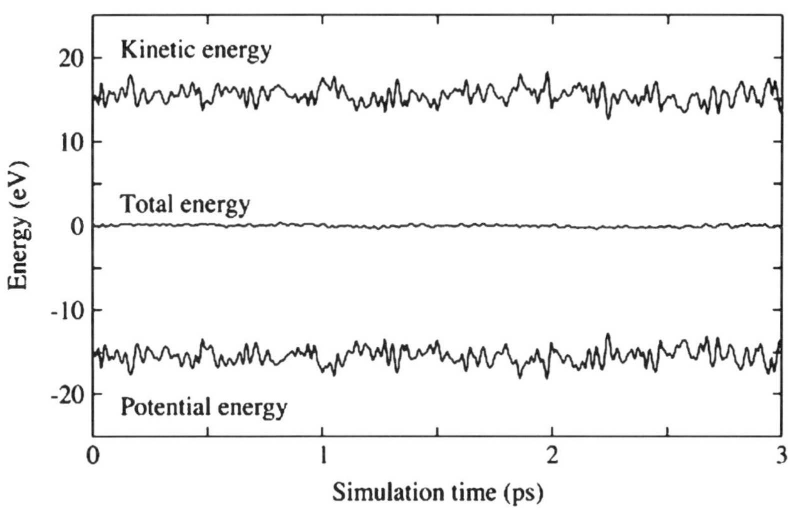

Fig. 4. Time courses of the kinetic, potential and total energies of liquid silicon during our microcanonical real-space MD simulation. The equations of motion of the ions were integrated for a time step of 165 a.u. using the Beeman algorithm [96]. No velocity rescaling was performed. The potential and total energies have been shifted by a constant so that the total energy averages to zero.

eigenstates, which implies that the procedure used to discretize and solve the Kohn–Sham is also very accurate.

Figure 5 shows results on the static structure of liquid silicon obtained from the simulation and by experiment [47]. The two sets of data for the radial distribution

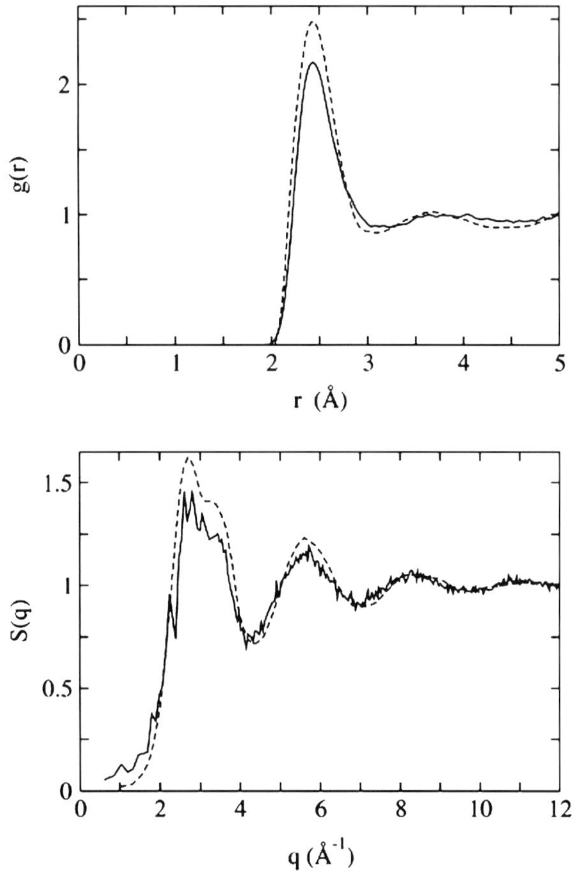

Fig. 5. Radial distribution function (upper panel) and static structure factor (lower panel) of liquid silicon. The continuous curves represent real-space MD results, the dashed curves experimental data from Waseda and Suzuki [47].

function $g(r)$ agree well (upper panel). The radial distribution represents the average spatial distribution of atoms around a given atom. The structure factor is related to this quantity by Fourier transform. The simulation results give the correct position of the principal peak of the function, 2.43 Å, and exhibits the correct trends at greater distances. Integrating $g(r)$ up to the position of the first minimum determined by experiment, 3.10 Å, affords a value for the average coordination number that is nearly identical to experimental estimates for the coordination number of ~ 6.4 [47].

Comparison between the results of simulation and experiment for the static structure factor $S(q)$ shown in Fig. 5 (lower panel) is a stronger test than a comparison of the radial distributions because it is $S(q)$ that is obtained directly from experimental measurements. Since $g(r)$ is obtained by a Fourier transform of $S(q)$, it is more susceptible to numerical errors.

$S(q)$ can be calculated from the simulation results using the expression [48]:

$$S(q) = \frac{1}{N_a}\left\langle \sum_i \sum_j \exp[-i\mathbf{q} \cdot (\mathbf{R}_i - \mathbf{R}_j)] \right\rangle \tag{19}$$

where the sums are taken over the ions in the unit cell and the angle brackets denote averaging over both the trajectories of the particles during the microcanonical simulation run and over all the wave vectors $\mathbf{q}$ with the same modulus q. As is typical in such analyses, one assumes that the liquid is macroscopically isotropic. In terms of $S(q)$, the agreement between simulation and experiment is very good. For example, the simulation results correctly predicting all the successive maxima and minima of the function. Note, in particular, the successful prediction of the shoulder to the right of the first peak, which does not appear in the $S(q)$ functions of simple liquid metals [49,50].

Molecular dynamical simulations of a liquid can also be used to examine the kinetic processes of the melt, e.g., the diffusion of atomic species in the liquid. Two quantities can be used for this purpose. One is the mean square displacement of an atom. The mean square displacement is obtained from

$$\langle \Delta R^2(t) \rangle = \frac{1}{N_a}\left\langle \sum_i [\mathbf{R}_i(t) - \mathbf{R}_i(0)]^2 \right\rangle \tag{20}$$

The other quantity is the velocity–velocity autocorrelation function, $Z(t)$:

$$Z(t) = \frac{\langle \sum_i \mathbf{v}_i(t) \cdot \mathbf{v}_i(0) \rangle}{\langle \sum_i \mathbf{v}_i(0) \cdot \mathbf{v}_i(0) \rangle} \tag{21}$$

where the sums are taken over the ions in the unit cell, $\mathbf{v}_i$ is the velocity of ion i, and the angle brackets on the right-hand sides denote averaging over time origins during the microcanonical simulation. These quantities are shown in Fig. 6.

Except for very small values of t, where the atoms are in a ballistic regime, $\langle \Delta R^2(t) \rangle$ is perfectly linear, in keeping with the theoretical expression [48]

$$\langle \Delta R^2(t) \rangle \sim 6Dt + c, \quad \text{as } t \to \infty \tag{22}$$

in which D is the self-diffusion coefficient and c a constant.

The autocorrelation function tends to zero with increasing time due to the absence of correlation between the velocity of each particle and its initial value. It is interesting that $Z(t)$ does not take negative values. This is in contrast with the typical behavior for a simple liquid metal where near the melting point a "cage effect" can occur. In a cage effect, the metal atoms surrounding any given atom allow the latter to move over short distances, but then reflect it, causing $Z(t)$ to change sign [51,52]. The absence of such backscattering in liquid silicon is likely related to its smaller average coordination number.

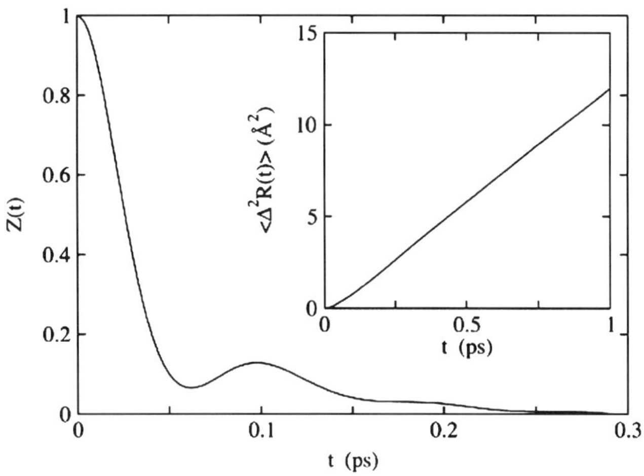

Fig. 6. Normalized velocity autocorrelation function $Z(t)$ and mean square displacement $\langle \Delta R^2(t) \rangle$ (inset) of liquid silicon as obtained from our real-space MD simulation.

Like $\langle \Delta R^2(t) \rangle$, the velocity autocorrelation function is related to the self-diffusion coefficient [48]:

$$D = \frac{k_B T}{M} \int_0^\infty Z(t)\,dt \tag{23}$$

where M is the atomic mass. The values of D obtained from the simulation data via Eqs (22) and (23) are both the same, 2.1 Å^2/ps. In principle, the numbers should be identical as the mean-square displacement and the velocity–velocity autocorrelation function are directly related by a time integration. Of course, in practice there are numerical differences.

This value for the self-diffusion constant is consistent with those afforded by previous *ab initio* calculations using plane wave representations: Born–Oppenheimer molecular dynamics results for $\langle \Delta R^2(t) \rangle$ have given a value of $D =$ 1.9 Å^2/ps, [53,54] and a Car–Parrinello study has yielded values of 2.3 Å^2/ps [from $\langle \Delta R^2(t) \rangle$] and 2.0 Å^2/ps [from $Z(t)$] [55]. A simulation based on interatomic potentials gave $D = 1.0$ Å^2/ps, [56] a value significantly smaller than the quantum mechanical estimates. The failure of classical, or interatomic potentials, may be due to the lack of the extra electronic degrees of freedom present in the quantum mechanical treatments.

This is a key advantage of quantum mechanical treatments versus classical descriptions. Namely, quantum methods are able to quantify the electronic properties of the liquid. For example, in Fig. 7 we illustrate the electronic density of states distribution for liquid silicon. This distribution can be obtained by constructing a histogram of the eigenvalues for every few time steps during the microcanonical simulation and averaging these results. For a comparison, the density of states

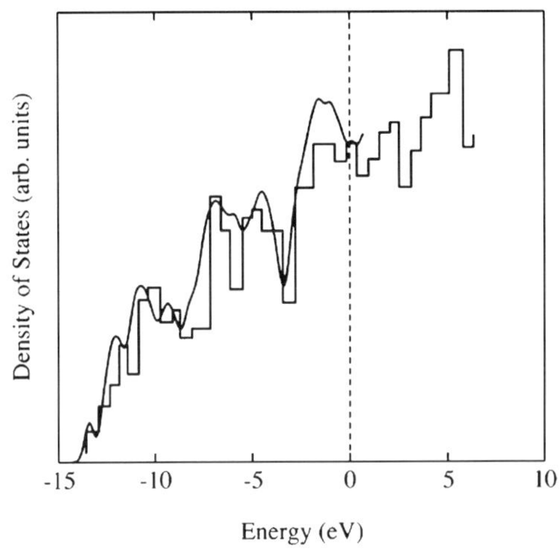

Fig. 7. Electronic density of states of liquid silicon as obtained from our real-space molecular dynamics simulation. The histogram is from a previous plane-wave molecular dynamics simulation [55]. The Fermi levels are set to zero.

obtained in an earlier Car–Parrinello study [55] of the same thermodynamic state is also shown. The agreement between the two approaches is quite good, bearing in mind the technical differences between the two calculations. These studies clearly show liquid silicon to be a good metal, i.e., a large density of states is present at the Fermi level.

It is now possible to examine a variety of liquid state properties using the standard model. For example, compound semiconductors are now routinely examined [35]. However, liquid–solid interfaces remain more problematic as does the possibility of simulating crystal growth. The use of accelerated dynamics is a promising development in this area [57].

5. PROPERTIES OF CONFINED SYSTEMS: CLUSTERS

The electronic and structural properties of atomic clusters stand as one of the outstanding problems in materials physics. Clusters often possess properties that are characteristic of neither the atomic nor solid state. For example, the energy levels in atoms may be discrete and well separated in energy relative to kT. In contrast, solids have continuum of states (energy bands). Clusters may reside between these limits, i.e., the energy levels may be discrete, but with a separation much less than kT.

5.1. Structure

The most fundamental issue in dealing with clusters is determining their structure. Before any accurate theoretical calculations can be performed for a cluster, the atomic geometry of a system must be defined. However, this can be a formidable exercise. Serious problems arise from the existence of multiple local minima in the potential energy surface of these systems; many similar structures can exist with vanishingly small energy differences. A complicating issue is the transcription of interatomic forces into tractable classical force fields. This transcription is especially difficult for clusters such as those involving semiconducting species. In these clusters, strong many body forces can exist that preclude the use of pairwise forces.

A convenient method to determine the structure of small or moderate sized clusters is *simulated annealing* [58,45]. Within this technique, atoms are randomly placed within a large cell and allowed to interact at a high (usually fictitious) temperature. The atoms will sample a large number of configurations. As the system is cooled, the number of high-energy configurations sampled is restricted. If the anneal is done slowly enough, the procedure should quench out structural candidates for the ground state structures. One can use Langevin dynamics for this purpose as presented in Eq. (17).

To illustrate the simulated annealing procedure, we consider a silicon cluster. With respect to the technical details for this example, the initial temperature of the simulation was taken to be about 3000 K; the final temperature was taken to be 300 K. The annealing schedule lowered the temperature of 500 K at each 50 time step. The time step was taken to be 5 fs. The friction coefficient in the Langevin equation was taken to be 6×10^{-4} a.u. After the clusters reached a temperature of 300 K, they were quenched to 0 K. The ground-state structure was found through a direct minimization by a steepest descent procedure.

Choosing an initial atomic configuration for the simulation takes some care. If the atoms are too far apart, they will exhibit Brownian motion, which is appropriate for Langevin dynamics with the interatomic forces zeroed. In this case, the atoms may not form a stable cluster as the simulation proceeds. Conversely, if the atoms are too close together, they may form a metastable cluster from which the ground state may be kinetically inaccessible even at the initial high temperature. Often the initial cluster is formed by a random placement of the atoms with a constraint that any given atom must reside within, say 1.05 and 1.3 times, the bond length from at least one atom where the bond length is defined by the crystalline environment. The cluster in question is placed in a spherical domain. Outside of this domain, the wave function is required to vanish. The radius of the sphere is such that the outermost atom is at least 6 a.u. from the boundary. Initially, the grid spacing was 0.8 a.u. For the final quench to a ground state structure, the grid spacing was reduced to 0.5 a.u. As a rough estimate, one can compare this grid spacing with a plane wave cutoff of $(\pi/h)^2$ or about 40 Ry for $h = 0.5$ a.u.

In Fig. 8, we illustrate the simulated anneal for the Si_7 cluster. The initial cluster contains several incipient bonds, but the structure is far removed from the ground state by approximately 1 eV/atom. In this simulation, at about 100 time steps a tetramer and a trimer form. These units come together and precipitate a large drop in the binding energy (BE). After another $\sim$ 100 time steps, the ground state structure is essentially formed. The ground state of Si_7 is a bicapped pentagon, as is the corresponding structure for the Ge_7 cluster. The BE shown in Fig. 8 is relative to

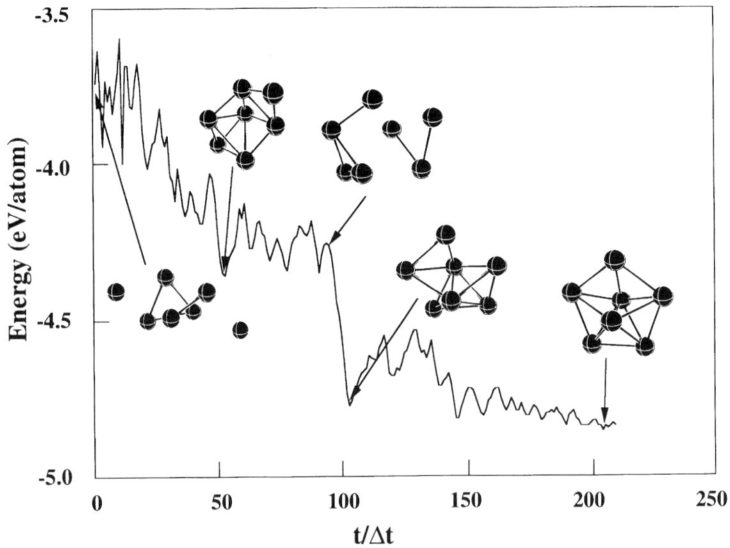

Fig. 8. Binding energy of Si_7 during a Langevin simulation. The initial temperature is 3000 K; the final temperature is 300 K. Bonds are drawn for interatomic distances of less than 2.5 Å. The time step is 5 fs.

that of an isolated Si atom. Gradient corrections [59], or spin polarization [60] have not been included in this example. Therefore, the binding energies indicated in the figure are likely to be overestimated by ~ 20% or so.

In Fig. 9, we present the ground state structures for Si_n for $n \leqslant 7$. The structures for Ge_n are very similar to Si_n. The primary difference resides in the bond lengths. The Si bond length in the crystal is 2.35 Å, whereas in Ge the bond length is 2.44 Å. This difference is reflected in the bond lengths for the corresponding clusters; Si_n bond lengths are typically a few percent shorter than the corresponding Ge_n clusters.

Simulating annealing simulation is just one example of an optimization procedure. As such, other optimization procedures may be used to extract the minimum energy structures. For example, a genetic algorithm has been used to examine carbon clusters [61]. In this algorithm, an initial set of clusters is "mated" with the lowest energy offspring "surviving". By examining several thousand generations, it is possible to extract a reasonable structure for the ground state. The genetic algorithm has some advantages over a simulated anneal, especially for clusters which contain more than ~ 20 atoms. One of these advantages is that kinetic barriers are more easily overcome. However, the implementation of the genetic algorithm is more involved than an annealing simulation, e.g., in some cases "mutations," or *ad hoc* structural rearrangements, must be introduced to obtain the correct ground state. Recently another simulation procedure has been used with some success [62]. In this technique, a random cluster is artificially compressed and then allowed to expand rapidly in a "big bang." The cluster is then quenched using steepest decent

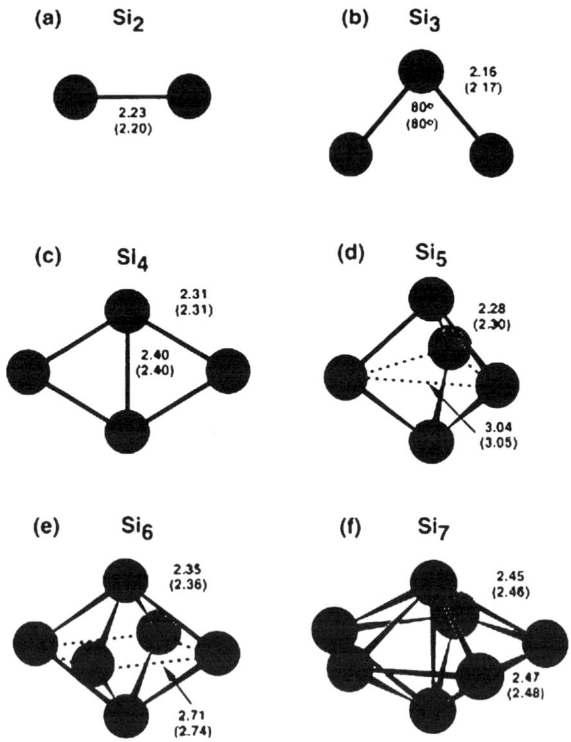

Fig. 9. Ground state geometries and some low-energy isomers of Si_n ($n \leqslant 7$) clusters. Interatomic distances are in Å. The values in parentheses are from quantum chemistry methods [76,97,98].

methods. Typically, millions of candidate clusters can be explored using this technique, provided the structural energies can be quickly evaluated using approximate methods such as tight binding [63].

5.2. Photoemission Spectra

A very useful probe of condensed matter involves the photoemission process. Incident photons are used to eject electrons from a solid. If the energy and spatial distributions of the electrons are known, then information can be obtained about the electronic structure of the materials of interest. For crystalline matter, the photoemission spectra can be related to the electronic density of states. For confined systems, the interpretation is not as straightforward.

One of the earliest experiments performed to examine the electronic structures of small semiconductor clusters examined negatively charged Si_n and Ge_n ($n \leqslant 12$) clusters [64]. The photoemission spectra obtained in this work were used to gauge the energy gap between the highest-occupied state and the lowest unoccupied state.

Large gaps were assigned to the "magic number" clusters, while other clusters appeared to have vanishing gaps. Unfortunately, the first theoretical estimates [65] for these gaps showed substantial disagreements with the measured values. It was proposed by Cheshnovsky et al. [64], that sophisticated calculations including transition cross sections and final states were necessary to identify the cluster geometry from the photoemission data. The data were first interpreted in terms of the gaps obtained for *neutral* clusters; it was later demonstrated that *atomic relaxations* within the *charged* cluster are important in analyzing the photoemission data [42,66]. In particular, atomic relaxations as a result of charging may change dramatically the electronic spectra of certain clusters. These charge-induced changes in the gap were found to yield very good agreement with the experiment.

The photoemission spectrum of Ge_{10}^- illustrates some of the key issues. Unlike Si_{10}^- , the experimental spectra for Ge_{10}^- does not exhibit a gap. Cheshnovsky et al. interpreted this to mean that Ge_{10}^- does not exist in the same structure as Si_{10}^-. This is a strange result. Si and Ge are chemically similar and the calculated structures for both *neutral* structures are similar. The lowest energy structure for both 10 atom clusters is the tetracapped trigonal prism (labeled by *I* in Fig. 10). The photoemission spectra for these clusters can be simulated by using Langevin dynamics. The clusters are immersed in a fictitious heat bath, and subjected to stochastic forces. If one maintains the temperature of the heat bath and averages over the eigenvalue spectra, a density of states for the cluster can be obtained. The heat bath resembles a buffer gas as in the experimental setup, but the time intervals for collisions are not similar to the true collision processes in the atomic beam. The simulated photoemission spectrum for Si_{10}^- is in very good agreement with the experimental results, reproducing both the threshold peak and other features in the spectrum. If a simulation is repeated for Ge_{10}^- using the tetracapped trigonal prism structure, the resulting photoemission spectrum is *not* in good agreement with experiment. Moreover, the calculated electron affinity is 2.0 eV in contrast to the experimental value of 2.6 eV. However, there is no reason to believe that the tetracapped trigonal prism structure is correct for Ge_{10} when this cluster is *negatively charged*. In fact, we find that the bicapped antiprism structure is lower in energy for Ge_{10}^-. The resulting spectra using both structures (*I* and *II* in Fig. 10) are presented in Fig. 11, and

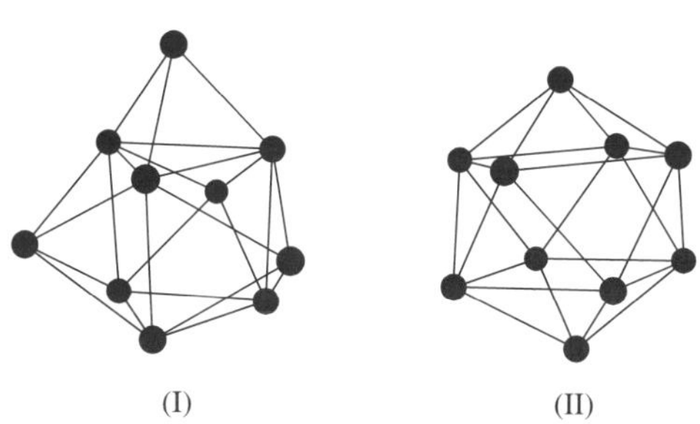

Fig. 10. Two possible isomers for Si_{10} or Ge_{10} clusters. (I) is a tri-capped trigonal prism cluster and (II) is a bi-capped anti-prism cluster.

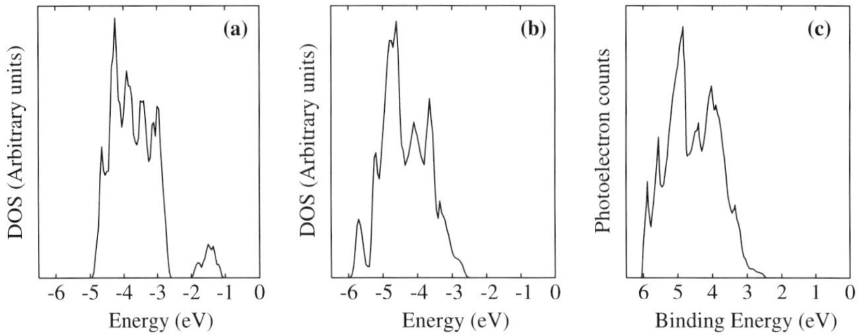

Fig. 11. Calculated "density of states" for Ge_{10}^{-} in (a) the tetracapped trigonal prism structure, (b) the bicapped antiprism structure and (c) experimental photoemission spectra from Cheshnovsky et al. [64].

compared to the photoemission experiment. The calculated spectrum using the bicapped antiprism structure is in very good agreement with the photoemission. The presence of a gap is indicated by a small peak removed from the density of states (Fig. 11(a)). This feature is absent in the bicapped antiprism structure (Fig. 11(b)) and consistent with experiment. For Ge_{10}, charging the structure reverses the relative stability of the two structures. This accounts for the major differences between the photoemission spectra.

While we have focused on semiconductor clusters in this section, the techniques we have are also appropriate for clusters composed of other materials. We can illustrate this for a transition metal clusters such as Cu. There are several significant technical differences compared to semiconductor clusters such as those composed of Si. For example, the Cu *d*-states are localized compared to the valence states of Si. This requires a finer grid mesh to describe these states in Cu. For the example shown here, we use $h = 0.35$ a.u. or the equivalent of a plane wave cutoff of roughly 80 Ry. Since Cu has 11 valence electrons, if we include the 3*d*-shell as part of the valence states, the computational load for a Cu cluster, can be an order of magnitude greater than for a Si cluster without including the role of *spin*. Since transition metals clusters can be *magnetic*, this requires the use of spin-polarized density-functional methods and doubles the size of the Hamiltonian matrix. Moreover, this complicates finding the correct ground state geometry, which is now a function of the spin state of the cluster.

Yet another difficulty centers on electron–hole correlations, which can be more pronounced in transition metal clusters. Early comparisons between the eigenvalue spectra and photoemission spectra for transition metal clusters were met with limited success using different density functional approaches [67–69]. Methods that explicitly include electron–hole relaxations have met with better success. For example, Massobrio et al. [67,68] proposed a simple method for calculating single-particle excitation energies. They obtained promising results when applied to the study of the photoemission spectra of Cu clusters. We illustrate approach here, albeit with a different implementation.

Within their method, the excitation energy of each of the Kohn–Sham eigenstates is calculated as the difference of the total energy of the neutral cluster obtained after the removal of the electron in the original anionic cluster, and the total energy of the anionic cluster. When determining the final excited state for the neutral cluster, the electronic cloud surrounding the hole is allowed to relax, but the geometric structure of the cluster is held fixed, i.e., a vertical excitation process is considered. The relaxation of the electronic degrees of freedom requires a self-consistent procedure, which is computationally intensive. In our example here, both hole and valence states are allowed to relax.

In Fig. 12, we present the ground state structures calculated for several Cu_n^- clusters ($n = 3, 5, 8$ and 9). All these structures possess C_{2v} symmetry, and are largely consistent with previous DFT-based calculations [67,68,70,71]. The most stable structure for Cu_3^- is a linear chain, which is lower by $\sim 0.2\,\mathrm{eV/atom}$ than an isosceles triangle. A planar trapezoidal structure represents the ground-state geometry for Cu_5^-; this structure is more stable than a trigonal bipyramid by

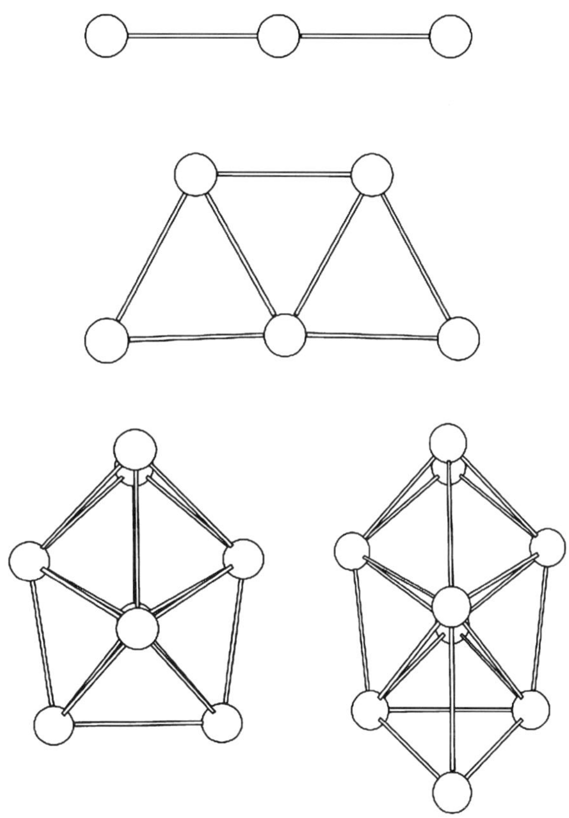

Fig. 12. Ground-state structures for Cu_n^- ($n = 3, 5, 8$ and 9) as obtained from a real-space pseudopotential approach.

$\sim 0.06\,\text{eV/atom}$, which is the second most stable structure for $n = 5$. The lowest energy structures for Cu_8^- are a bicapped octahedron and a capped pentagonal bipyramid (separated in energy by about $0.02\,\text{eV/atom}$). By adding an atom capped at the bottom of the bicapped octahedron we obtained the most favorable structure for Cu_9^-. A similar structure was found for $n = 9$ by Jug et al. [70] (the atom is added at the top instead than at the bottom). However, Massobrio et al. [67,68] obtained as ground state structure a bicapped pentagonal bipyramid; in the calculations illustrated here this structure is only $\sim 0.02\,\text{eV/atom}$ higher in energy than that plotted in Fig. 12. The ground-state structures obtained for $n = 3, 5$ and 9 have zero magnetic moment; the structure obtained for $n = 8$ has magnetic moment one μ_B.

In Fig. 13, we compare the density of states and excitation energy obtained for the Cu anionic structures plotted in Fig. 12 with available experimental data [71]. The excitation energy calculated using a similar method [67,68], but with a plane-wave approach for clusters with $n = 3, 5$ and 9 is also shown. The theoretical values shown here were broaden using a Gaussian distribution function with width at half-maximum of 0.1 eV. The energy scale of the theoretical results was shifted by a constant so the highest occupied molecular orbital of each cluster was aligned to the corresponding peak in the photoemission spectra. In making comparisons between theory and experiment, one must focus on the positions of the peaks in the spectra and not in their relative intensities. The intensities are affected by angular-momentum-dependent photodetachment cross-sections which are not included in theory.

The different nature of the $3d$ and $4s$ bands in Cu is reflected in the experimental spectra of the Cu_n^- clusters. The core-like character of the $3d$ electrons give rise to a "band" in the photoemission spectra that is quite localized. The position of this band shifts upwards monotonically in energy when the size of the cluster increases [72], and its onset is determined to be roughly 2 eV below the highest occupied state, at a BE of $\sim 4\,\text{eV}$ for the clusters studied in this work. The features that occur between the highest occupied states and the onset of the $3d$ band are mostly s like. Owing the delocalized character of the $4s$ valence electrons, these features strongly depend on the surface size and geometry of the clusters, and do not show any evident pattern with increasing cluster size.

As one can see from Fig. 13, the density of states as calculated from the single-particle eigenvalues give a very poor description of the photoemission spectra for the smaller clusters studied, Cu_3^- and Cu_5^-; even a qualitative assignment of most of the theoretical peaks to corresponding peaks in the photoemission spectra is not possible (e.g., for Cu_3^-, experiment clearly shows the existence of two distinct features before the onset of the d band, which exhibits five closely spaced peaks with binding energies less than 4 eV). For Cu_8^- and Cu_9^-, the two first major peaks discernible in the spectra can be identified in the density of states, although their relative position is not perfectly matched, but beyond the second peak the experimental features are not properly reproduced, i.e., theory tends to place peaks very close in energy.

The limitations showed by the single-particle eigenvalues in reproducing the photoemission spectra are overcome when final state effects are taken into account.

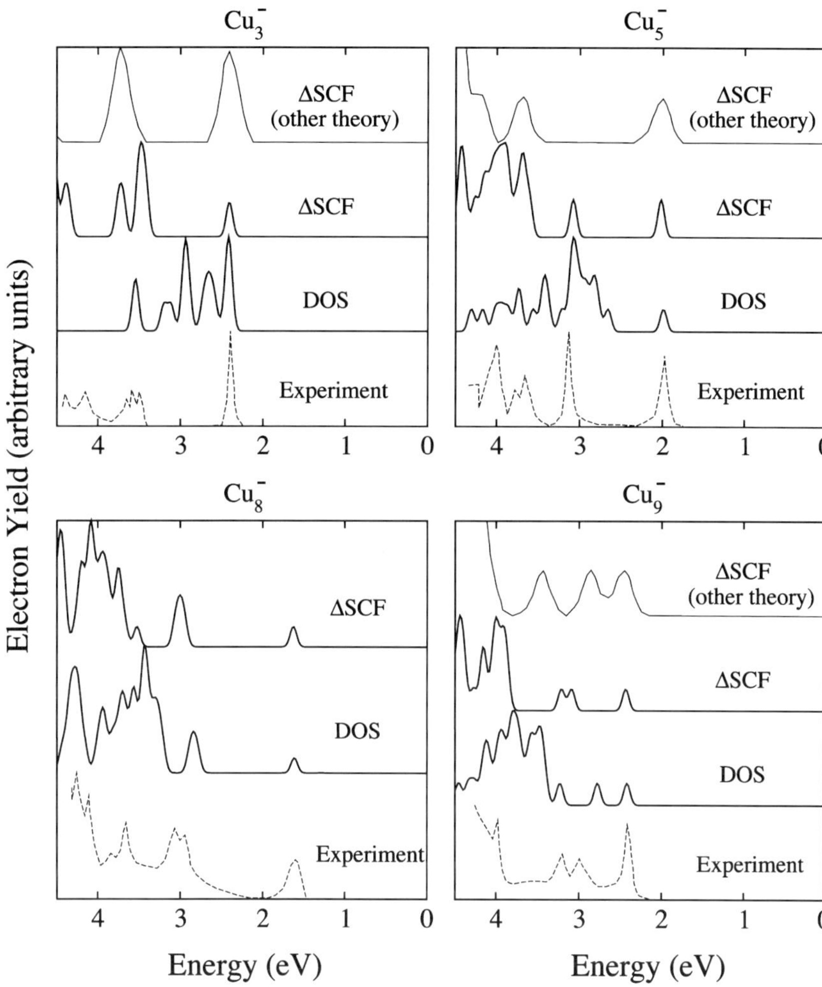

Fig. 13. DOS and excitation energies (ΔSCF) as obtained from a real-space pseudopotential approach for the Cu_n^- ground-state structures represented in Fig. 12. Results for the excitation energies as obtained from a plane-wave pseudopotential approach by Massobrio et al. [67,68] for $n = 3, 5$ and 9 are also shown ("other theory"; upper lines). The theoretical results are compared with available experimental data from C.-Y. Cha et al. [71].

As we see in Fig. 13, one can identify all the main features of the photoemission spectra (PES) through the theoretical excitation energies. Also, the relative position of the experimental features is accurately reproduced. The most-important mismatch between theory and experiment occurs for Cu_3^-, where theory yields a doublet peaks in contrast to the experimental feature at $\sim 3.5\,eV$. The experimental feature is divided in several subpeaks. The real space pseudopotential results for the excitation energy also match the experiment better than the results obtained in previous work [67,68], especially for Cu_5^- and Cu_9^-. The spectra for these two clusters clearly show the existence of three distinguishable peaks before the onset of the $3d$ band at $\sim 4\,\text{eV}$. Previous calculations of the spectra for Cu_5^- only gives two peaks, with a separation between them that is off by $\sim 0.5\,\text{eV}$ when compared with experiment. For Cu_9^-, three distinct peaks also are discernible from the previous results, but their distribution along the 0–4 eV BE range differs greatly from that observed in experiment. The differences between the current results and previous work for Cu_9^- can be attributed, at least in part, to geometry differences between the ground state structures for which the excitation spectra was calculated.

5.3. Vibrational Modes

Experiments on the vibrational spectra of clusters can provide us with very important information about their physical properties. Recently, Raman experiments have been performed on clusters which have been deposited on inert substrates [73]. Since different structural configurations of a given cluster can possess different vibrational spectra, it is possible to compare the vibrational modes calculated for a particular structure with the Raman experiment in a manner similar to the previous example with photoemission. If the agreement between experiment and theory is good, this is a necessary condition for the validity of the theoretically predicted structure.

There are two common approaches for determining the vibrational spectra of clusters. One approach is to calculate the dynamical matrix, $\mathcal{M}_{i\alpha,j\beta}$, for the ground state structure of the cluster

$$\mathcal{M}_{i\alpha,j\beta} = \frac{1}{M}\frac{\partial^2 E}{\partial R_i^\alpha \partial R_j^\beta} = -\frac{1}{M}\frac{\partial F_i^\beta}{\partial R_j^\alpha} \tag{24}$$

where M is the mass of the atom, E the total energy of the system, F_i^α the force on atom i in the direction α, R_i^α the α component of coordinate for atom i. One can calculate the dynamical matrix elements by calculating the first-order derivative of force versus atom displacement numerically. From the eigenvalues and eigenmodes of the dynamical matrix, one can obtain the vibrational frequencies and modes for the cluster of interest [74].

The other approach to determine the vibrational modes is to perform a molecular dynamics simulation. The cluster in question is excited by small random displacements. By recording the kinetic (or binding) energy of the cluster as a function of the simulation time, it is possible to extract the power spectrum of the cluster and determine the vibrational modes. This approach has an advantage for large clusters in that one never has to do a mode analysis explicitly. Another advantage is that anaharmonic mode couplings can be examined. It has the disadvantage in that the simulation must be performed over a long time to extract accurate values for all the modes.

As a specific example, consider the vibrational modes for a small silicon cluster: Si_4. The starting geometry was taken to be a planar structure for this cluster as established from a higher-order finite difference calculation [74].

It is straightforward to determine the dynamical matrix and eigenmodes for this cluster. In Fig. 14, the fundamental vibrational modes are illustrated. In Table 1, the frequency of these modes are presented. One can also determine the modes via a simulation. To initiate the simulation, one can perform a Langevin simulation [42] with a fixed temperature at 300 K. After a few dozen time steps, the Langevin simulation is turned off, and the simulation proceeds following Newtonian dynamics with "quantum" forces. This procedure allows a stochastic element to be introduced and establish initial conditions for the simulation without bias toward a particular mode. For this example, time step in the molecular dynamics simulation was taken to be 3.7 fs, or approximately 150 a.u. The simulation was allowed to proceed for 1000 time steps or roughly 4 ps. The variation of the kinetic and binding energies is given in Fig. 15 as a function of the simulation time. Although some fluctuations of the total energy occur, these fluctuations are relatively small, i.e., less than ~ 1 meV, and there is no noticeable drift of the total energy. Such fluctuations arise, in part, because of discretization errors. As the grid size is reduced, such errors are minimized [74]. Similar errors can occur in plane wave descriptions using supercells, i.e., the artificial periodicity of the supercell can introduce erroneous

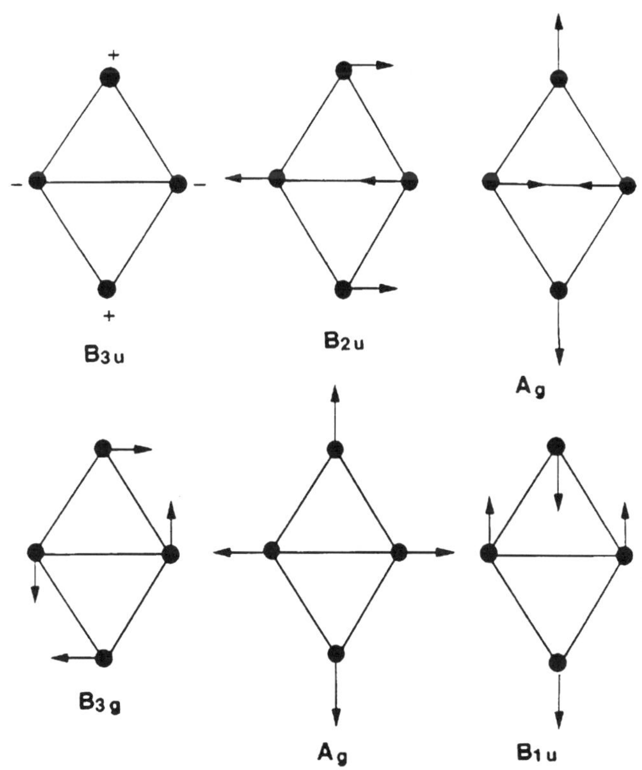

Fig. 14. Normal modes for a Si_4 cluster. The + and − signs indicate motion in and out of the plane, respectively.

Table 1. Calculated and experimental vibrational frequencies in an Si_4 cluster.

	B_{3u}	B_{2u}	A_g	B_{3g}	A_g	B_{1u}
Experiment [73]			345		470	
Dynamical matrix [74]	160	280	340	460	480	500
MD simulation [74]	150	250	340	440	490	500
HF [76]	117	305	357	465	489	529
LCAO [75]	55	248	348	436	464	495

Note: See Fig. 14 for an illustration of the normal modes. The frequencies are given in cm^{-1}.

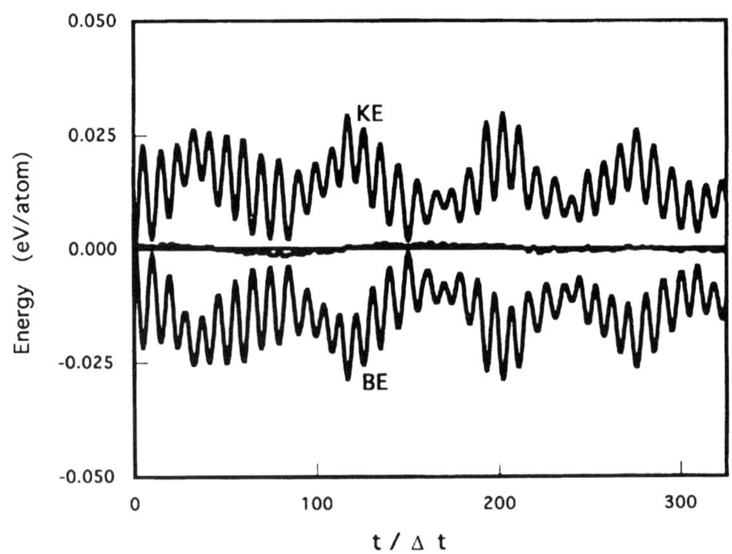

Fig. 15. Simulation for a Si_4 cluster. The kinetic energy (KE) and binding energy (BE) are shown as a function of simulation time. The total energy (KE + BE) is also shown with the zero of energy taken as the average of the total energy. The time step, Δt, is 3.7 fs.

forces on the cluster. By taking the power spectrum of either the KE or BE over this simulation time, the vibrational modes can be determined. These modes can be identified with the observed peaks in the power spectrum as illustrated in Fig. 16.

A comparison of the calculated vibrational modes from the molecular dynamics simulation and from a dynamical matrix calculation are listed in Table 1. Overall, the agreement between the simulation and the dynamical matrix analysis is quite satisfactory. In particular, the softest mode, i.e., the B_{3u} mode, and the splitting between the (A_g, B_{1u}) modes are well replicated in the power spectrum. The splitting of the (A_g, B_{1u}) modes is less than $10\,cm^{-1}$, or about 1 meV, which is probably at the resolution limit of any *ab initio* method.

The theoretical values are also compared to experiment. The predicted frequencies for the two A_g modes are surprisingly close to Raman experiments on silicon

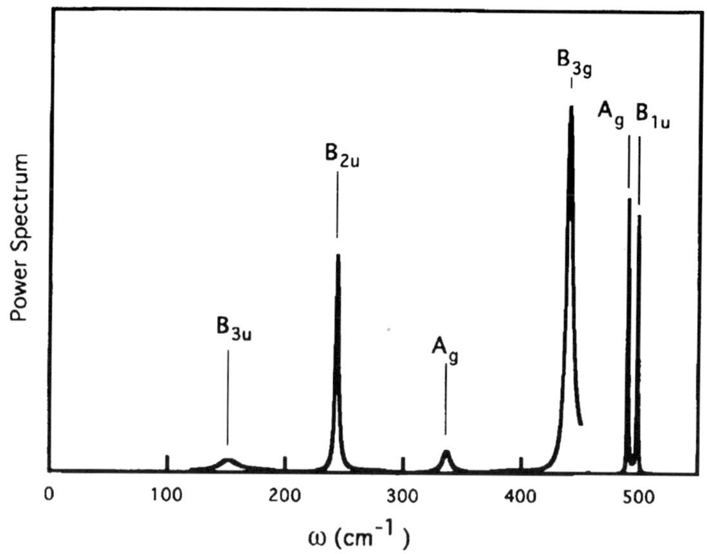

Fig. 16. Power spectrum of the vibrational modes of the Si_4 cluster. The simulation time was taken to be 4 ps. The intensity of the B_{3g} and (A_g, B_{1u}) peaks has been scaled by 10^{-2}.

clusters [73]. The other allowed Raman line of mode B_{3g} is expected to have a lower intensity and has not been observed experimentally.

The theoretical modes using the formalism outlined here are in good accord (except the lowest mode) with other theoretical calculations given in Table 1: an linear combination of atomic orbits (LCAO) calculation [75] and a Hartree–Fock (HF) calculation [76]. The calculated frequency of the lowest mode, i.e., the B_{3u} mode, is problematic. The general agreement of the B_{3u} mode as calculated by the simulation and from the dynamical matrix is reassuring. Moreover, the real-space calculations agree with the HF value to within $\sim$ 20–30 cm^{-1}. On the other hand, the LCAO method yields a value which is 50–70% smaller than either the real space or HF calculations. The origin of this difference is not apparent. For a poorly converged basis, vibrational frequencies are often overestimated as opposed to the LCAO result which underestimates the value, at least when compared to other theoretical techniques. Setting aside the issue of the B_{3u} mode, the agreement between the measured Raman modes and theory for Si_4 suggests that Raman spectroscopy can provide a key test for the structures predicted by theory.

5.4. Polarizabilities

Recently polarizability measurements [77] have been performed for small semiconductor clusters. These measurements will allow one to compare computed values with experiment.

The polarizability tensor, α_{ij}, is defined as the second derivative of the energy with respect to electric field components. For a noninteracting quantum mechanical system, the expression for the polarizability can be easily obtained by using second-order perturbation theory where the external electric field, $\mathscr{E}$, is treated as a weak perturbation.

Within the DFT, since the total energy is not the sum of individual eigenvalues, the calculation of polarizability becomes a nontrivial task. One approach is to use density-functional perturbation theory, which has been developed recently in Green's function and variational formulations [78,79].

Another approach, which is very convenient for handling the problem for *confined* systems, like clusters, is to solve the full problem exactly within the one-electron approximation. In this approach, the external ionic potential $V_{\text{ion}}(\mathbf{r})$ experienced by the electrons is modified to have an additional term given by $-e\mathscr{E} \cdot \mathbf{r}$. The Kohn–Sham equations are solved with the full external potential $V_{\text{ion}}(\mathbf{r}) - e\mathscr{E} \cdot \mathbf{r}$. For quantities like polarizability, which are derivatives of the total energy, one can compute the energy at a few field values, and differentiate numerically. Real-space methods are very suitable for such calculations on confined systems, since the position operator $\mathbf{r}$ is not ill defined, as is the case for supercell geometries in plane wave calculations. The next section of this text elaborates on this point.

There is another point that should be emphasized. It is difficult to determine the polarizability of a cluster or molecule owing to the need for a complete basis in the presence of an electric field. Often polarization functions are added to complete a basis and the response of the system to the field can be sensitive to the basis required. In both real-space and plane wave methods, the lack of a "prejudice" with respect to the basis are considerable assets. The real-space method implemented with a uniform grid possesses a nearly "isotropic" environment with respect to the applied field. Moreover, the response can be easily checked with respect to the grid size by varying the grid spacing. Typically, the calculated electronic response of a cluster is not sensitive to the magnitude of the field over several orders of magnitude.

In Table 2, we present some recent calculations for the polarizability of small Si and Ge clusters and illustrate their polarizability as a function of cluster size in Fig. 17. It is interesting to note that some of these clusters have permanent dipoles. For example, Si_6 and Ge_6 both have nearly degenerate isomers. One of these isomers possesses a permanent dipole, the other does not. Hence, in principle, one might be able to separate the one isomer from the other via an inhomogeneous electric field.

5.5. Optical Spectra

While the theoretical background for calculating ground state properties of many-electron systems is now well established, excited state properties such as optical spectra present a challenge for DFT. Linear response theory within the time-dependent density-functional formalism provides a powerful tool for calculating

Table 2. Static dipole moments and average polarizabilities of small silicon and germanium clusters.

Silicon			Germanium		
Cluster	$\|\mu\|$ (D)	$\langle\alpha\rangle$ (Å^3/atom)	Cluster	$\|\mu\|$ (D)	$\langle\alpha\rangle$ (Å^3/atom)
Si_2	0	6.29	Ge_2	0	6.67
Si_3	0.33	5.22	Ge_3	0.43	5.89
Si_4	0	5.07	Ge_4	0	5.45
Si_5	0	4.81	Ge_5	0	5.15
Si_6 (I)	0	4.46	Ge_6 (I)	0	4.87
Si_6 (II)	0.19	4.48	Ge_6 (II)	0.14	4.88
Si_7	0	4.37	Ge_7	0	4.70

excited states properties [80–82]. This method, known as the time-dependent local density approximation, allows one to compute the true excitation energies from the conventional, time independent Kohn–Sham transition energies and wave functions.

Within the time-dependent local density approximation (TDLDA), the electronic-transition energies Ω_n are obtained from the solution of the following eigenvalue problem [81,82]:

$$[\omega_{ij\sigma}^2 \delta_{ik}\delta_{jl}\delta_{\sigma\tau} + 2\sqrt{f_{ij\sigma}\omega_{ij\sigma}} K_{ij\sigma,kl\tau} \sqrt{f_{kl\tau}\omega_{kl\tau}}]\mathbf{F}_n = \Omega_n^2 \mathbf{F}_n \tag{25}$$

where $\omega_{ij\sigma} = \varepsilon_{j\sigma} - \varepsilon_{i\sigma}$ are the Kohn–Sham transition energies, $f_{ij\sigma} = n_{i\sigma} - n_{j\sigma}$ are the differences between the occupation numbers of the ith and jth states, the eigenvectors $\mathbf{F}_n$ are related to the transition oscillator strengths, and $K_{ij\sigma,kl\tau}$ is a coupling matrix given by

$$K_{ij\sigma,kl\tau} = \int\int \phi_{i\sigma}^*(\mathbf{r})\phi_{j\sigma}(\mathbf{r}) \left(\frac{1}{|\mathbf{r} - \mathbf{r}'|} + \frac{\partial v_\sigma^{\mathrm{xc}}(\mathbf{r})}{\partial \rho_\tau(\mathbf{r}')} \right) \phi_{k\tau}(\mathbf{r}')\phi_{l\tau}^*(\mathbf{r}')\, \mathrm{d}\mathbf{r}\, \mathrm{d}\mathbf{r}' \tag{26}$$

where i, j, σ are the occupied state, unoccupied state and spin indices respectively; $\phi(\mathbf{r})$ the Kohn–Sham wave functions and $v^{\mathrm{xc}}(\mathbf{r})$ the LDA exchange-correlation potential.

The TDLDA formalism is easy to implement in real space within the higher-order finite difference pseudopotential method [19,83]. The real-space pseudopotential code represents a natural choice for implementing TDLDA due to the real-space formulation of the general theory. With other methods, such as the plane wave approach, TDLDA calculations typically require an intermediate real-space basis. After the original plane wave calculation has been completed, all functions are transferred into that basis, and the TDLDA response is computed in real space [84]. The additional basis complicates calculations and introduces an extra error. The real-space approach simplifies implementation and allows us to perform the complete TDLDA response calculation in a single step.

We illustrate the TDLDA technique by calculating the absorption spectra of a sodium cluster. Sodium clusters were chosen as well-studied objects, for which

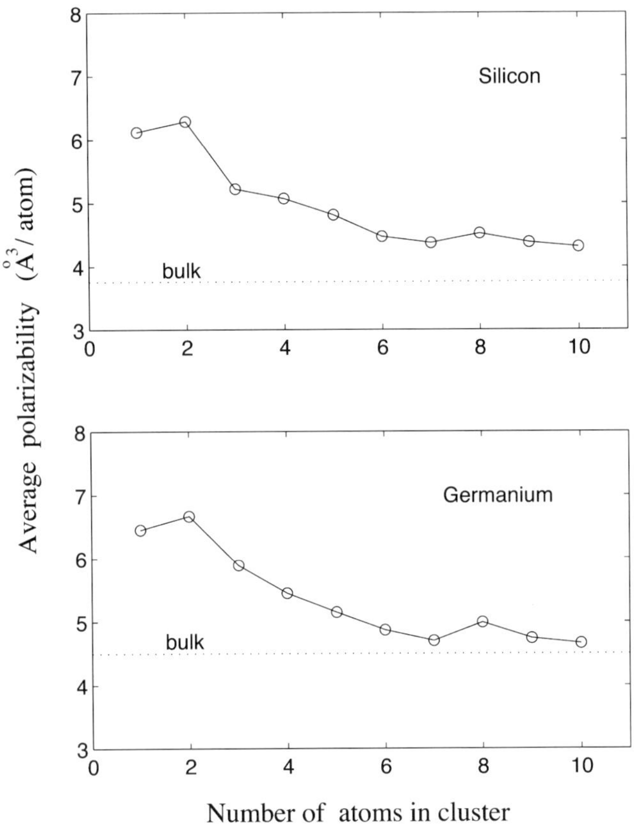

Fig. 17. The average polarizability for clusters of Si and Ge as a function of cluster size. Note that the polarizability exceeds that of the bulk value owing to dangling bonds on the cluster surface.

accurate experimental measurements of the absorption spectra are available [85]. The ground-state structures of the clusters were determined by simulated annealing [42]. In all cases the obtained cluster geometries agreed well with the structures reported in other works [86]. Since the wave functions for the unoccupied electron states are very sensitive to the boundary conditions, TDLDA calculations need to be performed within a relatively large boundary domain. For sodium clusters, we used a spherical domain with a radius of 25 a.u. and a grid spacing of 0.9 a.u.

The calculated absorption spectrum for small sodium clusters is shown in Fig. 18 along with experiment. In addition, we illustrate the spectrum generated by considering transitions between the LDA eigenvalues. The agreement between TDLDA and experiment is remarkable, especially when contrasted with the LDA spectrum. TDLDA correctly reproduces the experimental spectral shape, and the calculated peak positions agree with experiment within 0.1–0.2 eV. The comparison with other theoretical work demonstrates that our TDLDA absorption spectrum is as accurate

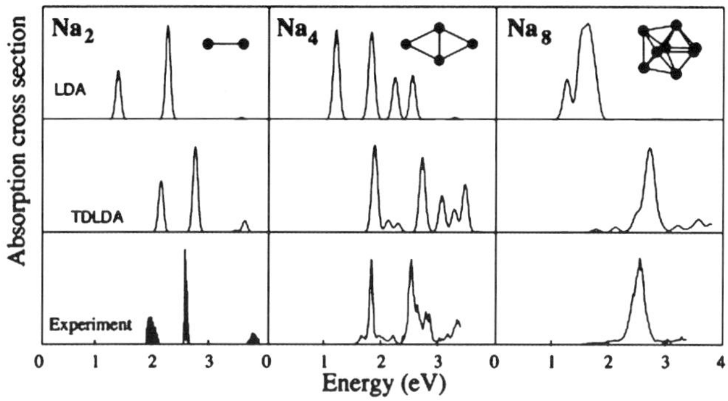

Fig. 18. The calculated and experimental absorption spectrum for Na_2, Na_4 and Na_8. The top panel shows a LDA to the spectrum using Kohn–Sham eigenvalues. The middle panel shows a TDLDA calculation. The bottom panel is experiment [85,99].

as the available CI spectra [87,88]. Furthermore, the TDLDA spectrum for the Na_4 cluster seems to be in better agreement with experiment than the GW absorption spectrum calculated using quantum chemistry methods [89].

6. QUANTUM CONFINEMENT IN NANOCRYSTALS AND DOTS

At the atomic scale, electronic states are typically characterized by discrete-energy levels that are often separated by electron volts; the spatial distribution of these states is highly localized. In contrast, electronic states in crystalline matter are typically characterized by energy bands: the energy states form a quasi-continuous spectrum with delocalized spatial distributions. At the nanoscale, the distribution of energy states resides between these limits. This evolution of energy levels from an atom to large clusters to "quantum dots" is illustrated in Fig. 19. A large cluster in this scheme may possess hundreds or thousands of atoms. When its surface is electronically passivated, such large clusters are called "quantum dots." While the energy spacing between electronic levels in a "quantum dot" is not quasi-continuous, it is usually too small to be measured. For example, the spacing may be much smaller than the thermal energy associated with a vibrational mode.

Understanding the evolution of the electronic structure of matter at the nanoscale is a challenging task. Matter at this length scale is not especially amenable to techniques based on atomic or macroscopic descriptions, as the spatial and energetic distribution of electron states are quite disparate. Methods for calculating the electronic properties of these systems usually start from either the atomic or molecular limit or from the solid state limit. Atomic or molecular methods, are often not equipped to handle the large number of atoms or electrons in nano-scale materials whereas solid state methods often require translational symmetry, which is not present for confined systems. Moreover, the electronic interactions and spatial

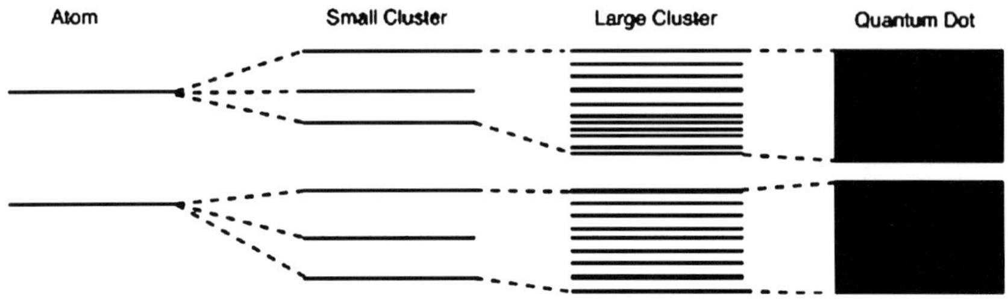

Fig. 19. Schematic energy levels for an atom, a cluster and a quantum dot.

extent of wave functions at the nanoscale can be remarkably different than either the atomic or macroscopic limit.

This electronic effect can be illustrated by considering an electron confined by rigid walls to a one-dimensional box of size a. The lowest energy level of this system is given in elementary textbooks as

$$E = \frac{\pi^2 \hbar^2}{2ma^2} \tag{27}$$

where $\hbar$ is the Planck constant divided by 2π, and m the mass of the particle, e.g., the mass of an electron. If the size of the box is reduced, then E increases.

This phenomenon, i.e., the energy level spacing increasing with reduced dimensionality, is called *quantum confinement*. Quantum confinement can be readily understood from the Heisenberg uncertainty principle. The uncertainty principle states that the uncertainty in momentum and position must be such that the product exceeds $\hbar/2$:

$$\Delta p \cdot \Delta x \geqslant \hbar/2 \tag{28}$$

where Δp is the uncertainty in the momentum and Δx the uncertainty in the position of the electron. Consider the energy of a free electron with momentum p:

$$E = \frac{p^2}{2m} \tag{29}$$

Since the uncertainty in the momentum cannot exceed the momentum itself, one can write $p > \Delta p$. As such, one has the following inequality:

$$E > \frac{\hbar^2}{8ma^2} \tag{30}$$

If one tries to localize the position of an electron by reducing the box size, its energy must increase and diverge as the confining region vanishes.

Although this very simple picture of quantum confinement is illustrated for one dimension. It is also true for an electron confined in three dimensions. For example, if a particle is confined within a sphere whose radius is R, one might expect an energy level to vary with the radius according to

$$E(R) = E_\infty + \frac{\alpha}{R^2} \tag{31}$$

where α is a constant and E_∞ the energy as $R \to \infty$. Suppose one considers an optical gap, E_{gap}, as the difference between the high-filled and lowest empty states for a system confined in a sphere of size R. In the simplest description of the gap we might expect E_{gap} to scale as

$$E_{gap}(R) = E_{gap} + \frac{\beta}{R^2} \tag{32}$$

where β is a constant.

Quantum confinement has been observed experimentally for nanostructures such as quantum dots of Si and CdSe. In Fig. 20 a quantum dot of hydrogenated silicon is illustrated. The interior of the dot consists of silicon atoms in the bulk diamond structure; the surface of the dot is hydrogenated. Hydrogen removes any dangling bonds on the surface. As such, all the atoms in the system should be fully coordinated and should not contribute to the electronic properties of the quantum dot. Typically, a quantum dot is a few nm to tens of nm in size. The smallest dots contain hundreds of atoms, whereas a large dot may contain hundreds of thousands of atoms.

Also illustrated in Fig. 20 is the corresponding optical gap as function of dot size. The quantum dot gap with this size range is approximately twice the gap of crystalline silicon; silicon exhibits weak optical absorption near 1.1 eV. The absorption is weak owing to the nature of the gap in silicon. It is an indirect gap because the optical excitation couples an electron with a hole of different crystalline momentum or wave vector. Conservation of crystal momentum would require the electron and hole to have the same momentum. As such, this process can only occur by the participation of the quantized vibrational modes of the lattice or phonons, which provide the required momentum to satisfy the conservation rules. Unlike a two-body process (electron and hole), a three-body process (electron, hole and phonon) results in a very weak interaction between light and matter.

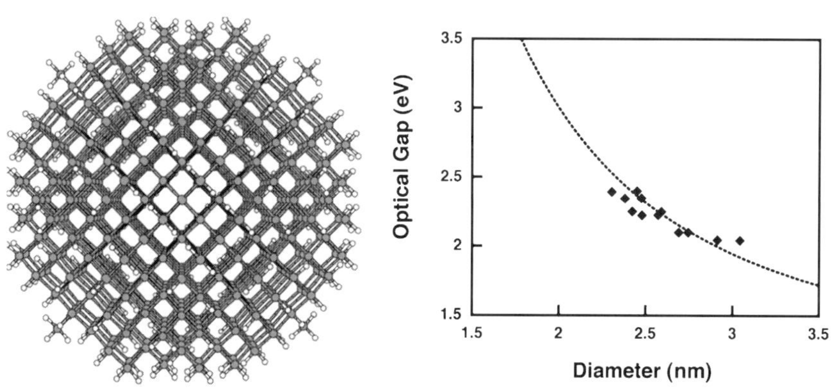

Fig. 20. Ball and stick model for a quantum dot of hydrogenated silicon (left). The gray balls represent the silicon atoms, the white balls represent the hydrogens. Also shown is the corresponding optical absorption [101]. Note the increase in gap size from the known bulk gap of silicon, which is 1.1 eV.

In finite systems, crystal momentum is not well defined, owing to the lack of translation symmetry. As the system approaches nanoscale sizes, the nature of an indirect gap or direct gap becomes moot. Consequently, the optical absorption becomes stronger in quantum dots and the size of the gap increases owing to quantum confinement. An approximate fit of Eq. (32) is consistent with the experimental data shown in Fig. 20. This example shows the striking effect of size on the electronic structure of nanostructures, i.e., silicon at this length scale is transformed from an optical inactive material to an optically active one.

Solutions of the Kohn–Sham equation can be used to describe the electronic structure of a variety of nanostructures, such as nanowires, nanotubes or quantum dots. In the case of nanowires or nanotubes, the structure is localized in two dimension and often periodic in the third. In this case, Bloch wave functions can be used for the periodic structure. For quantum dots, supercells or localized domains can be utilized along with plane waves or grid methods.

In Fig. 21, the evolution of energy levels are illustrated for GaAs quantum dots. The stoichiometry of the dots are $(GaAs)_nH^*_m$. The H^*_m notation refers to fictitious hydrogen-like atoms used to passivate any dangling or partially occupied bonds. In this example, the structure of the dots is taken to be that of a fragment of the GaAs crystal. For small dots such as $(GaAs)_2H^*_6$, the energy levels are well separated and the gap between filled and empty states is on the order of $\sim 6\,eV$. This gap cannot be compared to observed gap as these energy levels do not correspond to those created by an electron–hole pair. The Kohn–Sham eigenvalues have meaning only

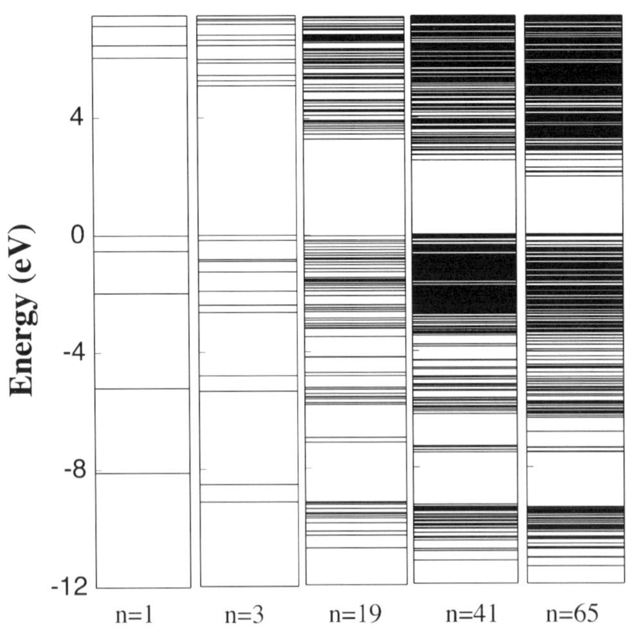

Fig. 21. Energy levels of $(GaAs)_nH^*_m$ quantum dots. The number of GaAs units is indicated in the figure. The highest-occupied level is taken as the zero energy reference.

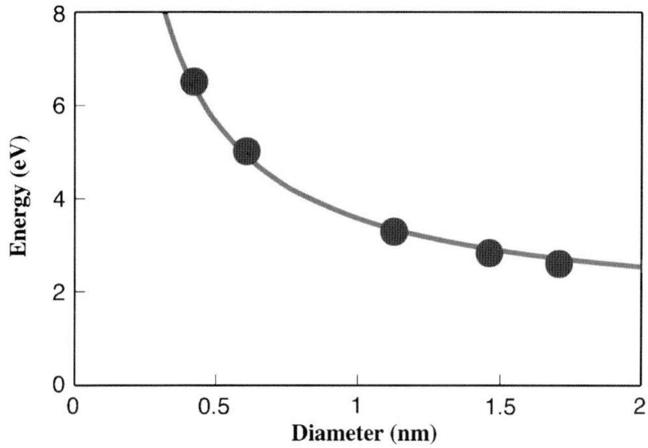

Fig. 22. The energy gap in quantum dots of $(GaAs)_n H_m^*$ quantum dots versus the size of the dot. The curve is after Eq. (32) with different scaling than R^{-2} (See text).

in solving for the total energy. However, more realistic approaches to excited state properties have validated the use of Kohn–Sham eigenvalues as a qualitative or in some case semiquantitative method of determining the optical gap.

As a function of cluster size the size of the gap between empty and filled states decreases as the size of the cluster increases. This is illustrated in Fig. 22 where the gap size is plotted versus the cube root of the number of GaAs units present. The cube root should scale with diameter of the quantum dot, provided a large number of atoms are present. The general trend of the gap with the size of the dot is consistent with Eq. (32) with one notable exception. The scaling is not consistent with R^{-2} behavior, which would only be expected for a particle contained by an infinite well. Owing to the finite size of the quantum well in realizable systems, the scaling obtained from accurate calculations yield gaps that scale closer to R^{-1} as indicated in this figure.

A notable feature of the energy level spacings is the distribution of these states as a function of the dot size. For large systems, the distribution of the eigenvalues should approach the crystalline state. In Fig. 23, we compare the crystalline density of states, i.e., the number of states per unit energy, to the number of eigenvalues per unit of time for a large dot. Structure in the density of states can be attributed to the topology of energy bands. When the energy band is flat, i.e., the derivative of the energy with respect to wave function vanishes, the density of states possesses a identifiable structure. As a function of size, one would expect large clusters to possess similar structural features. The number of atoms in a dot required to reproduce bulk features of the density of states can be assessed through the direct calculation of the eigenvalue spectrum. In Fig. 23, a comparison between the crystal and dot density of states for the dot: $(GaAs)_{342}H_{192}^*$ is illustrated. The distribution of eigenvalue states matches the bulk density of states very well with one notable exception. Some states are associated with the fictitious hydrogens; these states have

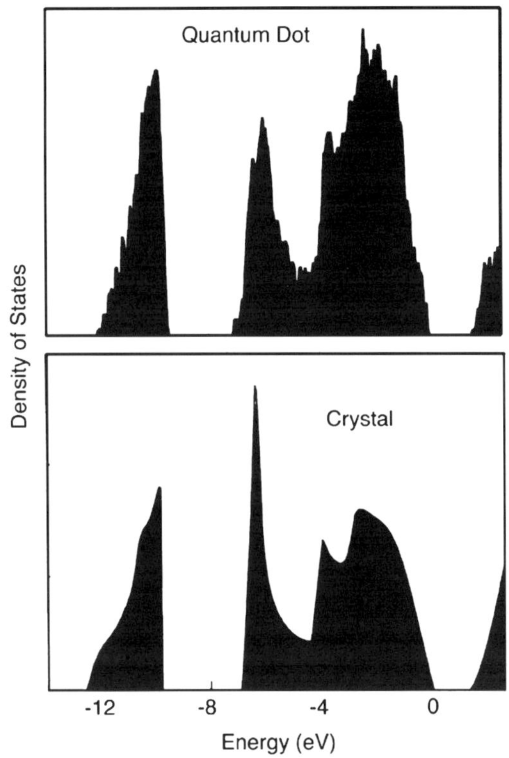

Fig. 23. Densities of states for crystal gallium arsenide (bottom) and the eigenvalue distribution for $(GaAs)_{342}H^*_{192}$.

been removed from Fig. 23. More complex nanostructures can be examined using similar approaches.

Of course, the gap between occupied and empty states as calculated from DFT is not directly comparable to the measure optical gap, although this gap may scale as the optical gap. This issue was discussed earlier in Chapter 2, where methods for determining optical excitations were discussed in some detail. Here we illustrate one such approach: the application of time-dependent DFT to quantum dots. This method is known to work well for systems of less than a 1000 atoms or so, but is expected to fail for extended systems using the formalism resulting in Eq. (25).

The variation of the optical absorption gaps as a function of cluster size is shown in Fig. 24. Along with the TDLDA values, we include optical gaps calculated by the Bethe–Salpeter (BS) technique [90]. For very small clusters, SiH_4, Si_2H_6 and Si_5H_{12}, the gaps computed by the TDLDA method are close to the Bethe-Salpeter values, although for $Si_{10}H_{16}$ and $Si_{14}H_{20}$ the TDLDA gaps are considerably smaller than the Bethe-Salpeter gaps. At the same time, the TDLDA gaps for clusters in the size range from 5 to 71 silicon atoms are larger by ~ 1 eV than the gaps calculated by the HF technique with the correlation correction included through the configuration-interaction approximation (HF–CI) [91].

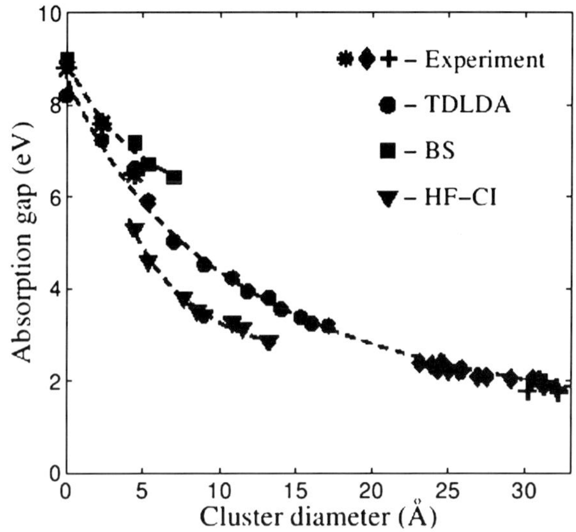

Fig. 24. Variation of optical absorption gaps as a function of cluster diameter. Theoretical values shown in the plot include the gaps calculated by the TDLDA method (this work), by the Bethe–Salpeter technique (BS) [90] and by the Hartree–Fock method with the correlation included through the configuration–interaction approximation (HF–CI) [91]. Experimental values are taken from Refs. [92,100–102]. The dashed lines are a guide to the eye.

These differences are consistent with the fact that the BS calculations systematically overestimate and the HF–CI calculations of Ref. [91] underestimate the experimental absorption gaps. For example, for the optical absorption gap of Si_5H_{12} the BS, TDLDA and HF–CI methods predict the values of 7.2, 6.6 and 5.3 eV, respectively, compared to the experimental value of 6.5 eV. However, it is not clear whether the gaps of Ref. [91] refer to the optically allowed or optically forbidden transitions, which may offer a possible explanation for the observed discrepancy. For large clusters, we find the TDLDA optical gaps to be in generally good agreement with the photoabsorption gaps evaluated by the majority of self-energy corrected LDA [66,92] and empirical techniques [93–95]. At present, the full TDLDA calculations for clusters larger than a few nm can exceed the capabilities of most computational platforms. Nevertheless, the extrapolation of the TDLDA curve in the limit of large clusters comes very close to the experimental values for the photoabsorption gaps. Software and hardware advances should make a direct verification of this possible in the near future.

ACKNOWLEDGMENTS

The author acknowledges support from the National Science Foundation, the United States Department of Energy, the Minnesota Supercomputing Institute and the Institute for Computational Engineering and Sciences.

REFERENCES

[1] J.R. Chelikowsky and S.G. Louie (Eds), *Quantum Theory of Materials* (Kluwer, Boston, 1996).
[2] J.R. Chelikowsky and M.L. Cohen, *Ab initio* Pseudopotentials for Semiconductors, in *Handbook of Semiconductors*, 2nd ed., vol. 1, edited by P.T. Landsberg (Elsevier, Amsterdam, 1992), p. 219.
[3] P. Hohenberg and W. Kohn, Inhomogeneous electron gas, *Phys. Rev.* **136**, B864 (1964).
[4] W. Kohn and L.J. Sham, Self-consistent equations including exchange and correlation effects, *Phys. Rev.* **140**, A1133 (1965).
[5] W. Pickett, Pseudopotential methods in condensed matter applications, *Comput. Phys. Rep.* **9**, 115 (1989).
[6] M.C. Payne, M.P. Teter, D.C. Allan, T.A. Arias and J.D. Joannopoulos, Iterative minimization techniques for *ab initio* total-energy calculations: Molecular dynamics and conjugate gradients, *Rev. Mod. Phys.* **64**, 1045 (1992).
[7] M.L. Cohen and J.R. Chelikowsky, *Electronic Structure and Optical Properties of Semiconductors*, 2nd Ed. (Springer, Berlin, 1989).
[8] J.R. Chelikowsky and S.G. Louie, First-principles linear combination of atomic orbitals method for the cohesive and structural properties of solids: Application to diamond, *Phys. Rev. B* **29**, 3470 (1984).
[9] S.C. Erwin, M.R. Pederson and W.E. Pickett, First-principles, general-potential local-orbital calculations for bulk crystals, *Phys. Rev. B* **41**, 10437 (1990).
[10] T.L. Beck, Real-space mesh techniques in density functional theory, *Rev. Mod. Phys.* **74**, 1041 (2000).
[11] J.E. Pask, B.M. Klein, P.A. Sterne and C.Y. Fong, Finite-element methods in electronic-structure theory, *Comp. Phys. Commun.* **135**, 1 (2001).
[12] J.R. Chelikowsky, L. Kronik, I. Vasiliev, M. Jain and Y. Saad, Real space pseudopotentials for the electronic structure problem, in *Handbook of Numerical Methods: Computational Chemistry*, edited by C.L. Bris (Elsevier, Amsterdam, 2003), p. 613.
[13] T. Ono and K. Hirose, Timesaving double-grid method for real-space electronic structure calculations, *Phys. Rev. Lett.* **82**, 5016 (1999).
[14] E.L. Briggs, D.J. Sullivan and J. Bernholc, Large-scale electronic-structure calculations with multigrid acceleration, *Phys. Rev. B* **52**, R5471 (1995).
[15] J.-L. Fattebert and J. Bernholc, Towards grid-based $O(N)$ density-functional theory methods: Optimized nonorthogonal orbitals and multigrid acceleration, *Phys. Rev. B* **62**, 1713 (2000).
[16] G. Zumbach, N.A. Modine and E. Kaxiras, Adaptive-coordinate real-space electronic-structure calculations for atoms, molecules, and solids, *Solid State Commun.* **99**, 57 (1996).
[17] F. Gygi and G. Galli, Real-space adaptive-coordinate electronic-structure calculations, *Phys. Rev. B* **52**, R2229 (1995).
[18] A. Stathopoulos, S. Öğüt, Y. Saad, J. Chelikowsky and H. Kim, Parallel methods and tools for predicting material properties, *Comput. Sci. Eng.* **2**, 19 (2000).
[19] J.R. Chelikowsky, The pseudopotential-density functional method applied to nanostructures, *J. Phys. D: Appl. Phys.* **33**, R33 (2000).
[20] G. Smith, *Numerical Solutions of Partial Differential Equation: Finite Difference Methods*, 2nd Ed. (Oxford University Press, Oxford, 1978).
[21] B. Fornberg and D.M. Sloan, A review of pseudospectral methods for solving partial differential equations, *Acta Numer.* **94**, 203 (1994).
[22] J.R. Chelikowsky, N. Troullier and Y. Saad, The finite-difference-pseudopotential method: Electronic structure calculations without a basis, *Phys. Rev. Lett.* **72**, 1240 (1994).
[23] J.R. Chelikowsky, N. Troullier, K. Wu and Y. Saad, Higher order finite difference pseudopotential method: An application to diatomic molecules, *Phys. Rev. B* **50**, 11135 (1994).
[24] N. Troullier and J.L. Martins, Efficient pseudopotentials for plane-wave calculations, *Phys. Rev. B* **43**, 1993 (1991).
[25] L. Kleinman and D.M. Bylander, Efficacious form for model pseudopotentials, *Phys. Rev. Lett.* **48**, 1425 (1982).

[26] J. Martins and M. Cohen, Diagonalization of large matrices in pseudopotential band-structure calculations: Dual space formalism, *Phys. Rev. B* **37**, 6134 (1988).

[27] Y. Saad, *Iterative Methods for Sparse Linear Systems* (PWS Publishing, Philadelphia, 1996).

[28] R. Lehoucq, D.C. Sorensen and C. Yang, *ARPACK User's Guide: Solution of Large Scale Eigenvalue Problems with Implicitly Restarted Arnoldi Methods* (SIAM, Philadelphia, 1998).

[29] J. Ihm, A. Zunger and M.L. Cohen, Momentum–space formalism for the total energy of solids, *J. Phys. C* **12**, 4409 (1979).

[30] R.D. King-Smith, M.C. Payne and J.S. Lin, Real-space implementation of nonlocal pseudopotentials for first-principles total-energy calculations, *Phys. Rev. B* **44**, 13063 (1991).

[31] P. Pulay, *Ab initio* calculation of force constants and equilibrium geometries, I. Theory, *Mol. Phys.* **17**, 197 (1969).

[32] J.R. Chelikowsky, Silicon in all its forms, *MRS Bull.* **27**, 951 (2002).

[33] T.E. Faber, *Introduction to the Theory of Liquid Metals* (Cambridge University Press, Cambridge, 1972).

[34] R. Car and M. Parrinello, Unified approach for molecular dynamics and density-functional theory, *Phys. Rev. Lett.* **55**, 2471 (1985).

[35] J.R. Chelikowsky, J.J. Derby, V. Godlevsky, M. Jain and J. Raty, *Ab initio* simulations of liquid semiconductors, *J. Phys. Cond. Matt.* **13**, R817 (2001).

[36] D.M. Ceperley and B.J. Alder, Ground state of the electron gas by a stochastic method, *Phys. Rev. Lett.* **45**, 566 (1980).

[37] J.P. Perdew and A. Zunger, Self-interaction correction to density-functional approximations for many-electron systems, *Phys. Rev. B* **23**, 5048 (1981).

[38] S. Nosé, A molecular dynamics method for simulations in the canonical ensemble, *Mol. Phys.* **52**, 255 (1984).

[39] S. Nosé, A unified formulation of constant temperature molecular dynamics, *J. Chem. Phys.* **81**, 511 (1984).

[40] N.G. van Kampen, *Stochastic Processes in Physics and Chemistry* (North-Holland, Amsterdam, 1981).

[41] J.C. Tully, G. Gilmer and M. Shugart, Molecular dynamics of surface diffusion. I. The motion of adatoms and clusters, *J. Chem. Phys.* **71**, 1630 (1979).

[42] N. Binggeli and J.R. Chelikowsky, Langevin molecular dynamics with quantum forces: Application to Silicon clusters, *Phys. Rev. B* **50**, 11764 (1994).

[43] R. Kubo, The fluctuation-dissipation theorem, *Rep. Prog. Theor. Phys.* **29**, 255 (1966).

[44] H. Risken, *The Fokker–Planck Equation* (Springer, New York, 1996).

[45] N. Binggeli, J.L. Martins and J.R. Chelikowsky, Simulation of Si clusters via Langevin molecular dynamics with quantum forces, *Phys. Rev. Lett.* **68**, 2956 (1992).

[46] R.M. Wentzcovitch and J.L. Martins, First principles molecular dynamics of Li: Test of a new algorithm, *Solid State Commun.* **78**, 831 (1991).

[47] Y. Waseda and K. Suzuki, Structure of molten silicon and germanium by X-ray diffraction (and calculation of resistivity and thermoelectric power), *Z. Phys. B* **20**, 339 (1975).

[48] J.P. Hanson and I.R. McDonald, *Theory of Simple Liquids* (Academic Press, London, 1986).

[49] W. Jank and J. Hafner, Structural and electronic properties of the liquid polyvalent elements: The group-IV elements Si, Ge, Sn, and Pb, *Phys. Rev. B* **41**, 1497 (1990).

[50] M.M.G. Alemany, O. Diéguez, C. Rey and L.J. Gallego, Molecular-dynamics study of the dynamic properties of fcc transition and simple metals in the liquid phase using the second-moment approximation to the tight-binding method, *Phys. Rev. B* **60**, 9208 (1999).

[51] U. Balucani, A. Torcini and R. Vallauri, Microscopic dynamics in liquid alkali metals, *Phys. Rev. A* **46**, 2159 (1992).

[52] M.M.G. Alemany, C. Rey and L.G. Gallego, Computer simulation study of the dynamic properties of liquid Ni using the embedded-atom model, *Phys. Rev. B* **58**, 685 (1998).

[53] J.R. Chelikowsky and N. Binggeli, First principles molecular dynamics simulations for liquid Silicon, *Solid State Commun.* **88**, 381 (1993).

[54] V. Godlevsky, J.R. Chelikowsky and N. Troullier, Simulations of liquid semiconductors using quantum forces, *Phys. Rev. B* **52**, 13281 (1995).
[55] I. Stich, R. Car and M. Parrinello, Structural, bonding, dynamical, and electronic properties of liquid silicon: An *ab initio* molecular-dynamics study, *Phys. Rev. B* **44**, 4262 (1991).
[56] P.B. Allen and J.Q. Broughton, Electrical conductivity and electronic properties of liquid silicon, *J. Phys. Chem.* **91**, 4964 (1987).
[57] A.F. Voter, Hyperdynamics: Accelerated molecular dynamics of infrequent events, *Phys. Rev. Lett.* **78**, 3908 (1997).
[58] R. Biswas and D.R. Hamann, Simulated annealing of silicon atom clusters in Langevin molecular dynamics, *Phys. Rev. B* **34**, 895 (1986).
[59] J.P. Perdew, K. Burke and Y. Wang, Generalized-gradient approximation for the exchange-correlation hole of a many-electron system, *Phys. Rev. B* **54**, 16533 (1996).
[60] F.W. Kutzler and G.S. Painter, First-row diatomics: Calculation of the geometry and energetics using self-consistent gradient-functional approximations, *Phys. Rev. B* **45**, 3236 (1992).
[61] D. Deaven and K.M. Ho, Molecular geometry optimization with a genetic algorithm, *Phys. Rev. Lett.* **75**, 288 (1995).
[62] K.A. Jackson, M. Horoi, I. Chaudhuri, T. Frauenheim and A.A. Shvartzburg, Unraveling the shape transformation in Silicon clusters, *Phys. Rev. Lett.* **93**, 013401 (2004).
[63] C.Z. Wang and K.M. Ho, Structure, dynamics, and electronic properties of diamondlike amorphous carbon, *Phys. Rev. Lett.* **71**, 1184 (1993).
[64] O. Cheshnovsky, S.H. Yang, C.L. Pettiet, M.J. Craycraft, Y. Liu and R.E. Smalley, Ultraviolet photoelectron spectroscopy of semiconductor clusters: Silicon and germanium, *Chem. Phys. Lett.* **138**, 119 (1987).
[65] D. Tomanek and M. Schluter, Calculation of magic numbers and the stability of small Si clusters, *Phys. Rev. Lett.* **56**, 1055 (1986).
[66] S. Ogut and J.R. Chelikowsky, Structural changes induced upon charging Ge clusters, *Phys. Rev. B* **55**, R4914 (1997).
[67] C. Massobrio, A. Pasquarello and R. Car, First principles study of photoelectron spectra of Cu_n^- clusters, *Phys. Rev. Lett.* **75**, 2104 (1995).
[68] C. Massobrio, A. Pasquarello and R. Car, Interpretation of photoelectron spectra in Cu_n^- clusters including thermal and final-state effects: The case of Cu_7^-, *Phys. Rev. B* **54**, 8913 (1996).
[69] H. Häkkinen, B. Yoon, U. Landman, X. Li, H.-J. Zhai and L.-S. Wang, On the electronic and atomic structures of small Au_n^- ($n = 4-14$) clusters: A photoelectron spectroscopy and density-functional study, *J. Phys. Chem. A* **107**, 6168 (2003).
[70] K. Jug and B. Zimmermann, Structure and stability of small copper clusters, *J. Chem. Phys.* **116**, 4497 (2002).
[71] C.-Y. Cha, G. Ganteför and W. Eberhardt, Photoelectron – spectroscopy of Cu_n^- clusters – comparison with jellium model predictions, *J. Chem. Phys.* **99**, 6308 (1993).
[72] O. Cheshnovsky, K.J. Taylor, J. Conceicao and R.E. Smalley, Ultraviolet photoelectron spectra of mass-selected copper clusters: Evolution of the 3*d* band, *Phys. Rev. Lett.* **64**, 1785 (1990).
[73] E.C. Honea, A. Ogura, C.A. Murray, K. Raghavachari, O. Sprenger, M.F. Jarrold and W.L. Brown, Raman spectra of size-selected silicon clusters and comparison with calculated structures, *Nature* **366**, 42 (1993).
[74] X. Jing, N. Troullier, J.R. Chelikowsky, K. Wu and Y. Saad, Vibrational modes of silicon nanostructures, *Solid State Commun.* **96**, 231 (1995).
[75] R. Fournier, S.B. Sinnott and A.E. DePristo, Density-functional study of the bonding in silicon clusters, *J. Chem. Phys.* **97**, 4149 (1992).
[76] C. Rohlfing and K. Raghavachari, Electronic structures and photoelectron spectra of Si_3^- and Si_4^-, *J. Chem. Phys.* **96**, 2114 (1992).
[77] Y. Saad, A. Stathopoulos, J.R. Chelikowsky, K. Wu and S. Ogut, Solution of large eigenvalue problems in electronic structure calculations, *BIT* **36**, 563 (1996).
[78] S. Baroni, P. Gianozzi and A. Testa, Green's-function approach to linear response in solids, *Phys. Rev. Lett.* **58**, 1861 (1987).

[79] X. Gonze, D.C. Allan and M.P. Teter, Dielectric tensor, effective charges, and phonons in α-quartz by variational density-functional perturbation theory, *Phys. Rev. Lett.* **68**, 3603 (1992).

[80] E.K.U. Gross and W. Kohn, Local density-functional theory of frequency-dependent linear response, *Phys. Rev. Lett.* **55**, 2850 (1985).

[81] M. Casida, Time-dependent density functional response theory for molecules, in *Recent Advances in Density-Functional Methods, Part I*, edited by D. Chong (World Scientific, Singapore, 1995), p. 155.

[82] M. Casida, Time-dependent density functional response theory of molecular systems: Theory, computational methods, and functionals, in *Recent Developments and Applications of Modern Density Functional Theory*, edited by J. Seminario (Elsevier, Amsterdam, 1996), p. 391.

[83] J.R. Chelikowsky, L. Kronik and I. Vasiliev, Time-dependent density functional calculations for the optical spectra of molecule, clusters and nanocrystals, *J. Phys. Cond. Matt.* **15**, R1517 (2003).

[84] X. Blase, A. Rubio, S.G. Louie and M.L. Cohen, Mixed-space formalism for the dielectric response in periodic systems, *Phys. Rev. B* **52**, R2225 (1995).

[85] C.R.C. Wang, S. Pollack, D. Cameron and M.M. Kappes, Optical absorption spectroscopy of sodium clusters as measured by collinear molecular beam photodepletion, *Chem. Phys. Lett.* **93**, 3787 (1990).

[86] I. Moullet, J.L. Martins, F. Reuse and J.B.J. Buttet, Static electric polarizabilities of sodium clusters, *Phys. Rev. B* **42**, 11598 (1990).

[87] V. Bonacic-Koutecky, P. Fantucci and J. Koutecky, Theoretical interpretation of the photoelectron detachment spectra of Na^-_{2-5} and of the absorption spectra of Na_3, Na_4 and Na_8, *J. Chem. Phys.* **93**, 3802 (1990).

[88] V. Bonacic-Koutecky, P. Fantucci and J. Koutecky, An *ab initio* configuration interaction study of the excited states of the Na_4 cluster: Assignment of the absorption spectrum, *Chem. Phys. Lett.* **166**, 32 (1990).

[89] G. Onida, L. Reining, R.W. Godby, R. del Sole and W. Andreoni, *Ab initio* calculations of the quasiparticle and absorption spectra of clusters: The Sodium tetramer, *Phys. Rev. Lett.* **75**, 818 (1995).

[90] M. Rohlfing and S.G. Louie, Excitonic effects and the optical absorption spectrum of hydrogenated Si clusters, *Phys. Rev. Lett.* **80**, 3320 (1998).

[91] R.J. Baierle, M.J. Caldas, E. Molinari and S. Ossicini, Optical emission from small Si particles, *Solid State Commun.* **102**, 545 (1997).

[92] B. Delley and E.F. Steigmeier, Quantum confinement in Si nanocrystals, *Phys. Rev. B* **47**, 1397 (1993).

[93] C. Delerue, G. Allan and M. Lannoo, Theoretical aspects of the luminescence of porous silicon, *Phys. Rev. B* **48**, 11024 (1993).

[94] L.W. Wang and A. Zunger, Electronic structure pseudopotential calculations of large (1000 atom) Si quantum dots, *J. Phys. Chem.* **98**, 2158 (1994).

[95] L.W. Wang and A. Zunger, Solving Schrodinger's equation around a desired energy: Application to silicon quantum dots, *J. Phys. Chem.* **100**, 2394 (1994).

[96] D. Beeman, Some multistep methods for use in molecular dynamics calculations, *J. Comput. Phys.* **20**, 130 (1976).

[97] K. Raghavachari and V. Logovinsky, Structure and bonding in small silicon clusters, *Phys. Rev. Lett.* **55**, 2853 (1985).

[98] K. Raghavachari and C. Rohlfing, Electronic structures of the negative ions $Si_2^- - Si_{10}^-$: Electron affinities of small silicon clusters, *J. Chem. Phys.* **94**, 3670 (1990).

[99] W.R. Fredrickson and W.W. Watson, The sodium and potassium absorption bands, *Phys. Rev.* **30**, 429 (1927).

[100] U. Itoh, Y. Toyoshima, H. Onuki, N. Washida and T. Ibuki, Vacuum ultraviolet absorption cross sections of SiH_4, GeH_4, Si_2H_6 and Si_3H_8, *J. Chem. Phys.* **85**, 4867 (1986).

[101] S. Furukawa and T. Miyasato, Quantum size effects on the optical band gap of microcrystalline Si:H, *Phys. Rev. B* **38**, 5726 (1988).

[102] D.J. Lockwood, A. Wang and B. Bryskiewicz, Optical absorption evidence for quantum confinement effects in porous silicon, *Solid State Commun.* **89**, 587 (1994).

Chapter 5
QUANTUM ELECTROSTATICS OF INSULATORS: POLARIZATION, WANNIER FUNCTIONS, AND ELECTRIC FIELDS

D. Vanderbilt and R. Resta

1. INTRODUCTION

The macroscopic polarization is a central concept in any phenomenological description of dielectric media [1]. Intuitively, it should be an intensive vector quantity carrying the meaning of electric dipole moment per unit volume. Most textbooks that attempt to give a microscopic definition of the polarization of a periodic crystal try to express it in terms of the dipole moment per unit cell [2,3], but such approaches have been shown to be deeply flawed because there is no unique choice of cell boundaries [4].

A new viewpoint emerged in the early 1990s and was instrumental to the development of a successful microscopic theory [5–7]. This new theoretical framework begins with a recognition that the bulk macroscopic polarization cannot be determined, even in principle, from a knowledge of the periodic charge distribution of the polarized crystalline dielectric. In this respect, the situation is fundamentally different for infinite periodic systems than for finite ones (e.g., molecules and clusters), for which the dipole moment can trivially be expressed in terms of the charge distribution. Instead, for periodic systems, one focuses on *differences* in polarization between two states of the crystal that can be connected by an adiabatic switching process [5]. The time-dependent Hamiltonian is assumed to remain periodic and insulating at all times, and the polarization difference is then equal to the integrated

Contemporary Concepts of Condensed Matter Science
Conceptual Foundations of Materials: A Standard Model for Ground- and Excited-State Properties

ISSN: 1572-0934/doi:10.1016/S1572-0934(06)02005-1

transient macroscopic current that flows through the insulating sample during this adiabatic switching process.

Thus, the macroscopic polarization of an extended system is – in the modern viewpoint – better understood as a dynamical property of the current in the adiabatic limit. While the density is a property connected with the square modulus of the wavefunction, the current also has an essential dependence on the phase. Indeed, it turns out that in the modern theory of polarization [6–9], the polarization difference is related to a special kind of phase, known as a "Berry phase" [10,11], defined over the manifold of Bloch orbitals. This theory, besides defining what polarization really is, also provides a powerful algorithm for computing macroscopic polarization from first principles that is implemented as a standard option in most crystalline electronic-structure codes.

In cases where the physical observable is a *derivative* of macroscopic polarization, an alternative approach is linear-response theory, developed over the years since the pioneering work of Adler and Wiser in the early 1960s [12–18]. In the modern context, this approach is also known as density-functional perturbation theory (DFPT). Among other things, the modern theory sheds new light on linear-response theory, and in particular, on the crucial role played by the current (as opposed to the charge) therein.

The modern theory can be equivalently reformulated – in a possibly more intuitive way – using localized Wannier functions (WFs) [19–22] instead of delocalized Bloch orbitals. The electronic contribution to the macroscopic polarization **P** is then expressed in terms of the dipole of the Wannier charge distribution associated with one unit cell. In this way, **P** is reformulated as a property of a localized charge distribution, apparently free of phase information. However, one has to bear in mind that the phases of the Bloch orbitals are essential for building the WFs, which are needed to specify how the periodic charge distribution is to be decomposed into localized ones. Indeed, a knowledge of the periodic charge distribution of the polarized dielectric is insufficient, in principle, to determine the WFs (or the polarization).

In its original formulation, the modern theory could only address the problem of computing electric polarization in a vanishing macroscopic electric field. While this ruled out the direct treatment of electronic dielectric response (i.e., field-induced polarization), it could nevertheless be used to investigate a wide variety of important effects including ferroelectricity (spontaneous polarization), piezoelectricity (strain-induced polarization), and lattice infrared activity and dielectric response (displacement-induced polarization). For these cases, the field could be set to zero, and the theory could continue to be framed in terms of the familiar lattice-periodical Hamiltonians and Bloch orbitals.

However, once a macroscopic electric field is applied, the Hamiltonian is not periodic, Bloch's theorem no longer applies, and the Hamiltonian eigenstates do not even qualitatively resemble Bloch functions [23]. Moreover, the potential operator is no longer bounded from below, and variational methods may, therefore, be expected to fail. For these reasons, practical–theoretical methods for treating an insulator in a finite electric field remained unavailable until the early 2000s. Not only was it impossible to compute the polarization, it was impossible to compute any property of an insulating crystal in a macroscopic electric field, except by perturbation methods. However, in the early 2000s it was realized that the zero-field Berry-phase polarization theory provides an avenue for the solution to this problem: one could retain a Bloch form for the occupied state wavefunctions, even though they are no longer Hamiltonian eigenstates, and treat the electric field through its coupling to the Berry-phase polarization [24–26]. This not only solves the problem of computing electric polarizabilities, but more generally opens up the possibility of computing a wide range of properties of materials subjected to electric fields.

This chapter is devoted to introducing and discussing the issues outlined above. In Section 2 we discuss the problems which long hampered a theoretical approach to electric polarization, together with the main phenomenological concepts and definitions. In Section 3 we introduce the basic concepts of linear-response theory (DFPT), showing how this theory provides a successful perturbative treatment of polarizations and electric fields. Section 4 is devoted to the formulation of the modern theory of polarization in terms of Berry phases, which are Brillouin-zone integrals of a certain form taken over the Bloch functions. The transition from continuous to discretized k space is discussed, with special attention to "gauge-invariance" (where the "gauge" is a choice of phase twist of the Bloch wavefunctions regarded as a function of wavevector). In Section 5 we present an alternative formulation of polarization in terms of WFs. In both the Bloch and Wannier representations, the modern theory expresses **P** in such a way that it is only well defined up to a "modulus" or "quantum", whose nature is discussed in Section 6. In Section 7 we explain how the modern theory of polarization also provides the solution to the long-standing problem of treating an insulator in a macroscopic electric field – and, in particular, of computing the polarization induced by such a field. Finally, in Section 8 we give a brief review of the main conclusions of this chapter.

2. THE POLARIZATION PROBLEM

The dipole moment of any *finite* N-electron system in its ground state is a simple and well-defined quantity. Given the single-particle density $n(\mathbf{r})$, the electronic contribution to the dipole is

$$\mathbf{d} = -e \int \mathrm{d}\mathbf{r}\, \mathbf{r}\, n(\mathbf{r}) \tag{1}$$

where $e > 0$ is the charge quantum. This looks trivial, but we are exploiting here an essential property of any finite N-electron system, namely, that the ground-state wavefunction is square-integrable and vanishes exponentially at infinity, so that the density decays exponentially as well.

Consider now a macroscopic solid. The corresponding quantity is the macroscopic polarization, which one is tempted to define as the dipole of a macroscopic sample divided by its volume. However, when using Eq. (1), one has to be careful about possible surface contributions. In condensed-matter physics, the standard approach for avoiding undesirable surface effects is to adopt periodic Born-von Kármán (BvK) boundary conditions. Indeed, the BvK choice is needed in order to introduce even the most elementary topics, such as the free-electron gas and its Fermi surface, or Bloch's theorem [2,3]. Normally, the BvK approach can be justified by taking the thermodynamic limit of a large but finite sample. However, this does *not* work for the case of the polarization. As we shall see, the surface contributions to the dipole moment do *not* vanish in the thermodynamic limit. Moreover, if one tries to evaluate Eq. (1) using wavefunctions obeying BvK boundary conditions, one finds that the integrals are ill defined due to the unbounded nature of the quantum mechanical position operator [8,9]. For these reasons, the formulation of a practical theory of macroscopic polarization persisted as a major challenge in electronic structure theory for many years.

The clue to understanding polarization comes from reconsidering a most fundamental experimental fact. Most measurements of bulk macroscopic polarization **P** of materials do not address its absolute value, but only its derivatives, which are expressed as Cartesian tensors. For example, the susceptibility

$$\chi_{\alpha\beta} = \frac{\mathrm{d}P_\alpha}{\mathrm{d}\mathscr{E}_\beta} \tag{2}$$

is defined as the derivative of polarization with respect to electric field. Here, as throughout this section, Greek subscripts indicate Cartesian coordinates. Similarly, the pyroelectric coefficient

$$\Pi_\alpha = \frac{\mathrm{d}P_\alpha}{\mathrm{d}T} \tag{3}$$

the piezoelectric tensor [27]

$$\gamma_{\alpha\beta\delta} = \frac{\partial P_\alpha}{\partial \varepsilon_{\beta\delta}} \tag{4}$$

and the dimensionless Born (or "dynamical" or "infrared") charge

$$Z^*_{\mathrm{s},\alpha\beta} = \frac{\Omega}{e} \frac{\partial P_\alpha}{\partial u_{\mathrm{s},\beta}} \tag{5}$$

are defined in terms of derivatives of **P** with respect to temperature T, strain $\varepsilon_{\beta\delta}$, and displacement $\mathbf{u}_\mathrm{s}$ of sublattice s, respectively, where Ω is the primitive cell volume. (In the above formulas, derivatives are to be taken at fixed electric field and fixed strain when these variables are not explicitly involved.)

Even "spontaneous polarization" is not experimentally accessible as an equilibrium property. Instead, one exploits the switchability of **P** in ferroelectric materials; the quantity actually measured is the *finite difference* ΔP between two different structures of the same material. In most crystalline ferroelectrics, these two structures are symmetry-equivalent (enantiomorphous), so that one speaks of polarization *reversal*. We illustrate a typical experiment by considering the case of the perovskite oxide $BaTiO_3$, whose equilibrium structure at room temperature is

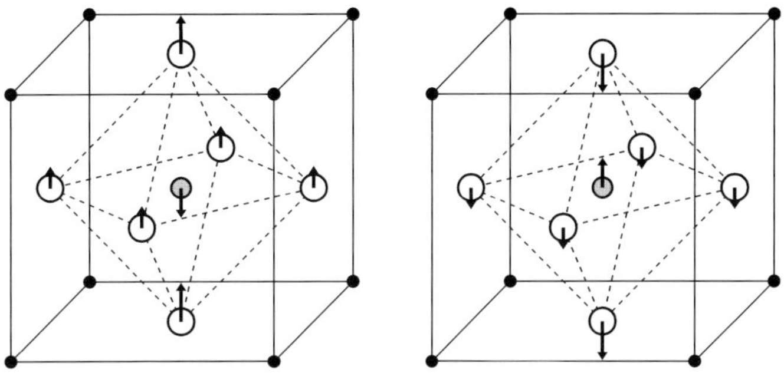

Fig. 1. Tetragonal ferroelectric structure of $BaTiO_3$. Solid, shaded, and empty circles represent Ba, Ti, and O atoms, respectively. The arrows indicate the atomic displacements (exagerated for clarity), where the origin has been kept at the Ba site. Two enantiomorphous structures, with polarization along [001], are shown. Application of a sufficiently large electric field causes the system to switch between the two structures, reversing the polarization.

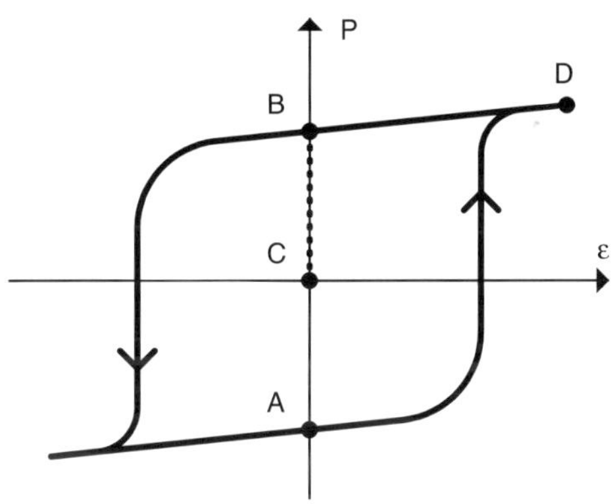

Fig. 2. A typical hysteresis loop for a ferroelectric material. The magnitude of the spontaneous polarization can be defined experimentally as half the polarization change along path ADB, or theoretically as a zero-field property by following the vertical dotted segment from C to B.

tetragonal. There are six enantiomorphous broken-symmetry structures; two of them, having opposite nuclear displacements, are shown in Fig. 1.

An experimental determination of the spontaneous polarization is normally extracted from a measurement of the hysteresis cycle [28] as shown schematically in Fig. 2. The transition between the two enantiomorphous structures A and B of Fig. 1 is driven by taking the system along path ADB by the application and subsequent removal of an electric field, and the spontaneous polarization is taken to be one-half of the integrated macroscopic current flowing through the sample during this process. We stress that the experiment measures neither $\mathbf{P}_A$ nor $\mathbf{P}_B$, but associates the spontaneous polarization with one-half of their difference $\mathbf{P}_B - \mathbf{P}_A$. It is only an additional symmetry argument that allows one to infer the value of $\mathbf{P}_A$ or $\mathbf{P}_B$ from the actual experimental data.

Notice that while a field is needed to induce the switching in the actual experiment, no such restriction applies to the theory. That is, one can imagine setting up a theoretical calculation in which the system is carried by hand from the centrosymmetric structure C in which the displacements are set to zero, to the spontaneously polarized one B, along the dotted path in Fig. 2. Since the macroscopic electric field is identically zero along path CB, one can formulate the theory entirely in the zero-field context. One can then use the zero-field version of DFPT [12–18] to compute the derivative of the polarization at each point along this path, and then integrate to obtain the total polarization change. (In case it is necessary to consider a path along which the macroscopic $\mathcal{E}$-field does not remain zero, the finite-field theory [24–26] to be described in Section 7 must be used.)

Of course, this trick of replacing the true physical path (solid curves in Fig. 2) by a theoretically more convenient one (dotted line in Fig. 2) is only permissible if the total change in polarization is guaranteed to be independent of path. The essential contribution of the modern theory of polarization, to be described in Section 4, is to show that such changes are indeed independent of path, and to provide a formula for evaluating $\Delta\mathbf{P}$ using only information from the end points of the path. This theory can then be used not only to evaluate the spontaneous polarization in a ferroelectric material, as above, but also for calculations of the piezoelectric tensor of Eq. (4), or the Born effective-charge tensor of Eq. (5), by a finite-difference calculation of the change in $\mathbf{P}$ resulting from a finite strain or atomic sublattice displacement.

Since this modern theory is built upon DFPT, we shall review this theory in the next section, with particular emphasis on derivatives of polarization.

3. OUTLINE OF DENSITY-FUNCTIONAL PERTURBATION THEORY

In this section we give an outline of the basic concepts of DFPT, also known as linear-response theory or, in the quantum chemistry context, "analytic derivative" methods. DFPT has an outstanding role in addressing many crystalline properties besides dielectric ones, including lattice-dynamical, elastic, and electron–phonon coupling properties. A comprehensive review of DFPT in crystalline solids is available in Ref. [18]. We will emphasize the formulation and application of DFTP for evaluating derivatives of the macroscopic polarization to compute such quantities as the Born effective charge from Eq. (5) [29–32] or the dielectric constant $\varepsilon_\infty = 1 + 4\pi\chi$ with χ from Eq. (2). Since the nuclear contribution to the macroscopic polarization is trivial, we focus on the electronic term only, indicating it with $\mathbf{P}$ for the sake of simplicity. We stress, however, that the nuclear term is essential to ensure charge neutrality and translational invariance.

Let us start with a large but finite insulating system having discrete single-particle orbitals $|\psi_i\rangle$ which vanish outside the sample. We will switch to the crystalline case, with periodic boundary conditions, only later in this section. We write the electron density as

$$n(\mathbf{r}) = \sum_i |\psi_i(\mathbf{r})|^2 \tag{6}$$

where the sum is over occupied ("valence") states i, and a factor of 2 may be inserted in all formulas for the spin-degenerate case. The electronic term in the macroscopic polarization is then, from Eqs (1) and (6),

$$\mathbf{P} = -\frac{e}{L^3}\sum_i \langle\psi_i|\mathbf{r}|\psi_i\rangle \tag{7}$$

where L^3 is the volume of the finite sample.

Suppose now that we switch on a given perturbation – e.g., a sublattice displacement $\mathbf{u}_s$ as in Eq. (5) – whose amplitude we measure by a dimensionless parameter λ. We expand all relevant quantities in powers of λ, e.g.,

$$V_{\text{ext}}(\mathbf{r}) = V_{\text{ext}}^{(0)}(\mathbf{r}) + V_{\text{ext}}^{(1)}(\mathbf{r}) + V_{\text{ext}}^{(2)}(\mathbf{r}) + \dots, \tag{8}$$

where $V_{\text{ext}}^{(n)}(\mathbf{r})$ contains the terms of order λ^n. Here $V_{\text{ext}}(\mathbf{r})$ is the bare (or unscreened) potential felt by the electrons, so that $V_{\text{ext}}^{(1)}(\mathbf{r})$ is the first-order perturbing term in the KS Hamiltonian [33]. We write similar expansions for the self-consistently screened total potential $V(\mathbf{r})$, the electron density $n(\mathbf{r})$, the wavefunctions $|\psi_i\rangle$, etc. Using Eq. (7), we wish to evaluate the corresponding first-order change in the polarization,

$$\mathbf{P}^{(1)} = -\frac{e}{L^3}\sum_i \langle\psi_i^{(0)}|\mathbf{r}|\psi_i^{(1)}\rangle + \text{c.c.} \tag{9}$$

where the sum is over occupied ("valence") states and "c.c." is the complex conjugate.

For a finite system, Eq. (9) is straightforward to evaluate as soon as $|\psi_i^{(1)}\rangle$ is available. The latter can be obtained by ordinary first-order perturbation theory

$$|\psi_i^{(1)}\rangle = \sum_{j\neq i} \frac{|\psi_j^{(0)}\rangle\langle\psi_j^{(0)}|}{E_i^{(0)} - E_j^{(0)}} V^{(1)}|\psi_i^{(0)}\rangle \tag{10}$$

involving a sum over all other states, including all unoccupied states [34]. Because such sums often converge very slowly, it is a common practice to obtain $|\psi_i^{(1)}\rangle$ instead by solving the equivalent implicit Sternheimer equation

$$(E_i^{(0)} - H)Q_i|\psi_i^{(1)}\rangle = Q_i V^{(1)}|\psi_i^{(0)}\rangle \tag{11}$$

for $|\psi_i^{(1)}\rangle$, where $Q_i = 1 - |\psi_i^{(0)}\rangle\langle\psi_i^{(0)}|$. This can be accomplished using standard iterative techniques.

Finally, since $V = V_{\text{ext}} + V_{\text{H}} + V_{\text{xc}}$, it is also necessary to iterate an outer self-consistent screening loop to solve for

$$n^{(1)}(\mathbf{r}) = \sum_i \psi_i^{(0)*}(\mathbf{r})\psi^{(1)}(\mathbf{r})_i + \text{c.c.} \tag{12}$$

and

$$V^{(1)} = V_{\rm ext}^{(1)} + f_{\rm H} \cdot n^{(1)} + f_{\rm xc} \cdot n^{(1)} \tag{13}$$

together with Eq. (11). Here the dot product indicates an integral over $\mathbf{r}'$; $f_{\rm H}$ is the (linear) Hartree potential kernel, and

$$f_{\rm xc}(\mathbf{r}, \mathbf{r}') = \frac{\delta V_{\rm xc}(\mathbf{r})}{\delta n(\mathbf{r}')} = \frac{\delta^2 E_{\rm xc}[n]}{\delta n(\mathbf{r})\delta n(\mathbf{r}')} \tag{14}$$

is the exchange-correlation kernel, where $E_{\rm xc}$ is the KS exchange-correlation energy functional [33]. This completes the solution of Eq. (9) in the case of a finite sample.

We now prepare to apply the thermodynamic limit and let the size of the sample become infinite. Then the occupied and empty states $|\psi_i\rangle$ will become Bloch states $|\psi_{n\mathbf{k}}\rangle$. However, it is not permissible to leave the position operator $\mathbf{r}$ in place in Eq. (9), because matrix elements of $\mathbf{r}$ between Bloch functions are ill defined. Therefore, before taking the thermodynamic limit, we first introduce an alternative method for obtaining $|\tilde{\psi}_{\alpha,i}\rangle = r_\alpha|\psi_i^{(0)}\rangle$ that avoids this problem. In particular, we compute it from

$$|\tilde{\psi}_{\alpha,i}\rangle = \sum_{j \neq i} \frac{|\psi_j^{(0)}\rangle\langle\psi_j^{(0)}|}{E_i^{(0)} - E_j^{(0)}} (\mathrm{i}\hbar v_\alpha)|\psi_i^{(0)}\rangle \tag{15}$$

as can easily be checked using the definition

$$\mathbf{v} = -\frac{i}{\hbar}[\mathbf{r}, H] \tag{16}$$

of the velocity operator [35]. This approach appears to reintroduce an undesirable sum over unoccupied states, but in view of the similarity of Eq. (15) to Eq. (10), we can once again replace it by a Sternheimer equation

$$(E_i^{(0)} - H)Q_i|\tilde{\psi}_{\alpha,i}\rangle = Q_i(\mathrm{i}\hbar v_\alpha)|\psi_i^{(0)}\rangle \tag{17}$$

which will be solved by iterative methods. Then Eq. (9) becomes

$$P_\alpha^{(1)} = -\frac{e}{L^3}\sum_i \langle\tilde{\psi}_{\alpha,i}|\psi_i^{(1)}\rangle + \text{c.c.} \tag{18}$$

A conceptually similar scheme was first proposed in the 1950s by Sternheimer [36–40] for evaluating atomic polarizabilities.

We emphasize that the replacement of $\mathbf{r}$ by $\mathbf{v}$ at this point essentially corresponds to switching from a formulation based on *charge* to one based on *current*. If we are actually dealing with a finite system, Eqs (9) and (18) are equally valid. But in taking the thermodynamic limit, matrix elements of $\mathbf{r}$ become ill defined, for essentially the same reason that Eq. (1) is ill defined. For an extended system in the thermodynamic limit, it is then mandatory to use the velocity formula, hence the current, as in Eq. (18). Focusing on currents has thus been crucial to the development of the modern theory of polarization, as illustrated further in the following section.

Let us now make the transition to the crystalline case, with periodic boundary conditions, and let the index i be identified with the band index n and the Bloch vector **k**. At this point we also drop the superscript "(0)" from unperturbed quantities, so that, e.g., $|\psi_i^{(0)}\rangle \to |\psi_{n\mathbf{k}}\rangle$. Then Eq. (18) becomes

$$\mathbf{P}^{(1)} = \frac{i\hbar e}{(2\pi)^3}\sum_n \sum_{m\neq n}\int_{\mathrm{BZ}} \mathrm{d}\mathbf{k}\,\frac{\langle\psi_{n\mathbf{k}}|\mathbf{v}|\psi_{m\mathbf{k}}\rangle\langle\psi_{m\mathbf{k}}|\psi_{n\mathbf{k}}^{(1)}\rangle}{E_{n\mathbf{k}}-E_{m\mathbf{k}}} + \text{c.c.} \tag{19}$$

where n runs over the occupied valence bands of the insulator and the integral is over the Brillouin zone (BZ). In combination with Eq. (10) or (11), this provides the solution to the problem of computing the first-order linear response of the polarization in response to a perturbation, such as a sublattice displacement, that preserves the crystal periodicity.

While such linear-response or DFPT methods have satisfactorily provided **P** derivatives over the years, the problem of evaluating the "polarization itself" remained an open and confusing one until the mid-1990s, when the advent of the modern theory of polarization provided a resolution. This is the subject of the next section.

4. THE BERRY-PHASE THEORY OF POLARIZATION

In this section, we provide a brief derivation of the central results that were uncovered in the early 1990s, which are often referred to as the "modern theory of polarization." The basic idea is to consider the change in polarization of a crystal as it undergoes some slow change, e.g., a slow displacement of one sublattice relative to the others, and relate this to the *current* that flows during this adiabatic evolution of the system. These considerations will allow us to arrive at an expression for the polarization that does not take the form of an expectation value of an operator, as is normally the case. Rather, it takes the form of a "Berry phase," which is a geometrical phase property of a closed manifold (the Brillouin zone) on which a set of vectors (occupied Bloch states) are defined.

Once again, we assume that the crystal Hamiltonian H_λ depends smoothly on parameter λ and has Bloch eigenvectors obeying $H_\lambda|\psi_{\lambda,n\mathbf{k}}\rangle = E_{\lambda,n\mathbf{k}}\,|\psi_{\lambda,n\mathbf{k}}\rangle$ (the λ subscripts will often be suppressed for clarity). We also assume that λ changes slowly with time, so that the adiabatic approximation is appropriate. With a slight change of notation (introducing $\partial_\lambda = \mathrm{d}/\mathrm{d}\lambda$), the principal result (19) of the previous section becomes

$$\partial_\lambda\mathbf{P} = \frac{i\hbar e}{(2\pi)^3}\sum_n \sum_{m\neq n}\int_{\mathrm{BZ}} \mathrm{d}\mathbf{k}\,\frac{\langle\psi_{n\mathbf{k}}|\mathbf{v}|\psi_{m\mathbf{k}}\rangle\langle\psi_{m\mathbf{k}}|\partial_\lambda\psi_{n\mathbf{k}}\rangle}{E_{n\mathbf{k}}-E_{m\mathbf{k}}} + \text{c.c.} \tag{20}$$

Since the spatially averaged current density is just $\mathbf{j} = \mathrm{d}\mathbf{P}/\mathrm{d}t = (\partial_\lambda\mathbf{P})\dot{\lambda}$, this equation can be converted into the statement that the instantaneous current is given by [41]

$$\mathbf{j} = \frac{i\hbar e\dot{\lambda}}{(2\pi)^3}\sum_n \sum_{m\neq n}\int_{\mathrm{BZ}} \mathrm{d}\mathbf{k}\,\frac{\langle\psi_{n\mathbf{k}}|\mathbf{v}|\psi_{m\mathbf{k}}\rangle\langle\psi_{m\mathbf{k}}|\partial_\lambda\psi_{n\mathbf{k}}\rangle}{E_{n\mathbf{k}}-E_{m\mathbf{k}}} + \text{c.c.} \tag{21}$$

We can then say that the change in polarization during some time interval is

$$\Delta\mathbf{P} = \int \mathbf{j}(t)\,\mathrm{d}t \tag{22}$$

where $\mathbf{j}(t)$ is given by Eq. (21). This formulation is particularly intuitive, since it is phrased in terms of the current density that is physically flowing through the crystal as the system traverses some adiabatic path. But since $\mathbf{j} = \mathrm{d}\mathbf{P}/\mathrm{d}t$ is proportional to $\dot{\lambda} = \mathrm{d}\lambda/\mathrm{d}t$ in Eq. (21), the $\mathrm{d}t$ can be factored out, and we can equivalently go back to integrating Eq. (20) directly to obtain

$$\Delta\mathbf{P} = \int (\partial_\lambda \mathbf{P})\,\mathrm{d}\lambda \tag{23}$$

where $\partial_\lambda \mathbf{P}$ is given by Eq. (20). This is the formally more direct path.

In the case that the Hamiltonian is just $p^2/2m_e$ plus a local potential, it follows that $\mathbf{v} = \mathbf{p}/m_e$, and it is possible to evaluate Eq. (20) directly (or, to avoid the sum over unoccupied states, to obtain $|\tilde{\psi}_{\alpha,n\mathbf{k}}\rangle$ by the method of Eq. (17)). However, we can proceed in a more general fashion, which also eliminates the energy denominator in Eq. (20), using ordinary $\mathbf{k}\cdot\mathbf{p}$ perturbation theory. Here we introduce the effective Schrödinger equation $H_\mathbf{k}|u_{n\mathbf{k}}\rangle = E_{n\mathbf{k}}\,|u_{n\mathbf{k}}\rangle$ where $u_{n\mathbf{k}}(\mathbf{r}) = \mathrm{e}^{-\mathrm{i}\mathbf{k}\cdot\mathbf{r}}\,\psi_{n\mathbf{k}}(\mathbf{r})$ and

$$H_\mathbf{k} = \mathrm{e}^{-\mathrm{i}\mathbf{k}\cdot\mathbf{r}} H \mathrm{e}^{\mathrm{i}\mathbf{k}\cdot\mathbf{r}} = \frac{1}{2m_e}(\mathbf{p} + \hbar\mathbf{k})^2 + \mathrm{e}^{-\mathrm{i}\mathbf{k}\cdot\mathbf{r}} V \mathrm{e}^{\mathrm{i}\mathbf{k}\cdot\mathbf{r}} \tag{24}$$

(The last term reduces back to V if the potential commutes with $\mathbf{r}$, but this is often not the case when dealing with modern pseudopotentials.) By elementary perturbation theory, the first-order change of $|u_{n\mathbf{k}}\rangle$ with wavevector is just

$$|\nabla_\mathbf{k} u_{n\mathbf{k}}\rangle = \sum_{m\neq n} \frac{(\nabla_\mathbf{k} H_\mathbf{k})|u_{m\mathbf{k}}\rangle}{E_{n\mathbf{k}} - E_{m\mathbf{k}}} \tag{25}$$

But using definition (16) of the velocity operator, and $\nabla_\mathbf{k} H_\mathbf{k} = -\mathrm{i}[\mathbf{r}, H_\mathbf{k}]$ (which follows immediately from $H_\mathbf{k} = \mathrm{e}^{-\mathrm{i}\mathbf{k}\cdot\mathbf{r}} H \mathrm{e}^{\mathrm{i}\mathbf{k}\cdot\mathbf{r}}$), the first matrix element in the numerator of Eq. (20) becomes

$$\langle \psi_{n\mathbf{k}}|\mathbf{v}|\psi_{m\mathbf{k}}\rangle = -\mathrm{i}\hbar^{-1}\langle u_{n\mathbf{k}}|[\mathbf{r}, H_\mathbf{k}]|u_{m\mathbf{k}}\rangle = \hbar^{-1}\langle u_{n\mathbf{k}}|(\nabla_\mathbf{k} H_\mathbf{k})|u_{m\mathbf{k}}\rangle \tag{26}$$

Then Eq. (20) turns into

$$\partial_\lambda \mathbf{P} = \frac{\mathrm{i}e}{(2\pi)^3} \sum_n \sum_{m\neq n} \int_{\mathrm{BZ}} \mathrm{d}\mathbf{k} \frac{\langle u_{n\mathbf{k}}|(\nabla_\mathbf{k} H_\mathbf{k})|u_{m\mathbf{k}}\rangle \langle u_{m\mathbf{k}}|\partial_\lambda u_{n\mathbf{k}}\rangle}{E_{n\mathbf{k}} - E_{m\mathbf{k}}} + \text{c.c.} \tag{27}$$

or, using Eq. (25), simply

$$\partial_\lambda \mathbf{P} = \frac{\mathrm{i}e}{(2\pi)^3} \sum_n \int_{\mathrm{BZ}} \mathrm{d}\mathbf{k} \langle \nabla_\mathbf{k} u_{n\mathbf{k}}|\partial_\lambda u_{n\mathbf{k}}\rangle + \text{c.c.} \tag{28}$$

Remarkably, the sum over empty states has been eliminated from this formula, showing that the rate of change of polarization with λ is a property of the occupied bands only, as one expects on physical grounds. This expression can be integrated with respect to λ to obtain [6]

$$\mathbf{P}(\lambda) = \frac{-ie}{(2\pi)^3} \sum_n \int_{\mathrm{BZ}} \mathrm{d}\mathbf{k} \langle u_{\lambda,n\mathbf{k}} | \nabla_{\mathbf{k}} | u_{\lambda,n\mathbf{k}} \rangle \tag{29}$$

as may be verified by taking the λ derivative of both sides of Eq. (29) and comparing with Eq. (28). The result is independent of the particular path of $\lambda(t)$ in time, and depends only on the final value of λ, as long as the change is slow in the adiabatic sense. We, therefore, associate $\mathbf{P}(\lambda)$ with the physical polarization of state λ, and henceforth drop the λ label. Since $\nabla_{\mathbf{k}}(\langle u_{\lambda,n\mathbf{k}} | u_{\lambda,n\mathbf{k}} \rangle) = 0$, the integrand is pure imaginary, and Eq. (29) can alternatively be written as

$$\mathbf{P} = \frac{e}{(2\pi)^3} \operatorname{Im} \sum_n \int_{\mathrm{BZ}} \mathrm{d}\mathbf{k} \langle u_{n\mathbf{k}} | \nabla_{\mathbf{k}} | u_{n\mathbf{k}} \rangle \tag{30}$$

To be more precise, this is the electronic contribution to the polarization; to this must be added the nuclear (or "ionic") contribution

$$\mathbf{P}_{\mathrm{ion}} = \Omega^{-1} \sum_s Z_s \mathbf{r}_s \tag{31}$$

where the sum is over atoms s having core charge Z_s and location $\mathbf{r}_s$ in the unit cell of volume Ω.

Equation (30) is the central result of the modern theory of polarization. It says that the electronic contribution to the polarization of a crystalline insulator may be expressed as a Brillouin-zone integral of an "operator" $i\nabla_{\mathbf{k}}$ that plays the role of an $\mathbf{r}$ operator in a heuristic sense. However, $i\nabla_{\mathbf{k}}$ is not a normal operator; it involves taking the derivative of the state vector $|u_{n\mathbf{k}}\rangle$ with respect to wavevector. In particular, the quantity $i\nabla_{\mathbf{k}}|u_{n\mathbf{k}}\rangle$ depends on the choice of relative phases of the Bloch functions at different $\mathbf{k}$; such a sensitivity is not expected for a "normal" quantum mechanical operator.

To explore this point, we specialize for the moment to a single band for a one-dimensional (1D) crystal of lattice constant a, so that $P = -e\phi/2\pi$, where

$$\phi = -\operatorname{Im} \int_0^{2\pi/a} \mathrm{d}k \langle u_k | \partial_k | u_k \rangle \tag{32}$$

Now suppose that a new set of Bloch functions is defined by

$$|\tilde{\psi}_k\rangle = \mathrm{e}^{-\mathrm{i}\beta(k)} |\psi_k\rangle \tag{33}$$

with an equivalent relation between $|\tilde{u}_k\rangle$ and $|u_k\rangle$, where $\beta(k)$ is some smooth real function of k. As mentioned in Section 1, such a twist of the phases is often referred to as a "change of gauge." Then

$$\langle \tilde{u}_k | \partial_k | \tilde{u}_k \rangle = \langle u_k | \partial_k | u_k \rangle - \mathrm{i} \frac{\mathrm{d}\beta}{\mathrm{d}k} \tag{34}$$

showing that the integrand of Eq. (32) depends on the choice of gauge. One may then wonder whether Eq. (30) is well defined at all.

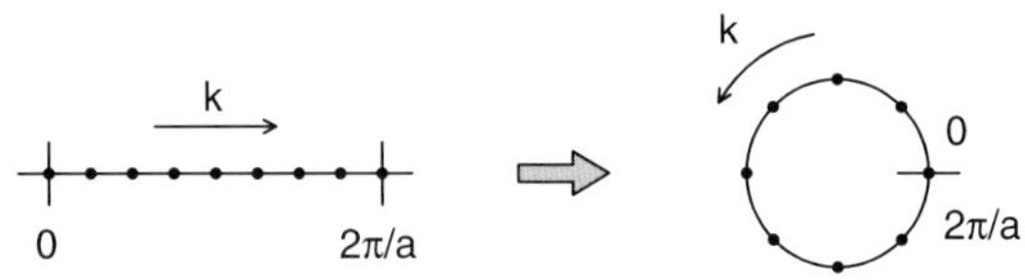

Fig. 3. At left, a sketch of the Brillouin zone of a 1D crystal, regarded as a segment of the real k axis. At right, a sketch of the same Brillouin zone regarded as a topologically closed loop.

However, it turns out that the entire integral in Eq. (32) is independent of gauge. The demonstration of this claim depends on the fact that the eigenvectors $|\psi_k\rangle$ are periodic functions of k with

$$|\psi_{2\pi/a}\rangle = |\psi_0\rangle \tag{35}$$

which is known as the "periodic gauge condition." Indeed, it is natural to regard the Brillouin zone not as an interval of the real axis, but as a closed space (i.e., a ring), as illustrated in Fig. 3. Thus, to be sensible, the gauge change should obey

$$\beta(2\pi/a) = \beta(0) + 2\pi m \tag{36}$$

where m is an integer, so that $\exp(-i\beta)$ will match seamlessly at the Brillouin zone boundary. We shall assume here that $m = 0$, returning to the possibility of $m \neq 0$ in Section 6. The integral of the last term of Eq. (34) over the entire Brillouin zone, therefore, vanishes, so that $\bar{\phi} = \phi$, and the polarization is indeed gauge invariant. The quantity ϕ is known a "Berry phase" [10]; it is a global-phase property of the Bloch bands as the wavevector k is carried around the Brillouin zone.

In order to perform practical calculations using Eq. (30), it is necessary to reformulate the theory on a discrete k mesh. We illustrate this again in the context of the 1D single-band case, where we note that Eq. (32) can be discretized as

$$\phi = -\text{Im} \ln \prod_{j=0}^{M-1} \langle u_{k_j} | u_{k_{j+1}} \rangle \tag{37}$$

where $k_j = 2\pi j/Ma$ is the jth k-point in the Brillouin zone. That this reproduces Eq. (32) can be checked by plugging the expansion $u_{k+\Delta k} = u_k + \Delta k\,(\partial_k u_k) + \mathcal{O}(\Delta k^2)$ into Eq. (37) and keeping the leading term as $\Delta k \to 0$. Equation (37) says to take the complex phase of the product of $\langle u_{k_1}|u_{k_2}\rangle$ with $\langle u_{k_2}|u_{k_3}\rangle$, etc., all the way around the Brillouin zone (as in the right panel of Fig. 3). The gauge invariance is manifest: changing the phase of one Bloch function $|u_k\rangle$ obviously has no effect, since it appears in the product once as a bra and once as a ket. (Note that, in evaluating Eq. (37), one should apply the periodic gauge condition of Eq. (35) in the form $u_{2\pi/a}(x) = e^{-2\pi i x/a}\, u_0(x)$ in evaluating the last inner product needed to close the loop.)

In three dimensions (3D), the Brillouin zone can be regarded as a closed 3 torus obtained by identifying boundary points $\psi_{n\mathbf{k}} = \psi_{n,\mathbf{k}+\mathbf{G}_j}$, where $\mathbf{G}_j$ are the three primitive reciprocal lattice vectors. Then the electronic polarization can be

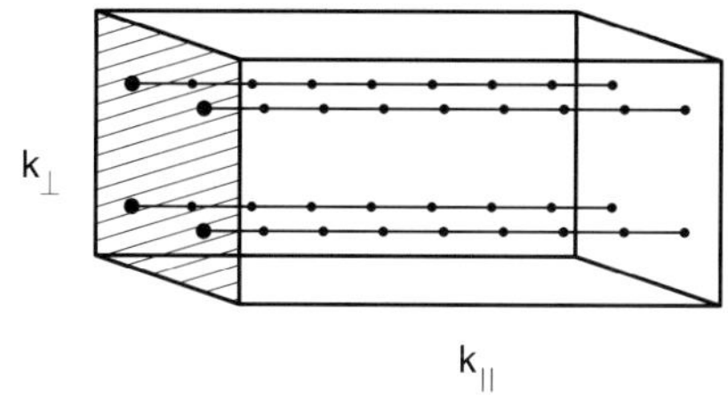

Fig. 4. Organization of the Brillouin zone needed to compute the component of **P** in direction $\mathbf{k}_{\parallel}$. A Berry-phase calculation is carried out for each k-point string, and a conventional average over $\mathbf{k}_{\perp}$ is then performed.

written as

$$\mathbf{P}_n = \frac{-e}{2\pi\Omega} \sum_j \phi_{nj} \mathbf{R}_j \tag{38}$$

where $\mathbf{R}_j$ is the real-space primitive translation corresponding to $\mathbf{G}_j$, and the Berry phase for band n in direction j is

$$\phi_{nj} = -\frac{\Omega}{(2\pi)^3} \mathrm{Im} \int_{\mathrm{BZ}} \mathrm{d}^3 k \langle u_{n\mathbf{k}} | \mathbf{G}_j \cdot \nabla_{\mathbf{k}} | u_{n\mathbf{k}} \rangle \tag{39}$$

To compute the ϕ_{nj} for a given direction j, the sampling of the Brillouin zone is arranged as in Fig. 4, where $\mathbf{k}_{\parallel}$ is the direction along $\mathbf{G}_j$ and $\mathbf{k}_{\perp}$ refers to the 2D space of wavevectors spanning the other two primitive reciprocal lattice vectors. For a given $\mathbf{k}_{\perp}$, the Berry phase $\phi_{nj}(\mathbf{k}_{\perp})$ is computed along the string of M_j k points extending along $\mathbf{k}_{\parallel}$ as in Eq. (37), and finally a conventional average is taken over the $\mathbf{k}_{\perp}$ via

$$\phi_{nj} = \frac{1}{N_{\mathbf{k}_{\perp}}} \sum_{\mathbf{k}_{\perp}} \phi_n(\mathbf{k}_{\perp}) \tag{40}$$

This is the form in which the Berry-phase theory of polarization is implemented in modern electronic structure codes. Further details and discussion, including the appropriate reformulation of Eqs (39) and (40) for the case of connected multiple bands (i.e., bands with symmetry-induced degeneracies at certain locations in the Brillouin zone) may be found in Refs. [6,7].

5. REFORMULATION IN TERMS OF WANNIER FUNCTIONS

An alternative, and in many ways more intuitive, way of thinking about the Berry-phase expression for the electronic polarization, Eq. (30), is in terms of WFs. The WFs are localized functions $w_{n\mathbf{R}}(\mathbf{r})$, labeled by band n and unit cell $\mathbf{R}$, that are constructed by carrying out a unitary transformation of the Bloch states $\psi_{n\mathbf{k}}$. Thus,

WFs and Bloch functions can be regarded as two different orthonormal representations of the same occupied Hilbert space. The construction is carried out via a Fourier transform of the form

$$|w_{n\mathbf{R}}\rangle = \frac{\Omega}{(2\pi)^3}\int_{\mathrm{BZ}} \mathrm{d}\mathbf{k}\, \mathrm{e}^{\mathrm{i}\mathbf{k}\cdot\mathbf{R}}|\psi_{n\mathbf{k}}\rangle \tag{41}$$

where the Bloch states are normalized in one unit cell. There is again some "gauge freedom" in the choice of these WFs; a set of Bloch functions

$$|\bar{\psi}_{n\mathbf{k}}\rangle = \mathrm{e}^{-\mathrm{i}\beta_n(\mathbf{k})}\,|\psi_{n\mathbf{k}}\rangle \tag{42}$$

results in WFs $|\bar{w}_{n\mathbf{R}}\rangle$ that are not identical to the $|w_{n\mathbf{R}}\rangle$. In practice, the gauge is often set by some criterion that keeps the WFs well localized in real space, such as the minimum quadratic spread criterion introduced by Marzari and Vanderbilt [22]. However, we should expect that any physical quantity, such as the electronic polarization arising from band n, should be invariant with respect to the phase twist $\beta_n(\mathbf{k})$. A typical WF constructed from the oxygen p-like valence bands of $BaTiO_3$ in its cubic centrosymmetric phase is shown in Fig. 5(a).

Once we have the WFs, we can locate the "Wannier centers" $\mathbf{r}_{n\mathbf{R}} = \langle w_{n\mathbf{R}}|\mathbf{r}|w_{n\mathbf{R}}\rangle$. Returning momentarily to one band in 1D, the Wannier center of the WF in the unit cell at the origin is just

$$x_0 = \langle w_0|x|w_0\rangle \tag{43}$$

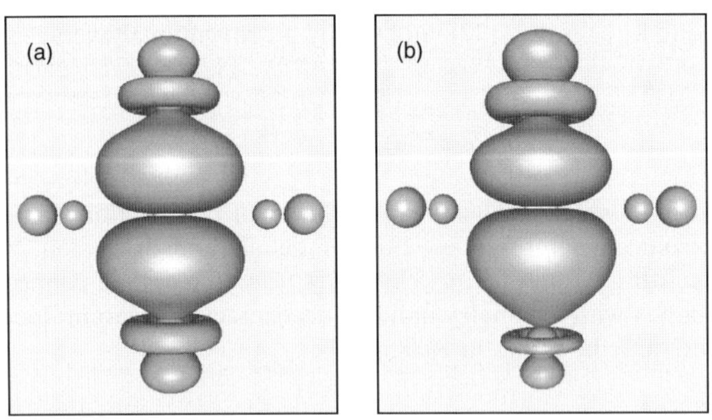

Fig. 5. Oxygen $2p$-like Wannier functions in $BaTiO_3$ as derived from the maximal localization algorithm of Ref. [22]. An isocontour of Wannier-function amplitude is shown, illustrating the hybridization of O $2p$ and Ti $3d$ orbital character in the Wannier function. Oxygen atom is at center, four second-neighbor Ba atoms also appear, and two first-neighbor Ti atoms are hidden under the d-like lobes of the Wannier function. Panel (a): Centrosymmetric paraelectric structure. Panel (b): Distorted ferroelectric structure, in which the Ti–O bond is shortened in the upper half of the figure and lengthened in the lower half, resulting in enhanced p–d hybridization in the upper portion of the figure and suppressed hybridization below. (See also Marzari and Vanderbilt [50].)

If Eq. (41) is rewritten as

$$|w_0\rangle = \frac{a}{2\pi}\int dk\, e^{ikx}|u_k\rangle \tag{44}$$

then it follows that

$$x|w_0\rangle = \frac{a}{2\pi}\int dk(-i\partial_k\, e^{ikx})|u_k\rangle = \frac{a}{2\pi}\int dk\, e^{ikx} i|\partial_k u_k\rangle \tag{45}$$

where an integration by parts has been used. Then

$$x_0 = -\frac{a}{2\pi}\,\mathrm{Im}\int_0^{2\pi/a} dk\langle u_k|\partial_k|u_k\rangle \tag{46}$$

Comparing with Eq. (32) of the previous section, we find that

$$x_0 = \frac{a\phi}{2\pi} \tag{47}$$

that is, *the Berry phase* ϕ *introduced earlier is nothing other than a measure of the location of the Wannier center in the unit cell.* The fact that ϕ was previously shown to be invariant with respect to choice of gauge implies that the same is true of the Wannier center x_0.

Similar arguments in 3D lead to the conclusion that

$$\mathbf{r}_{n\mathbf{R}} = \langle w_{n\mathbf{R}}|\mathbf{r}|w_{n\mathbf{R}}\rangle = \mathbf{R} + \sum_j \frac{\phi_{nj}}{2\pi}\mathbf{R}_j \tag{48}$$

where ϕ_{nj} is given by Eq. (39). That is, the location of the nth Wannier center in the unit cell is just given by the three Berry phases ϕ_{nj} of band n in the primitive lattice vector directions $\mathbf{R}_j$. In fact, the polarization is just related to the Wannier centers by

$$\mathbf{P} = -\frac{e}{\Omega}\sum_n \mathbf{r}_{n0} \tag{49}$$

This formula is very similar to the ionic one given in Eq. (31), but now the electron charge is taken to reside at the Wannier centers while the ionic charges reside at the nuclear positions.

The Berry-phase theory can thus be regarded as providing a mapping of the distributed quantum mechanical electronic charge density onto a lattice of negative point charges of charge $-e$, as illustrated in Fig. 6. Then, the change of polarization resulting from any physical change, such as the displacement of one atomic sublattice or the application of an electric field, can be related in a simple way to the displacements of the Wannier centers $\mathbf{r}_{n\mathbf{R}}$ occurring as a result of this change.

This viewpoint is illustrated for the case of $BaTiO_3$ by returning to Fig. 5, which shows the oxygen $2p$-like WF not only before (Panel (a)) but also after (Panel (b)) a displacement of the Ti sublattice by a small distance along the $-\hat{z}$ direction. We can think of this WF as being centered on the bottom oxygen atom shown in the left panel of Fig. 1. A calculation of the corresponding displacement of the Wannier

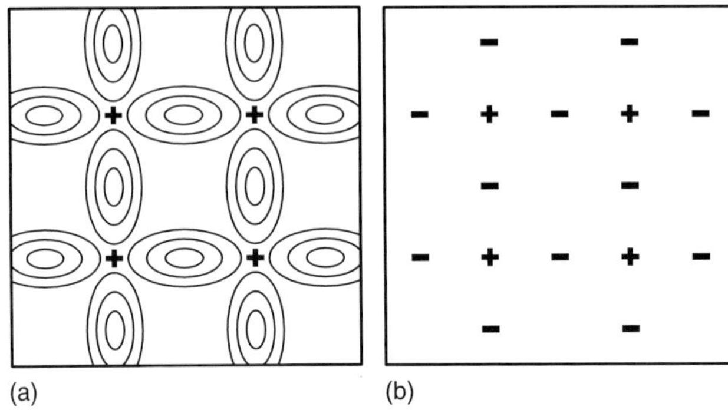

Fig. 6. Mapping of the distributed charge density (sketched in contour-plot fashion) onto the centers of charge of the Wannier functions.

center shows that it displaces strongly *upward* in response. This occurs because the hybridization between O 2p and Ti 3d orbitals is strengthened in the top lobe of the WF, and weakened in the bottom lobe, leading to the "swelling" and "shrinkage" of the d-like lobes that is evident in Fig. 5(b). This analysis provides an insightful microscopic explanation for the "anomalous dynamical effective charges" [$Z^*(\mathrm{O}) \simeq -5e$, $Z^*(\mathrm{Ti}) \simeq +7e$] that have been observed in this class of materials [42,43].

6. THE QUANTUM OF POLARIZATION AND THE SURFACE CHARGE THEOREM

The alert reader may have noticed that the formulas for the polarization given in Eqs (30), (38), and (49) have an arbitrariness modulo (e/Ω) times a lattice vector $\mathbf{R}$. This is perhaps most obvious in connections with Eq. (49), where the decision as to which of the periodic array of WFs is to be taken as belonging to the home unit cell (the one labeled $|w_{n\mathbf{0}}\rangle$) is arbitrary in Eq. (48). (Actually, a similar indeterminacy is also present in the ionic contribution, Eq. (31).) It is also fairly obvious if Eq. (38) is evaluated via Eqs (40) and (37), in which case each Berry phase ϕ_n is indeterminate modulo 2π.

In Eq. (30), or in the continuum evaluation (39) of Eq. (38), the difficulty is more subtle. Returning for the moment to the 1D, single-band case, we pointed out in Eq. (36) that, in general, a permissible gauge change (phase twist) of the Bloch functions allows for the phase to evolve by $2\pi m$ as k is transported around the Brillouin zone. Evaluating Eq. (32) using Eq. (34), it follows that $\bar{\phi} = \phi + 2\pi m$ and $\bar{P} = P - em$. The polarization is, therefore, only well-defined modulo an electron charge. In 3D, the corresponding statement is that one can apply a gauge twist

$$|\bar{u}_{n\mathbf{k}}\rangle = \mathrm{e}^{-\mathrm{i}\beta_n(\mathbf{k})}|u_{n\mathbf{k}}\rangle \tag{50}$$

obeying

$$\beta_n(\mathbf{k} + \mathbf{G}_j) = \beta_n(\mathbf{k}) + 2\pi m_{nj} \tag{51}$$

where $\mathbf{G}_j$ is the primitive reciprocal lattice vector in direction j. Then $\bar{\phi}_{nj} = \phi_{nj} + 2\pi m_{nj}$ in Eq. (39), so that

$$\bar{\mathbf{P}} = \mathbf{P} - \frac{e}{\Omega}\sum_{j=1}^{3} m_j \mathbf{R}_j \tag{52}$$

where $m_j = \sum_n m_{nj}$. Once again, the Berry-phase polarization is seen to be ill-defined modulo $e\mathbf{R}/\Omega$, that is, an electron charge times a lattice vector divided by cell volume. This uncertainty is sometimes known as the "quantum of polarization."

It is instructive to recall the argument leading to Eq. (29), which can be summarized by saying that

$$\mathbf{P}(\lambda_2) - \mathbf{P}(\lambda_1) := \int_{\lambda_1}^{\lambda_2} \mathrm{d}\lambda(\partial_\lambda \mathbf{P}) \tag{53}$$

where $\mathrm{d}\mathbf{P}/\mathrm{d}\lambda$ is given by Eq. (28). The symbol ":=" has been introduced here to emphasize that this equation needs a special interpretation, namely, that the two sides are equal *modulo* the quantum $e\mathbf{R}/\Omega$, where $\mathbf{R}$ is an arbitrary lattice vector. It is important to understand that while each quantity on the left-hand side is actually ill defined up to this modulus, the right-hand side has a definite and unambiguous value for a given evolution of the system. That is, the evaluation of the polarization change via Eq. (29) or (30) has a fundamental limitation; some of the information contained in the original definition, Eq. (23), is lost – namely, the information about the "choice of branch" of the polarization modulo $e\mathbf{R}/\Omega$.

Fortunately this limitation is rarely serious in practice. In most cases, the change in polarization that can be induced by a practical perturbation, such as a small sublattice displacement or electric field, is insufficient to cause $\mathbf{P}$ to change by a large fraction of $e\mathbf{R}/\Omega$. Where exceptions exist, as for the case of some strongly polarized ferroelectrics such as $PbTiO_3$, the ambiguity can be resolved by subdividing the adiabatic path into several shorter intervals, for each of which the change in $\mathbf{P}$ is unambiguous for practical purposes.

Nevertheless, the ambiguity inherent in Eq. (53) is an essential aspect of the theory. For example, for the case of a *closed cyclic* adiabatic evolution of the system, in which the parameter values λ_1 and λ_2 label the *same physical state* of the system, Eq. (53) becomes

$$\oint \mathrm{d}\lambda(\partial_\lambda \mathbf{P}) = \mathbf{0} \text{ modulo } \frac{e\mathbf{R}}{\Omega} \tag{54}$$

(As always, such equations are defined under the assumption that the system remains insulating everywhere along the path in λ space.) The modulus cannot be removed, because there are situations (e.g., sliding charge-density waves) in which the value of the integral is *not* zero [44]. In such cases, one says that there is "quantized adiabatic charge transport" and the cyclic evolution of λ acts as a charge pump that transfers an integral number of electrons from one side of the unit cell to the other in one cycle.

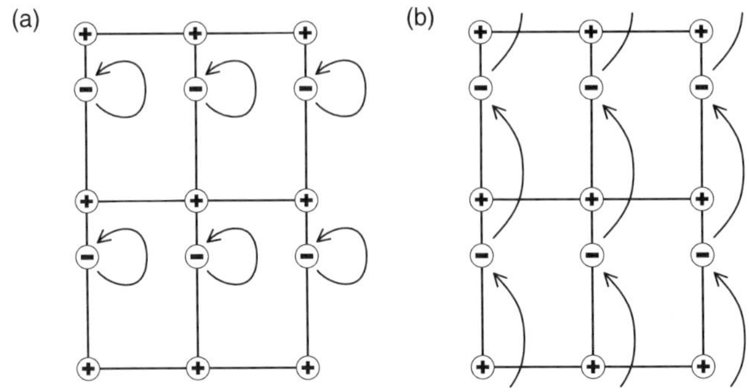

Fig. 7. Cyclic evolution in which quantum of polarization does not, or does, appear. Possible evolution of positions of Wannier centers (−), relative to lattice of ions (+), as Hamiltonian evolves adiabatically around a closed loop. Wannier functions must return to themselves, but can do so either (a) without, or (b) with, a coherent shift by a lattice vector.

It is amusing to consider the meaning of such a situation in the context of the WF picture introduced in the previous subsection. Because the initial and final states of the system are identical, the arrangement of WFs must be identical. However, if one follows individual Wannier centers during the evolution as illustrated in Fig. 7, they need not describe closed loops. If they do all describe closed loops, then the circuit integral in Eq. (54) does vanish. On the other hand, if the evolution results in the pumping of Wannier centers across the unit cell, then the system represents an example of adiabatic charge transport.

There is one more way in which the indeterminacy of $\mathbf{P}$ modulo $e\mathbf{R}/\Omega$ may be understood in a natural way. In elementary electrostatics, one learns that the macroscopic bound surface charge density σ_b residing on the surface of a sample is related to the polarization in the interior by $\sigma_\mathrm{b} = \hat{\mathbf{n}} \cdot \mathbf{P}$, where $\hat{\mathbf{n}}$ is the surface normal. One defines the bound charge σ_b by saying that no free charge is present, but what, precisely, does this mean? The surface must be insulating with the electron chemical potential lying in a gap that is common to both bulk and surface. But this is not a unique prescription. Consider, for example, the case of an insulating crystal having a surface band that lies entirely inside the bulk band gap. Then this surface band may either be completely occupied or completely empty, as indicated schematically in the density-of-states plots shown in Panels (a) and (d), respectively, of Fig. 8. From the point of view of the WF representation, this corresponds to the question of how many WFs exist at the surface, as illustrated in the remaining Panels (b–c) and (e–f). In either case, the condition of absence of bound charge is satisfied. But the surface charge densities σ clearly differ by an integer number of electron charges per primitive surface cell area A_surf, so we conclude that [45]

$$\sigma_\mathrm{b} = \hat{\mathbf{n}} \cdot \mathbf{P} \quad \text{modulo } \frac{e}{A_\mathrm{surf}} \tag{55}$$

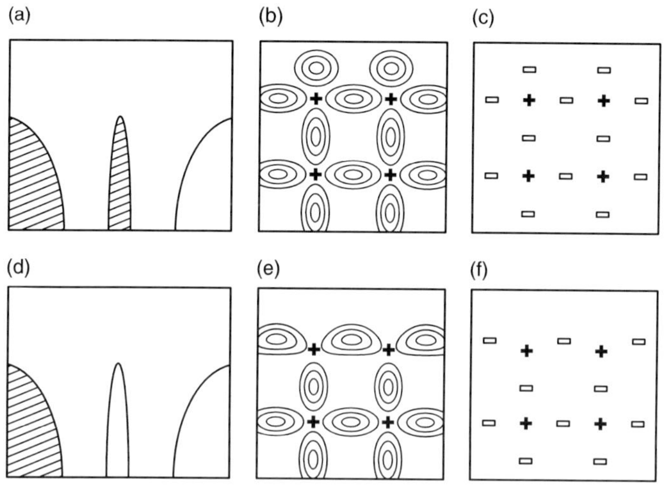

Fig. 8. Ambiguity of bound surface charge. Panels (a) and (d) illustrate the density of states of an insulating crystal having a full valence band (left) and empty conduction band (right), and a surface band lying entirely within the bulk gap (center) that may either be entirely (a) occupied, or (d) empty. Panels (b) and (e) show the corresponding charge densities, while (c) and (f) illustrate the mapping to Wannier centers in these two cases. As can be seen by comparing (c) and (f), the surface charges differ by precisely one electron charge (or two for spin) per surface unit cell area.

However, this is perfectly consistent with the fact that $\mathbf{P}$ is ill-defined modulo $e\mathbf{R}/\Omega$, since $\hat{\mathbf{n}} \cdot \mathbf{R} = mc = m\Omega/A_{\text{surf}}$, where m is an integer and c the lattice constant of the crystal in the surface-normal direction.

From arguments of this type, it should have been possible to anticipate this essential indeterminacy in the definition of crystal polarization. As a historical aside, it is interesting to note that the presence of this indeterminacy was not widely understood and appreciated until it was forced into the light by the efforts of the computational electronic-structure community to understand precisely how polarization should be computed in practice. For further discussion of the subtleties associated with the "quantum of polarization," the reader is referred to Refs [7,45,46].

7. TREATMENT OF FINITE ELECTRIC FIELDS

Earlier in this section, we have seen how the linear-response formula for the first-order change in polarization with respect to some perturbation (Section 3) can be manipulated into a theory of *finite* polarization changes (Section 4). An analogous situation applies to the case of applied electric fields. The problem of the first-order response of a system to an infinitesimal applied electric field can be treated by an essentially straightforward extension of the methods outlined in Section 3.

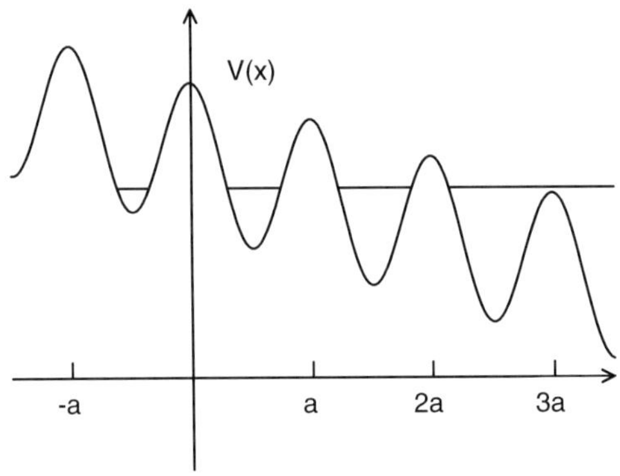

Fig. 9. Potential energy of an electron in a crystal in the presence of an electric field. The potential is neither periodic nor bounded from below. Horizontal line indicates regions of positive kinetic energy for a representative eigenstate of H.

However, the treatment of an insulator in a *finite* applied electric field introduces special difficulties. Luckily, these can be solved using methods based on the same modern theory of polarization introduced in Section 4.

Let us first appreciate the nature of the difficulty. The Hamiltonian per unit cell of an electron in a crystalline insulator in the presence of an electric field $\mathcal{E}$ is

$$H = \frac{p^2}{2m_e} + V_{\text{per}}(\mathbf{r}) + e\mathcal{E} \cdot \mathbf{r} \tag{56}$$

where $V_{\text{per}}(\mathbf{r})$ is a potential having the periodicity of the unit cell. Clearly the field-coupling term $e\mathcal{E} \cdot \mathbf{r}$ does not have this periodicity. The situation is illustrated in Fig. 9, where the total potential $V = V_{\text{per}} + e\mathcal{E} \cdot \mathbf{r}$ is sketched for the case of a 1D crystal. The crystal does not have the usual Born-von Kármán boundary conditions, $V(x) = V(x + a)$, and therefore Bloch's theorem *does not apply*. If an eigensolution of the Schrödinger equation is sought at some energy E, as shown by the horizontal line in Fig. 9, it will have a rather pathological form, decaying exponentially to the left of the pictured region, and oscillating with increasing rapidity to the right. Perhaps even more pathological is the fact that the potential energy $V(x)$ is not bound from below, so that the Hamiltonian of Eq. (56) *has no ground state even in principle*.

An alternative way to look at the problem is illustrated in Fig. 10, where the valence and conduction band energy ranges are shown as a function of spatial position in the presence of an electric field. The "state" that one has in mind is one in which all "valence" orbitals are occupied and all "conduction" orbitals are empty. However, for an insulator of gap E_g in a field $\mathcal{E}$, it is always possible to lower the energy of the system by transferring electrons from the valence band in one region to the conduction band in a region of distance $\gg L_t = E_g/\mathcal{E}$ down-field. This "Zener tunneling" is analogous to the auto-ionization that also occurs, in principle, for an atom or molecule in a finite electric field.

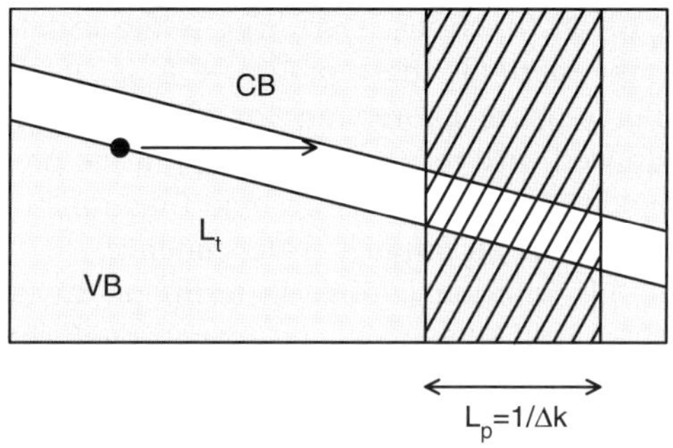

Fig. 10. Sketch of valence and conduction bands of an insulator in an electric field, showing two relevant length scales.

Despite these problems, the treatment of an infinitesimal electric field, as is done using linear-response (DFPT) methods, is rather straightforward (for a review, see [18]). In this case, the operator $V^{(1)}$ of Eq. (10) or (11) is just $e\mathcal{E}$ times $\mathbf{r}|\psi_i^{(0)}\rangle$, which can be rewritten as $|\tilde{\psi}_{\alpha,i}\rangle$ and solved via Eq. (15) or (17) as in Section 3. In short, a generic linear-response calculation requires one energy denominator, as in Eq. (10); an additional energy denominator becomes involved if the desired response is the polarization; and another one becomes involved if the perturbation is an electric field. Thus, in the case of computing the dielectric susceptibility, Eq. (2), three powers of the energy denominator enter in principle. However, all may be removed via the Sternheimer replacement, so that no infinite sum over unoccupied states is actually needed in practice [18].

Let us then turn to the harder case, that of an insulating crystal in a *finite* electric field. We expect that if we start with an insulating crystal in its ground state, and then adiabatically apply an electric field that remains small enough so that Zener tunneling is negligible, then a reasonably well-defined "state" should result. This should be a state in which the electronic charge density $n(\mathbf{r})$ remains periodic, even though the total potential of Fig. 9 would prefer to concentrate the electrons toward the right-hand side of the crystal.

We shall, therefore, *require* that our solution have this periodicity. In fact, if we work in the Wannier representation, we can do this just by insisting that each WF should remain identical to its periodic images, as happens automatically in the case of zero field. By requiring that the WFs also remain well localized, it is possible to develop an approach to the electric-field problem in which one solves for the distortion of each WF caused by the electric field [47]. Such an approach is not very convenient in practice, but it is possible to carry out a unitary transformation from the field-polarized WFs to a set of "field-polarized Bloch functions" $\psi_{n\mathbf{k}}(\mathbf{r})$. This results in a much more practical solution [25,26] in which one simply minimizes the electric enthalpy functional [24]

$$F = E_{\mathrm{KS}}(\{\psi_{n\mathbf{k}}\}) - \mathcal{E} \cdot \mathbf{P}(\{\psi_{n\mathbf{k}}\}) \tag{57}$$

with respect to these Bloch functions. Here E_{KS} is the usual Kohn–Sham energy functional involving the periodic part $p^2/2m_e + V_{per}$ of the Hamiltonian and $\mathbf{P}(\{\psi_{n\mathbf{k}}\})$ the usual Berry-phase expression for the polarization, Eq. (30). This equation is to be minimized with respect to all $\{\psi_{n\mathbf{k}}\}$ in the presence of a given field $\mathcal{E}$; the Bloch functions at minimum thus become functions of $\mathcal{E}$, so that the first term in Eq. (57) also acquires an implicit $\mathcal{E}$-dependence.

We emphasize that this solution is done in terms of wavefunctions $\psi_{n\mathbf{k}}(\mathbf{r})$ that obey Bloch symmetry, $\psi_{n\mathbf{k}}(\mathbf{r}+\mathbf{R}) = e^{i\mathbf{k}\cdot\mathbf{r}}\psi_{n\mathbf{k}}(\mathbf{r})$, even though the Hamiltonian of Eq. (56) does not have translational symmetry. The polarized Bloch functions *are not meant to be eigenstates* of the Hamiltonian. One way to justify this approach is to say that the Bloch functions $\psi_{n\mathbf{k}}(\mathbf{r})$ are only being introduced as a convenient representation

$$n(\mathbf{r},\mathbf{r}') = \sum_{n\mathbf{k}} \psi^*_{n\mathbf{k}}(\mathbf{r})\psi_{n\mathbf{k}}(\mathbf{r}') \tag{58}$$

for the one-particle density matrix $n(\mathbf{r},\mathbf{r}')$. This results in a density matrix that is periodic in the sense $n(\mathbf{r},\mathbf{r}') = n(\mathbf{r}+\mathbf{R},\mathbf{r}'+\mathbf{R})$ (where $\mathbf{R}$ is any lattice vector) as long as the $\psi_{n\mathbf{k}}(\mathbf{r})$ have Bloch symmetry.

Because the "state" of interest in the presence of a field is in principle not a true ground state, but only a long-lived resonance, it should not be surprising if the above theory has some corresponding limitation. Indeed, it fails to produce a perfectly well-defined solution, but in an unfamiliar way: the variational solution breaks down if the k-point sampling is taken to be too fine! In fact, when we find a minimum of Eq. (57), it is never a global minimum, but only a local minimum, and when the k-point sampling becomes too fine, even these local minima disappear.

We can understand this behavior heuristically by noting that there are two relevant length scales in the problem. One is the Zener tunneling distance $L_t = E_g/\mathcal{E}$ already discussed above and shown in the left portion of Fig. 10. The second length scale $L_p = 2\pi/\Delta k$ is proportional to the inverse of the k-point spacing Δk, and has the interpretation of the being the periodic repeat distance of the supercell to which the k-point mesh corresponds. In other words, L_p is the size of the "box" on which the periodic boundary conditions are imposed. This periodic repeat distance L_p has also been included in the sketch in Fig. 10. Now, it turns out that if $L_p \gg L_t$, then there is no solution. More precisely, for a given field $\mathcal{E}$, there is a critical L_c such that no solution exists for $L_p > L_c$, and L_c is on the order of (and scales as) $L_t = E_g/\mathcal{E}$. Intuitively, if the "size of the box" is larger than the Zener tunneling distance, then there is room for Zener tunneling to take place within the box, and no solution exists. Conversely, if the k-point sampling is not taken too fine, then L_p will remain smaller than the Zener tunneling length, and Zener tunneling will, therefore, be suppressed. The theory is thus limited to modest fields such that the potential drop across one primitive cell is much smaller than the band gap. In practice, this is usually not a severe restriction, since physical dielectric breakdown fields are also usually of this order or smaller. For more discussion and details, the reader is directed to Refs [25,26,48].

The minimization of Eq. (57) also raises some new issues in practice. The extra $\mathcal{E} \cdot \mathbf{P}$ term in the energy functional does not take the form of an expectation value of a quantum mechanical operator, so some care has to be exercised in evaluating its gradients. Moreover, because $\mathbf{P}$ is expressed in terms of inner products of the cell-periodic Bloch functions $|u_{n\mathbf{k}}\rangle$ on neighboring k points, as in Eq. (37), there is now effectively a coupling between neighboring k points, proportional to $\mathcal{E}$, that did not exist before. Thus, a self-consistent solution is required. This is usually not a serious problem in practice, since iterative methods are normally used in any case, and the self-consistency over k points can easily be incorporated in this context.

To summarize this section, it is possible to develop a fairly straightforward extension of standard electronic bandstructure methods to handle the case of an applied electric field, as long as the field is not too large. Once the solution has been obtained, the polarization in finite $\mathcal{E}$-field is trivially obtained as a by-product, and it is fairly straightforward to compute various secondary quantities (such as charge densities, total energies, forces, stresses, etc.) with little or no modification relative to the zero-field case. While calculations in finite $\mathcal{E}$-field are not yet as common as zero-field calculations of the polarization itself, we expect them to become more widespread as the capability to handle finite $\mathcal{E}$-fields is added to standard code packages.

8. CONCLUSIONS

In this chapter, we have reviewed the modern theory of polarization, which forms the basis for modern non-perturbative calculations of dielectric, piezoelectric, and related properties of insulating materials. The capability of computing polarization is available in almost all commonly used software packages for bulk electronic-structure calculations. While initially formulated in vanishing electric field, the case of finite field can be treated simply by letting the external electric field couple to the polarization while retaining the Bloch form of the wavefunctions. Because the Bloch functions are no longer eigenstates of the Hamiltonian, this approach requires some careful justification, but is not particularly difficult to implement in practice. Together, these methods provide a foundation for modern computational studies of the dielectric properties of insulating materials, including the coupling of the polarization and the electric fields with lattice vibrational, elastic, compositional, magnetic, and other degrees of freedom.

REFERENCES

[1] L.D. Landau and E.M. Lifshitz, *Electrodynamics of Continuous Media* (Pergamon Press, Oxford, 1984).
[2] C. Kittel, *Introduction to Solid State Physics*, 7th ed. (Wiley, New York, 1996).
[3] N.W. Ashcroft and N.D. Mermin, *Solid State Physics* (Holt, Reinhart and Winston, New York, 1976).
[4] R.M. Martin, Comment on calculations of electric polarization in crystals, *Phys. Rev. B* **9**, 1998 (1974).

[5] R. Resta, Theory of the electric polarization in crystals, *Ferroelectrics* **136**, 51 (1992).
[6] R.D. King-Smith and D. Vanderbilt, Theory of polarization of crystalline solids, *Phys. Rev. B* **47**, 1651 (1993).
[7] R. Resta, Macroscopic polarization in crystalline dielectrics: The geometric phase approach, *Rev. Mod. Phys.* **66**, 899 (1994).
[8] R. Resta, Quantum-mechanical position operator in extended systems, *Phys. Rev. Lett.* **80**, 1800 (1998).
[9] R. Resta, Macroscopic polarization from electronic wave functions, *J. Quantum Chem.* **75**, 599 (1999).
[10] A. Shapere and F. Wilczek (Eds), *Geometric Phases in Physics* (World Scientific, Singapore, 1989).
[11] R. Resta, Manifestations of Berry's phase in molecules and condensed matter, *J. Phys.: Condens. Matter* **12**, R107 (2000).
[12] S.L. Adler, Quantum theory of the dielectric constant in real solids, *Phys. Rev.* **126**, 413 (1962).
[13] N. Wiser, Dielectric constant with local field effects included, *Phys. Rev.* **129**, 62 (1963).
[14] V.N. Genkin and P.M. Mednis, *Contribution to the theory of nonlinear effects in crystals with account taken of partially filled bands, Sov. Phys. JETP* **27**, 609 (1968).
[15] S. Baroni and R. Resta, *Ab initio* calculation of the macroscopic dielectric constant in silicon, *Phys. Rev. B* **33**, 7017 (1986).
[16] S. Baroni, P. Giannozzi and A. Testa, Green's-function approach to linear response in solids, *Phys. Rev. Lett.* **58**, 1861 (1987).
[17] X. Gonze, D.C. Allan and M.P. Teter, Dielectric tensor, and phonons in α-quartz by variational density-functional perturbation theory, *Phys. Rev. Lett.* **68**, 3603 (1992).
[18] S. Baroni, S. de Gironcoli, A.D. Corso and P. Giannozzi, Phonons and related crystal properties from density-functional perturbation theory, *Rev. Mod. Phys.* **73**, 515 (2001).
[19] G.H. Wannier, The structure of electronic excitation levels in insulating crystals, *Phys. Rev.* **52**, 191 (1937).
[20] W. Kohn, Analytic properties of Bloch waves and Wannier functions, *Phys. Rev.* **115**, 809 (1959).
[21] E. Blount, Formalisms of band theory, *Solid State Phys.* **13**, 305 (1962).
[22] N. Marzari and D. Vanderbilt, Maximally localized generalized Wannier functions for composite energy bands, *Phys. Rev. B* **56**, 12847 (1997).
[23] R. Nenciu, Dynamics of band electrons in electric and magnetic fields: rigorous justification of the effective Hamiltonians, *Rev. Mod. Phys.* **63**, 91 (1991).
[24] R.W. Nunes and X. Gonze, Berry-phase treatment of the homogeneous electric field perturbation in insulators, *Phys. Rev. B* **63**, 155107 (2001).
[25] I. Souza, J. Iniguez and D. Vanderbilt, First-principles approach to insulators in finite electric fields, *Phys. Rev. Lett.* **89**, 117602 (2002).
[26] P. Umari and A. Pasquarello, *Ab initio* molecular dynamics in a finite homogeneous electric field, *Phys. Rev. Lett.* **89**, 157602 (2002).
[27] R.M. Martin, Piezoelectricity, *Phys. Rev. B* **5**, 1607 (1972).
[28] M.E. Lines and A.M. Glass, *Principles and Applications of Ferroelectrics and Related Materials* (Clarendon Press, Oxford, 1977).
[29] M. Born and K. Huang, *Dynamical Theory of Crystal Lattices* (Oxford University Press, Oxford, 1954).
[30] A.A. Maradudin, E.W. Montroll, G.H. Weiss and I.P. Ipatova, *Theory of Lattice Dynamics in the Harmonic Approximation*, 3rd ed. Solid State Phys. Suppl. (Academic, New York, 1971).
[31] R. Pick, M.H. Cohen and R.M. Martin, Microscopic theory of force constants in the adiabatic approximation, *Phys. Rev. B* **1**, 910 (1970).
[32] P. Brüesch, *Phonons: Theory and Experiment I* (Springer, Berlin, 1982).
[33] S. Lundqvist and N.H. March (Eds), *Theory of the Inhomogeneous Electron Gas* (Plenum, New York, 1983).
[34] It is permissible to replace the sum over $j \neq i$ in Eq. (10) by a sum over unoccupied states only, and also the projector Q_i in Eq. (11) by a projector onto unoccupied states only, for the purposes of

computing any physical change such as $\mathbf{P}^{(1)}$ of Eq. (9). This is almost always done in practice, but for simplicity we will not follow this path here.

[35] The velocity operator $\mathbf{v}$ coincides with $\mathbf{p}/m_e$ in the simple case where all potentials are local. The generalization to nonlocal pseudopotentials – either norm conserving or ultrasoft [49] – has been considered in Ref. [18].

[36] R.M. Sternheimer, Electronic polarizabilities of ions from the Hartree–Fock wave functions, *Phys. Rev.* **96**, 951 (1954).
[37] R.M. Sternheimer, Electronic polarizabilities of ions, *Phys. Rev.* **107**, 1565 (1957).
[38] R.M. Sternheimer, Electronic polarizabilities of ions, *Phys. Rev.* **115**, 1198 (1959).
[39] R.M. Sternheimer, Electronic polarizabilities of the alkali atoms. II, *Phys. Rev.* **183**, 112 (1969).
[40] R.M. Sternheimer, Quadrupole polarizabilities of various ions and the alkali atoms, *Phys. Rev. A* **1**, 321 (1970).

[41] This equation can also be derived from the formalism of adiabatic perturbation theory, by taking into account the first-order correction to the instantaneous wavefunction that appears at linear order in $\dot{\lambda}$ and using this in the evaluation of the expectation value of the current operator.

[42] R. Resta, M. Posternak and A. Baldereschi, Towards a quantum theory of polarization in ferroelectrics: The case of $KNbO_3$, *Phys. Rev. Lett.* **70**, 1010 (1993).
[43] W. Zhong, R.D. King-Smith and D. Vanderbilt, Giant LO-TO splittings in perovskite ferroelectrics, *Phys. Rev. Lett.* **72**, 3618 (1994).
[44] D.J. Thouless, Quantization of particle transport, *Phys. Rev. B* **27**, 6083 (1983).
[45] D. Vanderbilt and R. King-Smith, Electric polarization as a bulk quantity and its relation to surface charge, *Phys. Rev. B* **48**, 4442 (1993).
[46] D. Vanderbilt, Berry-phase theory of proper piezoelectric response, *J. Phys. Chem. Solids* **61**, 147 (2000).
[47] R.W. Nunes and D. Vanderbilt, Real-space approach to calculation of electric polarization and dielectric constants, *Phys. Rev. Lett.* **73**, 712 (1994).
[48] I. Souza, J. Iniguez and D. Vanderbilt, Dynamics of Berry-phase polarization in time-dependent electric fields, *Phys. Rev. B* **69**, 085106 (2004).
[49] D. Vanderbilt, Soft self-consistent pseudopotentials in a generalized eigenvalue formalism, *Phys. Rev. B* **41**, 7892 (1990).
[50] N. Marzari and D. Vanderbilt, Maximally-localized Wannier functions in perovskites: Cubic $BaTiO_3$, in First-Principles Calculations for Ferroelectrics: *Fifth Williamsburg Workshop*, edited by R.E. Cohen (AIP, Woodbury, New York, 1998), p. 146.

Chapter 6
ELECTRON TRANSPORT

P. B. Allen

1. INTRODUCTION

A voltage gradient ($-\nabla\Phi = \mathbf{E}$) drives an electron current ($\mathbf{j} = \sigma\mathbf{E}$, where σ is the conductivity). Many mechanisms of electron transport are known.

(1) Electronic quasiparticle propagation, the first known mechanism, studied by Drude [1], and improved by Sommerfeld [2,3], Bloch [4], Landau [5], and others. The current is given by

$$\mathbf{j} = -\frac{e}{V}\sum_k \mathbf{v}_k F(k) \tag{1}$$

where $F(k)$ is the distribution function (i.e. the non-equilibrium occupation) of the quasiparticle state k (short for: $\mathbf{k}, n, \sigma$) which has energy ε_k and group velocity $\mathbf{v}_k = \partial\varepsilon_k/\partial(\hbar\mathbf{k})$, and V the sample volume.

(2) Quasiparticle tunneling through barriers.

(3) Supercurrent flow in superconductors. The (Landau–Ginzburg) order parameter is $\psi = \sqrt{n_S}\exp(i\phi)$, and the current, with no magnetic field, is

$$j_S = -\frac{(2e)\hbar}{2mi}(\psi^*\nabla\psi - \psi\nabla\psi^*) = -2en_S\hbar\nabla\phi/m \tag{2}$$

where it is assumed that the superfluid density n_S is spatially uniform.

(4) Intrinsically diffusive currents in dirty alloys, metallic glasses, etc.

(5) Hopping currents in dilute electron systems where states are localized by (i) trapping at localized defects, (ii) Anderson localization at band tails in disordered media, or (iii) polaronic self-trapping in insulators.

Contemporary Concepts of Condensed Matter Science
Conceptual Foundations of Materials: A Standard Model for Ground- and Excited-State Properties

ISSN: 1572-0934/doi:10.1016/S1572-0934(06)02006-3

(6) Separate propagation of charge and spin in 1D metals ("Luttinger liquids").
(7) Collective sliding of charge-density waves when incommensurate distortions open a gap in a metal.

This article will focus on examples where mechanisms can be clearly identified. Two other constraints are imposed. First, magnetic field effects (Hall effect, magneto-resistance, etc.) are excluded, for no good reason except to allow room for other things. Second, if an effect (like Anderson localization, or "weak localization") occurs in 3d bulk materials, and occurs in lower dimension in altered form, then only the 3d version is included here, for the same reason.

Transport is a subject of greater breadth and depth than can be covered in any one volume, let alone chapter. Thanks first to "mesoscopic" physics, and now, to "nanoscience," the field of transport is rapidly expanding and evolving. This chapter has three aims. First, to illustrate the breadth and beauty of the subject. Second, to illustrate how techniques from electronic structure theory are used to explain or predict values of transport coefficients such as electrical resistivity. Third, to guide a reader through some of the theoretical notions currently evolving as nanoscale research changes the way we think about these things.

2. CONDUCTIVITY

The conductivity σ is the usual starting point. Let there be an electric field varying slowly in space and sinusoidally in time, $\mathbf{E}(t) = \mathbf{E}\cos(\omega t)$. Then a bulk solid will respond to first order with an electrical current

$$j_\alpha(t) = \sigma^{(1)}_{\alpha\beta} E_\beta \cos(\omega t) + \sigma^{(2)}_{\alpha\beta} E_\beta \sin(\omega t) \tag{3}$$

To simplify, the tensor notation $\sigma_{\alpha\beta}$ will often be condensed to σ with implicit rather than explicit tensor aspects. Equation (3) is equivalent to $j(t) = \mathrm{Re}\left[\sigma(\omega)E\mathrm{e}^{-\mathrm{i}\omega t}\right]$, with a complex conductivity $\sigma = \sigma^{(1)} + \mathrm{i}\sigma^{(2)}$ whose real and imaginary parts denote in-phase (dissipative) and out-of-phase (reactive) response to the E field. I will use the standard Fourier conventions

$$E(t) = \int_{-\infty}^{\infty} \frac{\mathrm{d}\omega}{2\pi} E(\omega)\mathrm{e}^{-\mathrm{i}\omega t} \quad \text{and} \quad E(\omega) = \int_{-\infty}^{\infty} \mathrm{d}t E(t)\mathrm{e}^{\mathrm{i}\omega t} \tag{4}$$

to relate time to frequency; $j(t)$ and $\sigma(t)$ have the same connection to $j(\omega)$ and $\sigma(\omega)$. Since $j(t)$ and $E(t)$ are real, and since $j(\omega) = \sigma(\omega)E(\omega)$, it is necessary that $E(\omega)$ obey $E(-\omega) = E^*(\omega)$ and similarly for $j(\omega)$ and $\sigma(\omega)$.

The time-domain relation between j and E is

$$j(t) = \int_{-\infty}^{\infty} \mathrm{d}t' \sigma(t - t')E(t') \tag{5}$$

Thus, $\sigma(t)$ has the meaning that it gives the current at time t in response to an impulsive unit E-field $E(t') = \delta(t')$. "Causality" is the statement that if $E(t)$ is zero until it is "turned on" at time $t = 0$, then $j(t)$ must also be zero for times $t<0$. Thus

$\sigma(t)$ must vanish for negative t, from which the Fourier relation Eq. (4) becomes $\sigma(\omega) = \int_0^\infty \mathrm{d}t\sigma(t)\exp(\mathrm{i}\omega t)$. Now let us regard $\sigma(\omega)$ as a function of a complex ω. When $\mathrm{Im}[\omega]>0$ (that is, in the upper half ω-plane) the integral converges absolutely, permitting interchange of derivative and integral from which follows $\partial\sigma_1/\partial\omega_1 = \partial\sigma_2/\partial\omega_2$ and $\partial\sigma_1/\partial\omega_2 = -\partial\sigma_2/\partial\omega_1$. These are the Cauchy–Riemann equations which confirm that $\sigma(\omega)$ is analytic in the upper-half ω-plane. This is the basis for the Kramers–Kronig integral relations between real and imaginary part of $\sigma(\omega)$ for real ω.

Consider the Drude formula [1], originally proposed on classical grounds, for the conductivity of metals:

$$\sigma_{\mathrm{Drude}}(\omega) = \frac{\mathrm{i}e^2(n/m)_{\mathrm{eff}}}{\omega + i/\tau} \tag{6}$$

The factor $(n/m)_{\mathrm{eff}}$ is discussed later, see Eqs. (26) and (55). For simple metals, $(n/m)_{\mathrm{eff}}$ is fairly close to n/m, where n is the valence electron density and m the free electron mass. The electron charge is $-e$. Unlike these parameters with fixed classical meaning, τ is a phenomenological parameter indicating the time between the collision events which allow electrons to reach a steady-state current. Note that $\sigma(\omega)$ has a pole at $\omega = -i/\tau$, that is, in the lower part of the complex ω-plane. By Fourier inversion we find

$$\sigma_{\mathrm{Drude}}(t) = e^2\left(\frac{n}{m}\right)_{\mathrm{eff}} e^{-t/\tau}\theta(t) \tag{7}$$

where $\theta(t)$ is the unit step function, zero for $t<0$ and 1 for $t>0$. The real part of $\sigma(\omega)$ is a Lorentzian, with conserved spectral weight

$$\int_{-\infty}^{\infty} \mathrm{d}\omega\sigma^{(1)}_{\mathrm{Drude}}(\omega) = \pi e^2\left(\frac{n}{m}\right)_{\mathrm{eff}} \quad \text{but} \quad \int_{-\infty}^{\infty} \mathrm{d}\omega\sigma^{(1)}_{\mathrm{exact}}(\omega) = \pi e^2\left(\frac{n_{\mathrm{total}}}{m}\right) \tag{8}$$

Both of these are unaffected by the phenomenological parameter τ. The second version, for the *exact* σ, is the "f-sum rule." It is equivalent to the statement that $\sigma(t=0^+)$, which is the current just after a unit pulse $E(t) = \delta(t)$ has been applied, is given by the classical constant $n_{\mathrm{total}}e^2/m$, where n_{total} includes core as well as valence electrons, and the integral includes ultraviolet and x-ray regions of the spectrum. In the limit of an infinitely sharp pulse and zero elapsed time, electrons have not yet responded to anything other than the E-field. They do not notice the positive nuclei or the Coulomb repulsion with each other. Thus the sum rule is true for any system, molecule or extended, insulator or metal, crystal or glass.

Now consider the experimental results [6] for copper, the prototype ordinary metal, shown in Fig. 1. The dielectric "constant" $\varepsilon(\omega)$ is defined by $\varepsilon = 1 + 4\pi P/E$ where the polarization P is related to current j by $j = \partial P/\partial t$. Therefore, we have $\varepsilon(\omega) = 1 + 4\pi\mathrm{i}\sigma(\omega)/\omega$, and $\varepsilon^{(2)} = 4\pi\sigma^{(1)}/\omega$. The Drude Lorentzian, centered at $\omega = 0$, has a strength measured in terms of the "Drude plasma frequency" $\omega_{\mathrm{P}}^2 = 4\pi e^2(n/m)_{\mathrm{eff}}$. Theory [7] gives $\hbar\omega_{\mathrm{P}} = 9.1\,\mathrm{eV}$. This can be compared with the free electron value, 10.8 eV, which is based on a simplified model of 1 rather than 11 valence electrons per Cu atom. The Drude part of $\sigma(\omega)$ becomes quite small for

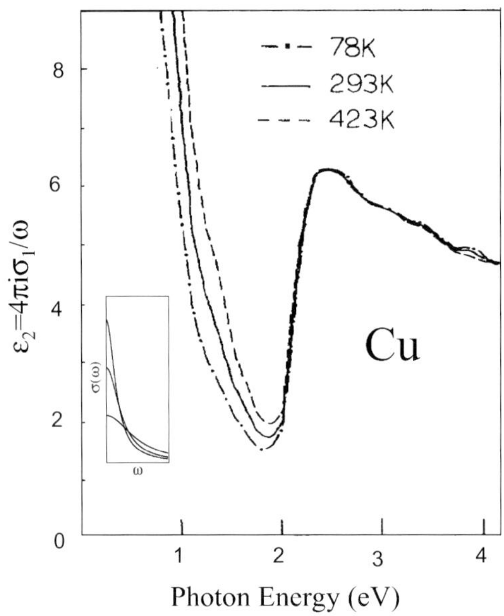

Fig. 1. Measured optical properties of Cu metal at three temperatures [6]. Interband transitions set in above 2 eV. At lower frequencies, the tail of the Drude part varies with temperature. The inset is a schematic of the real part of the Drude conductivity $\sigma(\omega)$, Eq. (6) at three temperatures.

photon energies $\hbar\omega$ as high as 2 eV. At higher photon energies, σ_1 increases, indicating additional non-Drude currents, which are caused by electrons making quantum transitions from d-states not far below the Fermi energy E_F into s-states above the Fermi energy, with a conserved wavevector **k**. The f-sum Eq. (8) is approximately divided into spectral weight in the Drude region below 2 eV, and spectral weight from interband transitions. As temperature T changes, the spectral weight in the Drude region is conserved. The formula for $(n/m)_{\text{eff}}$ needs a treatment of the energy bands, discussed in Section 4.2 from a quantum approach, and then discussed again from a semiclassical point of view in Section 6.2. Although the interband part of $\sigma(\omega)$ is normally classified as "optical properties" rather than "transport," nevertheless, there is no truly fundamental distinction, and understanding of low-ω transport requires some understanding of high-ω behavior.

3. CONDUCTANCE *VERSUS* CONDUCTIVITY: THE POINT CONTACT

For homogeneous bulk matter, the electrical properties are characterized by the intrinsic conductivity $\sigma = 1/\rho$ of the material. For other cases, conductance $G = 1/R$ is the appropriate parameter. Figure 2 illustrates the reason.

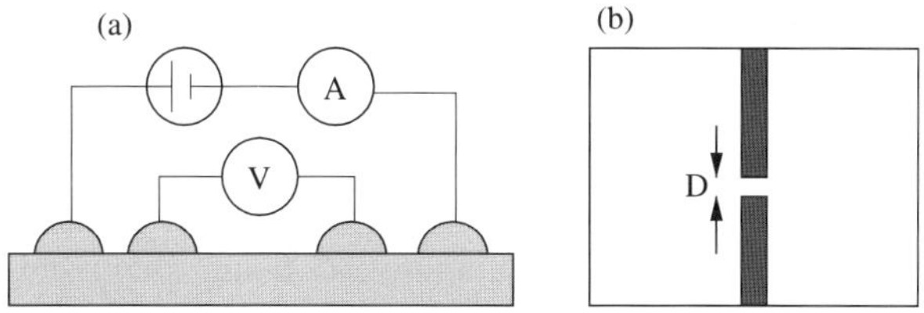

Fig. 2. (a) The usual four probe geometry for measuring resistivity of a bulk material. The voltmeter draws negligible current and thus does not disturb the electrical potential distribution. (b) An idealized orifice of diameter D in an otherwise bulk sample. For D small compared with the sample dimensions, the resistance is dominated by the alteration of electrical potential in the vicinity of the orifice. This is a model for a "point contact."

In the usual bulk case, the conductance is $G = \sigma A/L$ where σ is an intrinsic property, A the cross-sectional area of the conductor, and L the length. But in panel (b) of Fig. 2 (a model for a "point contact"), the conductance is limited by the orifice of diameter D. The spatial variation of the potential $V(\mathbf{r})$ must be solved self-consistently. Maxwell [8] solved this problem under the assumption of a local relation $\mathbf{j}(\mathbf{r}) = \sigma\mathbf{E}(\mathbf{r})$ between current and field, where σ is the conductivity of the homogeneous "bulk" material. He obtained $G_{\rm M} = D\sigma$. The resistance $R_{\rm M} = 1/D\sigma$ is dominated by the constriction; the remainder (the "electrodes") adds a series resistance $R = L/A\sigma \ll 1/D\sigma$) which vanishes in the limit of large dimensions $L \gg D$ and $A/L \gg D$. In such a case, the conductivity does not describe transport through the system and it is necessary to study conductance.

The local relation assumed by Maxwell holds when the mean electron free path ℓ (equal to $v_{\rm F}\tau$ where $v_{\rm F}$ is an average Fermi velocity) is small compared to the diameter D of the constriction. The problem becomes more interesting in the opposite limit. Because of the long mean-free path, the current at point $\mathbf{r}$ depends on the $\mathbf{E}$-field at points $\mathbf{r}'$ where the value of $\mathbf{E}$ is different. That is, current and field are related non-locally by $\mathbf{j}(\mathbf{r}) = \int d\mathbf{r}'\sigma(\mathbf{r}-\mathbf{r}')\mathbf{E}(\mathbf{r}')$ where the non-local conductivity $\sigma(\mathbf{r})$ has range ℓ. Sharvin [9] found when $\ell/D \gg 1$ the smaller conductance $G_{\rm S} = 1/R_{\rm S} = G_{\rm M}(3\pi D/16\ell)$. Sharvin's answer can be written

$$G = G_0[A_{\rm c}k_{\rm F}^2/4\pi] \quad \text{where } G_0 = 2e^2/h \tag{9}$$

where $A_{\rm c}$ is the area of the constriction. The factor $2e^2/h = G_0 = 0.775 \times 10^{-4}\,\Omega^{-1}$ is the "quantum unit of conductance," while the second factor is the number of quantum "channels" that can carry current through the orifice. Consider for example a square orifice of side d. The spacing of k-states (e.g. $\sin(k_x x)\sin(k_y y)$) whose nodes are on the boundaries of the orifice is π/d in the directions x and y in the plane of the orifice. Counting the number of such states which are occupied and lie in the quarter circle $|(k_x, k_y)| < k_{\rm F}$ yields the second factor $[A_{\rm c}k_F^2/4\pi]$. The crossover between the Maxwell and Sharvin limits was studied by Wexler [10] and Nikolic and Allen [11]. They found the interpolation formula,

$$R = R_{\rm S} + \gamma(D/\ell)R_{\rm M} \tag{10}$$

where the numerical correction $\gamma(x)$ is fitted by the Padé approximation $\gamma = (1 + 0.83x)/(1 + 1.33x)$.

Sharvin's answer is only valid if the number of channels is large. When the dimensions of the channel become comparable to the Fermi wavelength, so that the number of channels is of order one, the size quantization of wavefunctions transverse to the channel results in the conductance G being quantized in approximate integer units of G_0. This effect, shown in Fig. 3, was first seen by van Wees et al. [12,13] and Wharam et al. [14], using devices made from a "two-dimensional electron gas" (2deg). Long mean-free paths are available near interfaces in $GaAs/Al_{1-x}Ga_xAs$ "quantum well" structures. Metallic gate electrodes deposited above the 2deg provide a way to tune the width of a constriction, known as a "quantum point contact."

There is a very simple argument which gives the quantized conductance $G = nG_0$, and which will be used as a starting point for the discussion of the Landauer formula in the next chapter. Let the gate potential at the gate axis $y = 0$ of Fig. 3 be modeled as $V(x, y = 0) = V_0 + m\omega_0^2 x^2/2$. Near the constriction, the wavefunctions in effective mass approximation are $\exp(ik_y y)H_n(x)\exp(-x^2/2x_0^2)$, that is, propagating in the y direction and harmonically confined in the x direction. The energy levels are $eV_0 + (c + 1/2)\hbar\omega + \hbar k_y^2/2m^*$. The integer $c \geqslant 0$ is the "channel index." Inset (b) of Fig. 3 shows the occupied levels for a gate voltage near -1.8 V, where three sub-bands in the constriction are partly occupied. The inset also indicates that a small source-drain bias has been applied, $\mu_L - \mu_R = -eV_{SD}$, where L and R refer to $y > 0$ and $y < 0$. The current through the constriction is caused by ballistic propagation. We just have to count the imbalance between left- and

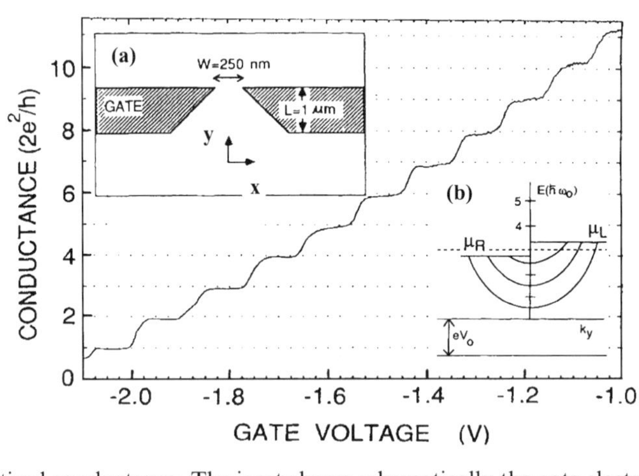

Fig. 3. Quantized conductance. The inset shows schematically the gate electrodes deposited in the insulating layer above a 2d electron gas (2deg, or GaAs quantum well). Varying the gate potential causes a variable width constriction for electrons of the 2deg. This device is called a "quantum point contact" (from van Wees et al. [12], [13]).

right-propagating states.

$$I = \frac{2}{L}\sum_{c}\sum_{k}(-ev_{kc})F(kc) \tag{11}$$

where the non-equilibrium occupancy is $F(kc) = f(\varepsilon_{kc} - \mu + \text{sign}(v_{kc})\Delta\mu/2)$ and $f(\varepsilon_{kc} - \mu)$ is the equilibrium Fermi–Dirac function of the state with longitudinal wavevector k and channel c. Left- and right-propagating states have Fermi levels shifted up and down by $\Delta\mu/2 = (\mu_L - \mu_R)/2$, and the factor of 2 is for spin degeneracy. For each channel, k is quantized in a fictitious box $-L/2 < k < L/2$ in the y direction with the orifice at the center. An occupied state transports charge $-e$ through the orifice in time L/v_{kc}. The box is long enough such that the spacing $\Delta k = 2\pi/L$ between k states is small. Convert the sum $\sum_k$ into an integral $(L/2\pi)\int \mathrm{d}k$, and convert the integration step $\mathrm{d}k$ into $\mathrm{d}\varepsilon_{kc}/\hbar v_{kc}$. The factors of velocity v_{kc} cancel, so each channel that intersects the Fermi level makes the same contribution. Thus we get

$$I = \frac{-2eN_c}{h}\int \mathrm{d}\varepsilon[f_L - f_R] = G_0 N_c V_{SD} \tag{12}$$

where N_c is the number of channels intersecting the Fermi level. Each occupied propagating sub-band gives a current $G_0 V_{SD}$ which increases ohmically with $|\mu_L - \mu_R|$. At gate voltages less than -2.2 V, the upper ($y>0$) and lower ($y<0$) halves of the 2deg are decoupled, and G is essentially zero. At higher gate voltages, more and more sub-bands become partly occupied, and the conductance rises in steps of height $G_0 = 2e^2/h$ each time a new sub-band dips below the Fermi level.

One may ask, what is the source of resistance $1/G$ in this case where there is no evident source of dissipation? Note that the conductance does not depend on the longitudinal size L which is the total path length of current flow, so that the dissipated heat per unit volume goes to zero in the large size limit. The non-zero value of $1/G$ should be considered a "contact resistance" between the narrow "channel" and the macroscopic electrodes. De Picciotto et al. [15] have verified that a 4-terminal resistance measurement on a ballistic quantum wire gives an inherent resistance of zero, so the quantum resistance $1/N_C G_0$ is safely assigned to the contacts which were outside their voltage probes.

The details of coherent quantum flow in a ballistic quantum point contact were imaged by Topinka et al. [16], giving a most beautiful visual verification of the coherent electron states in the different channels.

The point contact turns out to be a useful probe, as Sharvin anticipated. In common with other junction devices, it permits a significant bias ΔV across a short channel, and thus a much larger **E**-field than can be achieved in a bulk conductor. Further, the small transverse dimension makes less stringent demands on microfabrication than does a larger area "planar" junction. Section 5 gives an example.

4. KUBO AND OTHER FORMULAS

In this chapter, some of the basic formulas [17] of transport theory are summarized and discussed.

4.1. Kubo Formulas

We consider an applied field $\mathbf{E} = -(1/c)\partial\mathbf{A}/\partial t$, with monochromatic frequency-dependence $\mathbf{E} = (\mathrm{i}\omega/c)\mathbf{A}$. The current operator is

$$\hat{j}_{\mathrm{tot},\alpha} = (ine^2/m\omega)E_\alpha + \hat{j}_\alpha,$$
$$\hat{j}_\alpha = -\frac{e}{m}\sum_i p_{i\alpha} \tag{13}$$

where $\mathbf{p}_i$ is the momentum operator for the ith electron. The first term of $\hat{j}_{\mathrm{tot}}$ is the "diamagnetic" part related to the substitution $\mathbf{p}_i \to \mathbf{p}_i + e\mathbf{A}(\mathbf{r}_i, t)/c$. The Hamiltonian of the system is

$$H_{\mathrm{tot}} = H + H_{\mathrm{ext}},$$
$$H_{\mathrm{ext}} = -\hat{\mathbf{j}}\cdot\mathbf{A}/c = (i/\omega)\hat{\mathbf{j}}\cdot\mathbf{E} \tag{14}$$

where the "unperturbed" part H contains all the equilibrium properties of the interacting system. We want to calculate the current $\mathrm{tr}\hat{\rho}_{\mathrm{tot}}\hat{j}_{\mathrm{tot},\alpha}$ to first order in $\mathbf{E}$. The density matrix $\hat{\rho}_{\mathrm{tot}}$ is perturbed away from its equilibrium value $\rho = \exp(-\beta H)/Z$ by the field, and we need to know the correction $\delta\hat{\rho}$ to first order in $\mathbf{E}$. First-order time-dependent perturbation theory for $\delta\hat{\rho}$ can be written in operator language using standard field-theoretic methods [18,19] as

$$\delta\hat{\rho}(t) = -\frac{i}{\hbar}\mathrm{e}^{-\mathrm{i}Ht/\hbar}\int_{-\infty}^{t}\mathrm{d}t'[H_{\mathrm{ext}}(t'), \hat{\rho}]\mathrm{e}^{\mathrm{i}Ht/\hbar} \tag{15}$$

where the time dependence of $H_{\mathrm{ext}}(t')$ has two sources, first the fixed time dependence $\exp(-\mathrm{i}\omega t')$ of the classical field $\mathbf{A}(t)$, and second, the Heisenberg time-dependence, $\hat{j}(t) = \exp(\mathrm{i}Ht/\hbar)\hat{j}\exp(-\mathrm{i}Ht/\hbar)$ assigned to operators. The frequency ω is assigned a small positive imaginary part $\omega \to \omega + \mathrm{i}\eta$, which has the effect that at time $t = -\infty$, the perturbation vanishes and the density matrix has its equilibrium value. The Kubo formula [17–19] for the conductivity then follows by simple manipulation. It has at least two important versions,

$$\sigma_{\alpha\beta}(\omega) = \frac{ine^2}{m\omega}\delta_{\alpha\beta} + \frac{1}{\hbar\omega V}\int_0^\infty \mathrm{d}t\mathrm{e}^{\mathrm{i}\omega t}\langle[\hat{j}_\alpha(t), \hat{j}_\beta(0)]\rangle \tag{16}$$

$$\sigma_{\alpha\beta}(\omega) = \frac{1}{V}\int_0^\beta \mathrm{d}\lambda\int_0^\infty \mathrm{d}t\mathrm{e}^{\mathrm{i}\omega t}\langle\hat{j}_\alpha(-\mathrm{i}\hbar\lambda)\hat{j}_\beta(t)\rangle \tag{17}$$

The angular brackets $\langle\hat{Q}\rangle \equiv \mathrm{tr}(\hat{\rho}\hat{Q})$ mean a canonical ensemble average using the equilibrium density matrix. The first version, Eq. (16), has the usual form [18] from linear response theory of a "retarded" commutator correlation function. The word "retarded" refers to the fact that the commutator is set to zero for $t<0$, as is appropriate for a causal or retarded response. The first term of this expression, which comes from the diamagnetic part of the current operator, is singular, diverging as $\omega \to 0$. This divergence cancels against pieces from the second term.

The second version, Eq. (17) gets rid of the commutator and the singular first term, at the cost of introducing an imaginary time $-\mathrm{i}\hbar\lambda$. This is treated like a real time, with the replacement $t \to -\mathrm{i}\hbar\lambda$ in the Heisenberg time evolution. It can be derived from Eq. (16) using an operator identity for the density matrix,

$$[\hat{Q}, \hat{\rho}] = -\mathrm{i}\hbar\hat{\rho}\int_0^\beta \mathrm{d}\lambda \dot{\hat{Q}}(-\mathrm{i}\hbar\lambda) \tag{18}$$

where the time derivative of $\hat{Q}$ is $(i/\hbar)[H, \hat{Q}]$.

To analyze these formulas, let us pretend that we know a complete set of many-body eigenstates $H|n\rangle = E_n|n\rangle$ of the system before the field is applied. Then we get from Eqs. (16) and (17) the "spectral representations"

$$\sigma_{\alpha\beta}(\omega) = \frac{ine^2}{m\omega}\delta_{\alpha\beta} + \frac{i}{\omega V}\sum_{mn}\frac{\mathrm{e}^{-\beta E_n} - \mathrm{e}^{-\beta E_m}}{Z}\frac{\langle n|j_\alpha|m\rangle\langle m|j_\beta|n\rangle}{\hbar(\omega + \mathrm{i}\eta) - (E_m - E_n)} \tag{19}$$

$$\sigma_{\alpha\beta}(\omega) = \frac{\mathrm{i}\hbar}{V}\sum_{mn}\frac{\mathrm{e}^{-\beta E_n} - \mathrm{e}^{-\beta E_m}}{Z(E_m - E_n)}\frac{\langle n|j_\beta|m\rangle\langle m|j_\alpha|n\rangle}{\hbar(\omega + \mathrm{i}\eta) - (E_m - E_n)} \tag{20}$$

where the positive infinitesimal η has been added ($\omega \to \omega + \mathrm{i}\eta$) to make the time integrals well defined. Because the denominators have imaginary parts which are delta functions, it is easy to show that both expressions have the same real part. This real part can be written in spectral representation, and then resummed as a correlation function,

$$\mathrm{Re}\,\sigma_{\alpha\beta}(\omega) = \frac{\pi}{\omega V}\sum_{mn}\frac{\mathrm{e}^{-\beta E_n} - \mathrm{e}^{-\beta E_m}}{Z}\langle n|j_\alpha|m\rangle\langle m|j_\beta|n\rangle\delta(\hbar\omega - (E_m - E_n)) \tag{21}$$

$$\mathrm{Re}\,\sigma_{\alpha\beta}(\omega) = \frac{1 - \mathrm{e}^{-\beta\hbar\omega}}{2\hbar\omega V}\int_{-\infty}^{\infty}\mathrm{d}t\mathrm{e}^{\mathrm{i}\omega t}\langle j_\alpha(t)j_\beta(0)\rangle \tag{22}$$

This last version of the Kubo formula is known as the "fluctuation-dissipation" theorem, because it relates the random current fluctuations of the equilibrium system $\langle j(t)j(0)\rangle$ to the dissipative part $\mathrm{Re}\sigma$ of the conductivity. The spectral version Eq. (21) is the same as the Fermi golden rule for the rate of absorption of energy by the system from the electromagnetic field *via* the perturbation $H_{\mathrm{ext}} = -\mathbf{j}\cdot\mathbf{A}/c$. The Kubo formulas Eqs. (16) and (17) are the results of lowest-order time-dependent perturbation theory in powers of the external field. This formula is the starting point for many-body perturbation theory using the response of a non-interacting system as a reference.

4.2. Kubo–Greenwood Formula

Often materials are successfully approximated as non-interacting Fermi systems. The Hamiltonian is the sum $\sum_i H_0(i)$ of identical single-particle Hamiltonians for each electron. The single-particle eigenstates $H_0|n\rangle = \varepsilon_n|n\rangle$ are a convenient basis, and the current operator can be written as

$$\hat{j}_x(t) = \sum_{nm} \langle n|j_x|m\rangle c_n^\dagger c_m \mathrm{e}^{\mathrm{i}(\varepsilon_n - \varepsilon_m)t/\hbar} \tag{23}$$

For an interacting system the same basis is used, and the same operator expression works at $t = 0$, but the Heisenberg time dependence of the current operator can no longer be written out explicitly. To evaluate the Kubo formulas, Eqs. (16) and (17), in non-interacting approximation Eq. (23), we need

$$\begin{aligned} \mathrm{tr}\hat{\rho}([c_n^\dagger c_m, c_p^\dagger c_q]) &= \delta_{mp}\delta_{nq}(f_n - f_m) \\ \mathrm{tr}\hat{\rho}(c_n^\dagger c_m c_p^\dagger c_q) &= \delta_{mp}\delta_{nq} f_n(1 - f_m) + \delta_{nm}\delta_{pq} f_n f_p \end{aligned} \tag{24}$$

Here, $f_n = (\exp(\beta\varepsilon_n) + 1)^{-1}$ is the Fermi–Dirac occupation factor. Either procedure leads to

$$\sigma_{\alpha\beta}(\omega) = \frac{\mathrm{i}\hbar}{V} \sum_{mn} \left(\frac{f_n - f_m}{\varepsilon_m - \varepsilon_n}\right) \frac{\langle n|j_\alpha|m\rangle\langle m|j_\beta|n\rangle}{\hbar(\omega + \mathrm{i}\eta) + (\varepsilon_n - \varepsilon_m)} \tag{25}$$

To get this result from Eq. (19) requires separating denominators $1/\hbar\omega(\hbar\omega + \Delta)$ into $1/\Delta$ times $1/\Delta - 1/(\hbar\omega + \Delta)$ and then cancelling the first term against the diamagnetic part of Eq. (16) by use of operator relations $p_\alpha = (im/\hbar)[H, r_\alpha]$ and $[r_\alpha, p_\beta] = \mathrm{i}\hbar\delta_{\alpha\beta}$.

First, let us interpret Eq. (25) for the case of a perfect crystal. The single particle quantum number n is now $(\mathbf{k}n\sigma)$, the current matrix elements are $\mathbf{k}$-diagonal, and the sum over n, m goes over transitions between filled and empty band states. For a semiconductor or insulator, this is the usual band-theoretic optical interband conductivity. For a metal, we have additional diagonal elements $\langle n|j_\alpha|n\rangle$ which equal $-ev_{\mathbf{k}n\alpha}$. The $n = m$ diagonal part of $(f_n - f_m)/(\varepsilon_m - \varepsilon_n)$ must be interpreted as $-\partial f(\varepsilon_{\mathbf{k}n})/\partial\varepsilon_{\mathbf{k}n}$. This can be understood by remembering that the electric field in an optical experiment is not strictly homogeneous but has a wavevector $\mathbf{q}$ with $q = \omega/c$ small on the scale of the Brillouin zone. The current matrix elements are then not diagonal, but defined between states of wavevector $\mathbf{k}$ and $\mathbf{k} + \mathbf{q}$. The limit $q \to 0$ goes smoothly. The resulting intraband piece of Eq. (25) is

$$\sigma_{\mathrm{intraband},\alpha\beta}(\omega) = \left(\frac{\mathrm{i}e^2}{\omega + \mathrm{i}\eta}\right)\left[\frac{1}{V}\sum_{\mathbf{k}n}\left(-\frac{\partial f}{\partial\varepsilon_{\mathbf{k}n}}\right) v_{\mathbf{k}n\alpha} v_{\mathbf{k}n\beta}\right] \tag{26}$$

This is exactly the Drude result, Eq. (6), except that the true collision rate $1/\tau$ is replaced by the infinitesimal η, reflecting the absence of collisions in the non-interacting perfect crystal. The factor in square brackets above is the Drude $(n/m)_{\mathrm{eff}}$, written correctly as a tensor which becomes a scalar for high-symmetry solids.

To really derive the Drude conductivity including collisions from the Kubo formula, the correct route is through a Boltzmann equation. There are convincing derivations of the linearized Boltzmann equation starting from the Kubo formula [20–22]. The derivations are quite tedious. Amazingly, the correct (and not linearized) Boltzmann equation was guessed by Bloch long before the advanced perturbative techniques were available. The reasons for Bloch's success were clarified by Landau. Solving the Bloch–Boltzmann equation in all generality is also not easy. Simple approximations work well much of the time, and allow an easy alternate derivation of the Drude $(n/m)_{\rm eff}$. This will be shown in Section 6.2, especially Eq. (55) which gives further interpretation.

The Boltzmann approach works only if the crystal is pure enough, and has weak enough interactions, that the wavevector **k** is a fairly good quantum number. This fails in dirty alloys, yet a single-particle approximation may still be good. Therefore, Eq. (25) has a range application outside of Boltzmann theory. This is discussed in Section 8. Eq. (25) is often rewritten in terms of the single-particle Green's function

$$\hat{G}^{+}(E) = \frac{1}{E + i\eta - \hat{H}_0} \qquad \hat{G}^{-}(E) = \frac{1}{E - i\eta - \hat{H}_0}$$
$$\hat{G}^{\times}(E) = \frac{1}{2\pi i}(\hat{G}^{+} - \hat{G}^{-}) = \sum_n |n\rangle\delta(E - \varepsilon_n)\langle n|$$
$$\hat{G}^{\times}(x, x', E) = \langle x|\hat{G}^{\times}(E)|x'\rangle = \sum_n \psi_n(x)\psi_n^*(x')\delta(E - \varepsilon_n) \tag{27}$$

The notation is that of Ziman [23]. The label x denotes $(\mathbf{r}\sigma)$, the space and spin coordinates of the electron. The non-interacting Kubo formula in the dc limit can be written as

$$\sigma_{\alpha\beta} = \frac{\pi e^2 \hbar}{m^2 V}\int {\rm d}x\,{\rm d}x' \left(\frac{\hbar}{i}\nabla_\alpha G^{\times}(x, x', \varepsilon_{\rm F})\right)\left(\frac{\hbar}{i}\nabla'_\beta G^{\times}(x', x, \varepsilon_{\rm F})\right) \tag{28}$$

where $\nabla_\alpha = \partial/\partial r_\alpha$ and $\nabla'_\beta = \partial/\partial r'_\beta$. This is called the "Kubo–Greenwood" formula [17,24]. It is not any different from the previous version Eq. (25), but is useful if the disorder is treated perturbatively. One then averages over an ensemble of different representatives of the disorder. The correct procedure is to average the product of the two Green's functions. This is not the same as multiplying two separately averaged Green's functions. The latter procedure, for example, does not permit Anderson-insulating behavior in very disordered electron systems. The reason is that the averaged G loses off-diagonal information and cannot distinguish localized from delocalized states.

4.3. Conductance as Transmission

When coherent electron transmission plays a role, Kubo and Boltzmann formulations become less appropriate than an analysis starting from scattering theory [25–29]. Consider the system shown schematically in Fig. 4. On the left and right are

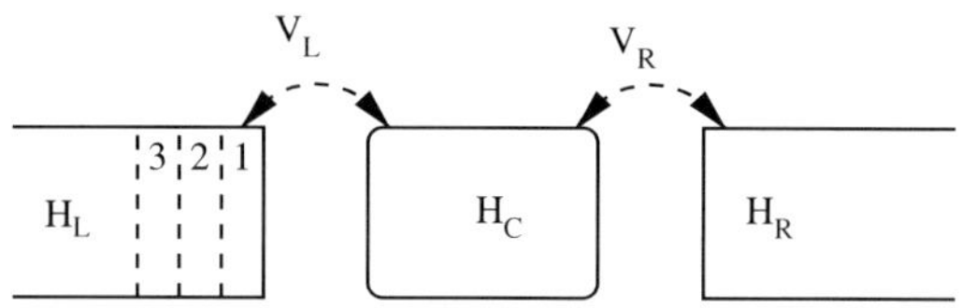

Fig. 4. Schematic diagram of a system with two ideal leads on left and right, modeled as semi-infinite and described by Hamiltonians H_L and H_R. In the center is an island which is coupled to the leads (perhaps weakly or perhaps strongly) by coupling matrix elements V_L and V_R. The material in the center (described by H_C) can be an oxide barrier, vacuum, a disordered metal, or a molecule.

perfect metals regarded as electrodes. The central section may have disorder, or an oxide barrier, or a vacuum break (modeling a scanning tunneling microscope, for example), or perhaps a break which has been filled with some molecular bridge material. The left and right ballistic leads each have their own chemical potential, μ_L and μ_R, set by distant reservoirs which are not modeled explicitly. Choosing each electrode sequentially as a source of incident waves, the Schrödinger equation is solved for the amplitude of transmission into the other electrode (or electrodes in a multi-terminal device). We may not wish to average over an ensemble of related systems, but instead might wish to study the interesting idiosycracies of some particular system.

The scattering state wavefunction for a state incident from the left is

$$\begin{aligned} |\psi_{knL}\rangle &= |knL\rangle + \sum_m r(k)_{mn}|-kmL\rangle \quad \text{on the left} \\ &= \sum_m t(k)_{nm}|kmR\rangle \quad \text{on the right} \end{aligned} \tag{29}$$

where r and t are reflection and transmission coefficients connecting the incident state $|knL\rangle$ to reflected $|-kmL\rangle$ and transmitted $|kmR\rangle$. Here k is the 1D wavevector for propagation along the electrode; it is not necessarily the same on left (L) and right (R). Each electrode has possibly many "channels" or tranverse quantum states n, m near the Fermi level. At fixed energy each channel has a different k. When left and right electrodes are inequivalent, the states $|knL,R\rangle$ are normalized with a factor of $\sqrt{(v_{knL,R})}$ so that unitary transmission matrices t_{mn} are required by current conservation. The conductance of this system was already discussed in Section 3 for the ballistic case $t = \delta_{mn}$ and $r = 0$. When there is non-zero scattering, Eq. (12) generalizes to [25,26]

$$I = \frac{2e}{h}\int d\varepsilon[f_L(\varepsilon) - f_R(\varepsilon)]\mathrm{tr}[t^\dagger t(\varepsilon)] \tag{30}$$

This is a Landauer formula [25,29]. In the Ohmic limit $I = GV$ we get

$$G = G_0\mathrm{tr}(t^\dagger t) \tag{31}$$

The 2 is for spin, and the Fermi–Dirac functions $f_{L,R}(\varepsilon)$ are $(\exp(\beta(\varepsilon - \mu_{L,R}) + 1)^{-1}$. Equation (31) follows from (30) in the limit where the source-drain bias

$V_{SD} = (\mu_L - \mu_R)/e$ is smaller than the characteristic energies of the leads or of the system between the leads. Equation (30) goes beyond the Kubo approach in allowing bias voltages outside the regime of linear response. However, inelastic scattering effects do not easily get incorporated.

There is a more general approach using non-equilibrium Green's function (NEGF) theory, discussed by Meir and Wingreen [30,31], which in principle permits a full treatment of interactions and inelastic scattering, but is not easy to implement. In the non-interacting limit, a result is found equivalent to Eq. (30), but more convenient for computation. The derivation was first given by Caroli et al. [31]. The total Hamiltonian is written

$$H_{tot} = H_L + [V_L + H_C + V_R] + H_R \tag{32}$$

Whereas the scattering method treats the three terms in square brackets as one scattering potential, now we think separately of the solutions to the three pieces described by H_L, H_R, and H_C. Long-range interactions between different subsystems are omitted. The parts of the Hamiltonian are usually expressed in a basis of single-particle local orbitals. The single-particle Schrödinger equation in schematic form is

$$\begin{pmatrix} H_L - \varepsilon & V_L & 0 \\ V_L^\dagger & H_C - \varepsilon & V_R \\ 0 & V_R^\dagger & H_R - \varepsilon \end{pmatrix} \begin{pmatrix} \psi_L \\ \psi_C \\ \psi_R \end{pmatrix} = 0 \tag{33}$$

Let us suppose that we can find the separate "partial" Green's functions

$$\begin{aligned} g_L(\varepsilon) &= (\varepsilon - H_L)^{-1} \\ g_C(\varepsilon) &= (\varepsilon - H_C)^{-1} \\ g_R(\varepsilon) &= (\varepsilon - H_R)^{-1} \end{aligned} \tag{34}$$

For the central region, perhaps the system is small enough that we can solve $H_C|C\rangle = E_C|C\rangle$ exactly by some method. The L and R electrodes are semi-infinite, and almost periodic except that the termination at the central region destroys translational invariance. We prefer not to find explicit solutions $H_L|L\rangle = E_L|L\rangle$ and $H_R|R\rangle = E_R|R\rangle$. It turns out that the partial Green's function $g_L(x, x')$ is only needed for x, x' lying in the surface region labeled 1 on Fig. 4, and similarly for g_R. If the surface part 1 couples by V_{12} to layer 2, and is decoupled from other interior layers, and if identical matrices V_{23}, etc. couple the other layers, then there is a closed equation which can be solved iteratively, with $g_L^{(n)}$ approaching g as $n \to \infty$, namely

$$g_L^{(n+1)} = (\varepsilon - h_1 - V_{12}^\dagger g_L^{(n)} V_{12})^{-1} \tag{35}$$

where h_1 is the part of H_L confined to layer 1, assumed identical in form to h_2, etc. There is an alternate method to calculate g_L due to Kalkstein and Soven [32]. Therefore all three partial Green's functions may be calculable. They can be used to find the transmission and the current

$$I = \frac{4\pi e}{\hbar}\int_{-\infty}^{\infty} \mathrm{d}\varepsilon \sum_{LR} |\langle L|\hat{T}|R\rangle|^2 \delta(\varepsilon - E_L)\delta(\varepsilon - E_R)[f_L(\varepsilon) - f_R(\varepsilon)] \tag{36}$$

without explicitly finding $|L\rangle$, or E_L, etc. The transmission matrix $\hat{T}$ is given by the Lippman–Schwinger equation,

$$\hat{T} = (V_L + V_R) + (V_L + V_R)^{\dagger} G(\varepsilon)(V_L + V_R) \tag{37}$$

where $G = (\varepsilon - H)^{-1}$ is the Green's function of the whole system. We assume the central region large enough that there are no direct couplings, $0 = \langle L|V_L + V_R|R\rangle$. Therefore the off-diagonal parts of the $\hat{T}$ matrix are

$$\langle L|\hat{T}|R\rangle = \sum_{MM'} \langle L|V_L^{\dagger}|M\rangle\langle M|G|M'\rangle\langle M'|V_R|R\rangle \tag{38}$$

From this we see that the system's Green's function matrix $G(x, x')$ is only needed for x, x' in the central region where the eigenstates are $|M\rangle, |M'\rangle$. Denoting this submatrix as G_C, and doing some matrix algebra, we find

$$G_C(\varepsilon) = (\varepsilon - H_C - \Sigma_L - \Sigma_R)^{-1} \tag{39}$$

where the self-energies $\Sigma_{L,R}$ contain the shift and broadening of the states of the central region which come from coupling to the leads,

$$\Sigma_L(\varepsilon) = V_L^{\dagger} g_L(\varepsilon) V_L \quad \Sigma_R(\varepsilon) = V_R^{\dagger} g_R(\varepsilon) V_R \tag{40}$$

Now we are ready to rewrite Eq. (36) in terms of this central part of the Green's function. The self-energies $\Sigma_{L,R}$ have imaginary parts

$$\begin{aligned} \langle N|\Gamma_L|M\rangle &= -2\mathrm{Im}\langle N|\Sigma_L(\varepsilon + \mathrm{i}\eta)|M\rangle \\ &= 2\pi \sum_L \delta(\varepsilon - E_L)\langle N|V_L|L\rangle\langle L|V_L^{\dagger}|M\rangle \end{aligned} \tag{41}$$

and similarly for Γ_R. Now we can express the current as

$$I = \frac{2e}{h}\int_{-\infty}^{\infty} \mathrm{d}\varepsilon \,\mathrm{tr}[\Gamma_L G_C(\varepsilon - \mathrm{i}\eta)\Gamma_R G_C(\varepsilon + \mathrm{i}\eta)][f_L(\varepsilon) - f_R(\varepsilon)] \tag{42}$$

where the trace goes over states of the central region only.

This equation is widely used to calculate the conductance $\mathrm{d}I/\mathrm{d}V_{\mathrm{SD}}$ both in the linear regime of small source-drain bias and beyond the linear regime, for small molecules or other small systems. For example, "exact" single-particle eigenstates of a model dirty alloy may be found for small samples, of order 100 atoms. Calculation of resistivity by direct application of the Kubo–Greenwood formula has some difficulties because of the discreteness of the spectrum. These difficulties are nicely smoothed over if the small system is attached to two ideal leads and Eq. (42) is used [33–35]. For molecular applications, theory and experiment for the non-linear $G(V)$ curves do not often agree closely. The subject is under rapid development [36–40].

5. SUPERCURRENT AND ANDREEV REFLECTION

Supercurrent j_S is a collective motion of the superconducting condensate, Eq. (2). The order parameter ψ is related to the pair amplitude $\langle c_{k\uparrow}c_{-k\downarrow}\rangle$, which gives the occupancy of a Cooper pair state built from the time-reversed pair of electrons $(k\uparrow, -k\downarrow)$. The pair amplitude acquires a current-carrying phase factor $\exp(i\mathbf{q}\cdot\mathbf{r})$ if all states $k'\uparrow$ and $-k'\downarrow$ are shifted by $\mathbf{q}/2$. Then the current is $\mathbf{j}_S = -2en_S\hbar\mathbf{q}/m$.

Andreev [41] explained how a current in a normal metal (N) can flow across the N/S boundary to the superconducting (S) side. An incident N quasiparticle $k\uparrow$ with current $-e\mathbf{v}_k$ can convert directly (with some amplitude for reflection as well as transmission) into a current-carrying S excited quasiparticle, but only if the N energy $\varepsilon_k - \mu$ exceeds the S gap Δ. If $|\varepsilon_k - \mu| < \Delta$, this is not possible, and the only route is for the incident N electron to bind to another N electron $-k\downarrow$, entering the S side as a Cooper pair. After this event, the N side lacks the $-k\downarrow$ electron, which means it has a hole with charge $+|e|$ and velocity $\mathbf{v}_{-k} = -\mathbf{v}_k$. This hole continues to carry N current $-e\mathbf{v}_k$, the same as the N current before the $k\uparrow$ state entered S. This process, called "Andreev reflection," doubles the current.

An elegant illustration [42] is shown in Fig. 5. A superconducting Nb wire was sharpened to make a point contact and carefully contacted to various clean normal metal surfaces (paramagnetic Cu and various ferromagnets). The conditions are not always reproducible, and only selected data were used for analysis. The selected

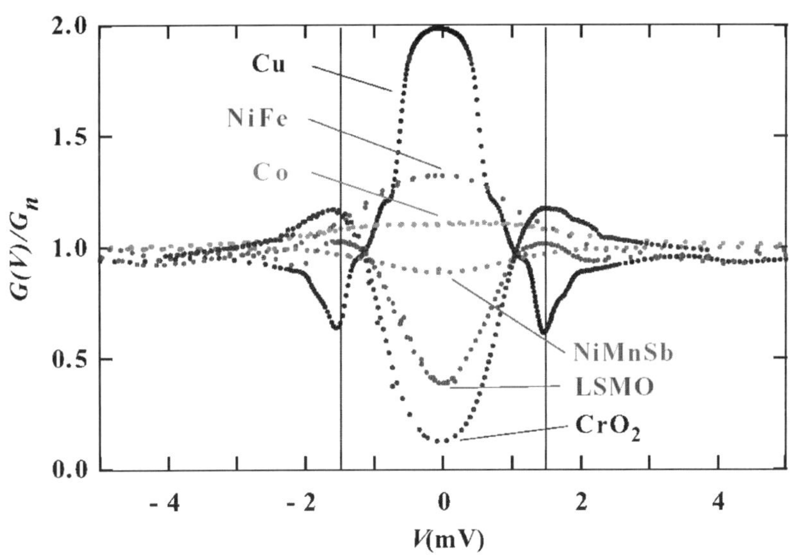

Fig. 5. Differential conductance dI/dV versus source-drain bias in a superconductor/normal point contact [42]. A superconducting Nb point is contacted to various metals. Cu shows the effect of Andreev reflection doubling the conductance at low bias, whereas the ferromagnetic samples with reduced minority spin population show suppression of the Andreev process.

data are believed to satisfy the criteria for Andreev reflection, cleanliness and transparency of the point contact. When the bias voltage V across the point contact is $\pm\Delta/e$ (the gap Δ of Nb is 1.5 meV, slightly higher than $3.52k_B T_c/2$), structure is seen in the differential conductance $G = dI/dV$, arising from the BCS peak $1/\sqrt{E^2 - \Delta^2}$ in the quasiparticle density of states. At larger biases, $V > 2\Delta/e$, G is independent of bias and approaches the conductance of the normal metal N. At low biases, $|V| < \Delta/2e$, the conductance doubles in the Cu point contact as expected from Andreev reflection [43]. But CrO_2 is different in that there are no down-spin electrons near the Fermi level. It is a "half-metallic" ferromagnet, meaning that the minority spins have an insulating gap while the majority spins are metallic. Therefore, in CrO_2, no Andreev process is possible, and the conductance is essentially zero at small bias. Other ferromagnets, which have reduced but non-zero numbers of minority carriers at the Fermi level, show an intermediate effect. Analysis using the theory of Blonder et al. [43] indicates a point contact area $\approx 10^4\,\text{Å}^2$, large compared to the area $< 100\,\text{Å}^2$ where quantized conductance is expected as in Fig. 3.

This experiment illustrates an important theme of present-day nanoscale physics. Use of nanoscale devices can enormously enhance the view of phenomena which are also present in bulk but hard to access experimentally. The cost is that data may have to be selectively sorted based on theoretically inspired criteria, since control over fabrication is so primitive. This enlarges the opportunity for ill-considered claims (not illustrated here!) or outright fraud [44].

6. BLOCH–BOLTZMANN THEORY

Transport theory of solids began with Bloch's [4] thesis of 1928 which explained metallic resistivity. Landau [5] clarified the meaning of Bloch's work. Around 1962 theoretical tools improved to the point that the rigorous basis for Bloch's ideas became clear [20–22]. Since around 1980 [45,46] it has been possible to compute with Bloch's theory for metals with non-trivial band structures, which fully tests the theory. It has also been possible to go outside the validity of Bloch's theory to calculate the resistivity of dirty alloys, liquid metals, and amorphous metals, using one-electron theory and neglecting inelastic scattering.

Following Landau [5], assume that there exist single-particle-like electron excitations, and that the occupation function or distribution function $F(k, \mathbf{r}, t)$ exists. F is an ensemble average non-equilibrium distribution. For large samples with homogeneous fields, the volume average equals the ensemble average, or the system is self-averaging. In small samples at low T, interesting deviations from self-averaging can be seen, requiring a more complete theory. Two effects cause F to evolve with t. First, scattering causes discontinuous changes of the quantum number k at some statistical rate. Second, there is smooth evolution arising from drift and acceleration of quasiparticles. Ignoring collisions, at a later time $t + \Delta t$, the new distribution $F(k, \mathbf{r}, t + \Delta t)$ will be the old distribution $F(k - \dot{k}\Delta t, \mathbf{r} - \dot{\mathbf{r}}\Delta t, t)$. This is expressed by the Boltzmann equation

$$\frac{\partial F}{\partial t} + \dot{\mathbf{k}} \cdot \frac{\partial F}{\partial \mathbf{k}} + \dot{\mathbf{r}} \cdot \frac{\partial F}{\partial \mathbf{r}} = \left.\frac{\partial F}{\partial t}\right|_{\text{coll}} \tag{43}$$

The left-hand side is the "co-moving" time derivative and the right-hand side takes account of collisions. Bloch [4] identified $\dot{\mathbf{r}}$ with $\mathbf{v}_k$ and $\hbar\dot{\mathbf{k}}$ with the force $-e(\mathbf{E} + \mathbf{v}_k \times \mathbf{B}/c)$ of applied fields. The collision term $(\partial F/\partial t)_{\text{coll}}$ was constructed using the probabilistic rules of t-dependent perturbation theory, and requires knowing the occupancies $F(k')$ of all other states k'. The theory is amazingly powerful and surprisingly accurate for a wide class of materials. It rests on assumptions of wide validity, although their truth is far from evident. The most fundamental underlying assumption was explained by Landau [5] – the assumption of the existence of "quasiparticles." A quasiparticle is an approximate excited state with charge $\pm|e|$, spin $\pm 1/2$, and a reasonably sharp wavevector $k \pm 1/\ell_k$ and energy $\varepsilon_k \pm \hbar/\tau_k$, where $\ell_k = v_k \tau_k$ is the mean free path and τ_k the time interval before the excitation loses its sharp definition in energy or momentum. In Section 6.6, the issue is discussed of whether band theory gives correct quasiparticle energies.

6.1. *Technical Definition of Quasiparticle*

The "temperature Green's function" $G(k,\sigma) = -\langle \hat{T} c_k(\sigma) c_k^\dagger(0)\rangle$ is the convenient object for perturbation theory. An imaginary time $\mathrm{i}t \to \sigma$ labels the electron destruction operators $c_k(\sigma) = \exp(\sigma H) c_k \exp(-\sigma H)$ and is ordered by the Wick operator $\hat{T}$. This Green's function is Fourier transformed into the imaginary Matsubara frequency $\mathrm{i}\omega_\nu$, giving $G(k, \mathrm{i}\omega_\nu)$. It is then analytically continued to the real frequency axis to give the retarded Green's function

$$G(k,\omega) = [\omega - \varepsilon_k - \Sigma(k, \omega + \mathrm{i}\eta)]^{-1} \tag{44}$$

where $\Sigma(k, \omega + \mathrm{i}\eta)$ is the complex self-energy which can be evaluated perturbatively. The imaginary part of G is the electron spectral function $A(k,\omega)$

$$A(k,\omega) = -\frac{1}{\pi} \mathrm{Im} G(k,\omega) \tag{45}$$

For fixed k, the ω-dependence of $A(k,\omega)$ is interpreted as the spectrum of excitation energy that results from insertion of an electron into the system in state k outside the Fermi surface, or from insertion of a hole into the system in a state k inside the Fermi surface. The latter process is experimentally accessed in a photoemission experiment, and the former in an "inverse" photoemission experiment, although there are complicating details that weaken these interpretations [47,48]. If the Green's function has a simple pole $1/(\omega - \omega_k)$ at complex frequency ω_k, below the real ω axis by an amount $-\mathrm{Im}\omega_k$, then $A(k,\omega)$ has a Lorentzian peak whose center defines the quasiparticle energy ε_k^* and whose width gives the relaxation rate $1/\tau_k^*$. We assume that the real part of the self-energy can be expanded for small ω as

$$\mathrm{Re}\Sigma(k, \omega + \mathrm{i}\eta) = \delta\varepsilon_k - \omega\lambda_k \tag{46}$$

Then we define

$$\varepsilon_k^* = \frac{\varepsilon_k + \delta\varepsilon_k}{(1+\lambda_k)}$$
$$\frac{1}{\tau_k^*} = \frac{-2}{(1+\lambda_k)}\mathrm{Im}\Sigma(k, \varepsilon_k^* + i\eta) \tag{47}$$

so that the spectral function has approximately the form of a Lorentzian lineshape,

$$A(k,\omega) = \frac{z_k}{\pi}\frac{\hbar/2\tau_k^*}{(\omega - \varepsilon_k^*)^2 + (\hbar/2\tau_k^*)^2} \tag{48}$$

Here $z_k = 1/(1+\lambda_k)$ is called the wavefunction renormalization.

If $\hbar/\tau_k^*$ is small compared to relevant energies like $|\varepsilon_k^* - \mu|$, then the Lorentzian is sharp as a function of ω. We may identify ε_k^* as the "quasiparticle energy." However, the quasiparticle is only well defined if the spectral function is sharply peaked also as a function of $|\mathbf{k}|$. Expanding ε_k^* as $\varepsilon_F + \hbar\mathbf{v}_k^* \cdot (\mathbf{k} - \mathbf{k}_F)$, we must also require that $\ell_k^* = |\mathbf{v}_k^*|\tau_k^*$ be large enough compared to relevant distances $1/k_F$ or the lattice constant a.

This is actually not the only way to make a connection between G and a quasiparticle. When deriving a Boltzmann equation from the non-equilibrium version of G [21] it is preferable to define F by integration of G over the component $k_\perp$ of the k-vector perpendicular to the Fermi surface. This permits a Boltzmann equation to persist even for certain cases where the quasiparticle defined above fails to be narrow enough to recognize.

6.2. Solution for Conductivity

Consider a bulk material with a homogeneous dc electric field. A steady-state current flows, derivable using Eq. (1) from a steady-state non-equilibrium distribution $F(k) = f(k) + \delta F(k)$, where $f(k)$ is the Fermi–Dirac distribution $1/[\exp(\varepsilon_k/k_B T) + 1]^{-1}$. To find δF to first order in $\mathbf{E}$ it is necessary to solve the linearized equation of motion for F, that is, the linearized version of Eq. (43),

$$-e\mathbf{E}\cdot\mathbf{v}_k\frac{\partial f}{\partial \varepsilon_k} = -\sum_{k'} I(k,k')\delta F(k'). \tag{49}$$

The left-hand side is the linearized version of $\dot{\mathbf{k}} \cdot \partial F/\partial k$, and the right-hand side is the linearized collision integral. After suitable manipulations (explained in Refs. [49,50]) an integral equation with a Hermitean non-negative kernel is obtained. The non-negativity is required by the second law of thermodynamics and guarantees that entropy (which can be defined [51] for near-equilibrium Fermi gases by $S/k_B = -\sum[F_k \ln F_k + (1-F_k)\ln(1-F_k)]$) increases steadily in time until equilibrium is reached. The Hermitean operator can be inverted by brute force in k-space [45] or using smaller matrices after expansion in a convenient set of orthogonal polynomials [52,53]. The solution can be guided by a variational principle (maximum entropy production) [49,50]. A general form of a variational *ansatz* is

$$F(k) = f(\mathbf{k} + e\mathbf{E}\tau_k/\hbar) \rightarrow f(k) + e\tau_k \mathbf{E} \cdot \mathbf{v}_k \left(\frac{\partial f}{\partial \varepsilon_k} \right) \tag{50}$$

which represents a Fermi–Dirac distribution pulled off-center by the E-field. The displacement of the Fermi surface is governed by the k-dependent parameter τ_k which is varied to optimize the solution of the exact linearized equation. One can pretend that this "trial" solution follows from a "relaxation-time" representation of the Boltzmann equation,

$$\left(\frac{\partial F}{\partial t} \right)_{\text{coll}} \rightarrow -\sum_{k'} I(k,k')\delta F(k') \rightarrow -\frac{\delta F(k)}{\tau_k} \tag{51}$$

However, the last form of this equation is *not* the true evolution equation. The τ_k which gives the exact solution of the real equation has no closed-form expression, and is definitely different from the quasiparticle lifetime τ_k^* defined above. One difference is that the "renormalization factor" $1 + \lambda_k$ in Eq. (47) is missing; all such factors cancel in the dc limit. Another difference is that a correction like $1 - \cos\theta_{kk'}$ needs to be included in the τ_k which solves the Boltzmann equation with a driving field. This factor suppresses the contribution of small angle scattering, because small angles do not much degrade the electrical current.

The formula for dc electrical conductivity is found from Eqs. (1) and (50)

$$\sigma_{xx} = \frac{e^2}{V} \sum_k v_{kx}^2 \tau_k \left(-\frac{\partial f}{\partial \varepsilon_k} \right) \tag{52}$$

where τ_k depends on the details of the scattering and on the temperature, and remains to be determined. There is one standard variational trial solution which works quite well and gives convenient closed-form answers. If we assume that the parameter τ_k in Eq. (50) is τ_{tr}, independent of k, the Boltzmann equation will specify the optimal value of τ_{tr} which gives σ closest to the exact solution. The answer is

$$1/\tau_{\text{tr}} = \frac{\sum_{k,k'} v_{kx} v_{k'x} I_{k,k'} (-\partial f/\partial \varepsilon'_k)}{\sum_k v_{kx}^2 (-\partial f/\partial \varepsilon_k)} \tag{53}$$

Within this variational approximation we now have a closed-form approximation for the electrical conductivity,

$$\sigma = \left(\frac{n}{m} \right)_{\text{eff}} e^2 \tau_{\text{tr}} \tag{54}$$

where τ is given by Eq. (53), and $(n/m)_{\text{eff}}$ is given by

$$\left(\frac{n}{m} \right)_{\text{eff}} = \frac{1}{V} \sum_k v_{kx}^2 \left(-\frac{\partial f}{\partial \varepsilon_k} \right) = \frac{1}{V} \sum_k \left(\frac{\partial^2 \varepsilon_k}{\partial (\hbar k_x)^2} \right) f_k \tag{55}$$

The first version of Eq. (55) is convenient for numerical computation from band theory [54]. It was already derived by a different method in Eq. (26). Using the replacement $\delta(\varepsilon_k) = -\partial f/\partial \varepsilon_k$, the k-sum is restricted to the Fermi surface. The second version is obtained from the first after integration by parts. In the second form, it is clear that $(n/m)_{\mathrm{eff}}$ is the sum over occupied states of the reciprocal band effective mass. The sum over a filled band has positive and negative contributions and gives zero.

6.3. Orders of Magnitude

A big part of transport theory concerns [55] the phenomenological relaxation rate $1/\tau$. Let us estimate the order of magnitude of various contributions. There are three interactions (impurities, electron–phonon interactions, and Coulomb scattering) which always affect the lifetime of quasiparticles in metals,

$$H_{\mathrm{imp}} = \sum_{kk'} V_{\mathrm{imp}}(kk')S_{\mathrm{imp}}(k-k')c_{k'}^{\dagger}c_k \tag{56}$$

$$H_{\mathrm{ep}} = \sum_{kk'Q} V_{\mathrm{ep}}(kk'Q)c_{k'}^{\dagger}c_k(a_Q^{\dagger}+a_{-Q}) \tag{57}$$

$$H_{\mathrm{C}} = \sum_{1234} V_{\mathrm{C}}(1234)c_1^{\dagger}c_2^{\dagger}c_3c_4 \tag{58}$$

For magnetic materials, spin waves [56] and spin disorder also scatter. Here $S_{\mathrm{imp}}(q)$ is the "impurity structure factor" $S_{\mathrm{imp}} = \sum_i \exp(\mathbf{iq}\cdot\mathbf{R}_i)$ where $\mathbf{R}_i$ is the position of the ith impurity. The operator a_Q destroys the phonon of wavevector Q and energy ω_Q. Discrete translation symmetry gives a crystal momentum selection rule requiring the matrix element $V_{\mathrm{ep}}(kk'Q)$ to vanish unless $\mathbf{k}'-\mathbf{k}=\mathbf{Q}+\mathbf{G}$ where $\mathbf{G}$ is a reciprocal lattice vector. The shorthand $(12\ldots)$ means $(k_1,k_2,\ldots)$, and crystal momentum conservation requires that $V_{\mathrm{C}}(1234)$ vanishes unless $(\mathbf{k}_1+\mathbf{k}_2-\mathbf{k}_3-\mathbf{k}_4)$ equals a reciprocal lattice vector.

One estimates the lifetimes from the Golden rule as follows:

$$\hbar/\tau_{\mathrm{imp}} = 2\pi n_{\mathrm{imp}}|V_{\mathrm{imp}}|^2N(0) \tag{59}$$

$$\hbar/\tau_{\mathrm{ep}} = 2\pi n_{\mathrm{ph}}|V_{\mathrm{ep}}|^2N(0) \tag{60}$$

$$\hbar/\tau_{\mathrm{C}} = 2\pi n_{\mathrm{pairs}}|V_{\mathrm{C}}|^2N(0) \tag{61}$$

where $N(0)$ is the density of states at the Fermi level. Here n_{imp} comes from averaging the impurity structure factor $\langle S(q_1)S(q_2)\rangle = n_{\mathrm{imp}}\delta(q_1+q_2)$. The factor n_{ph} is the number of phonons $\langle a^{\dagger}a\rangle$ which at high temperature is $k_{\mathrm{B}}T/\hbar\omega_{\mathrm{ph}}$. The factor n_{pairs} is the number of electron-hole pair states $(k_{\mathrm{B}}T/\varepsilon_{\mathrm{F}})^2$ available for an electron to create when it scatters by the Coulomb interaction. Thus the orders of magnitude can be written

$$\hbar/\tau_{\mathrm{imp}} \sim n_{\mathrm{imp}}\varepsilon_{\mathrm{F}} \tag{62}$$

$$\hbar/\tau_{\mathrm{ep}} \sim k_{\mathrm{B}}T \tag{63}$$

$$\hbar/\tau_{\mathrm{C}} \sim (k_{\mathrm{B}}T)^2/\varepsilon_{\mathrm{F}} \tag{64}$$

Here we use the fact that the typical impurity or Coulomb scattering matrix element is a few eV in magnitude, similar to the Fermi energy ε_F, while the electron–phonon matrix element $V_{ep} \sim \sqrt{(\hbar\omega_{ph}\varepsilon_F)}$ is smaller. Impurity scattering has $\hbar/\tau$ smaller than ε_F by the small parameter n_{imp}, electron–phonon scattering by the small parameter $(k_B T/\varepsilon_F)$, and Coulomb scattering by two factors of the same. At low T, impurity scattering always dominates (unless superconductivity intervenes and destroys the Fermi liquid). At high enough T, in principle, Coulomb scattering should dominate, but usually the required temperature is above the melting temperature. At intermediate temperatures, phonons dominate. At lower temperatures, the number of phonons decreases as the third rather than first power of $(k_B T/\hbar\omega_{ph})$, so there is a window at low T where the Coulomb interaction is larger than the electron–phonon interaction. Except for extremely clean metals, impurity scattering dominates Coulomb scattering in this window.

6.4. Computation of Resistivity

At the level of the variational solution Eqs. (53) and (54), the resistivity obeys Matthiessen's rule, being composed of additive parts from the various scattering mechanisms

$$1/\tau_{tr} = 1/\tau_{imp} + 1/\tau_{ep} + 1/\tau_{ee} \tag{65}$$

The dominant source of deviation from Mattheissen's rule is not because of higher than second-order effects where different scattering mechanisms mix, but rather in the fact that the true solution for the displacement τ_k of the Fermi distribution can reoptimize if the different scattering processes have differing anisotropies. The fact that measured deviations from Matthiessen's rule [57] are small indicates that the lowest order variational solution is generally quite good. Successful computation [58] of deviations from Matthiessen's rule shows that Bloch–Boltzmann theory is correct in many details.

Using general properties of the linearized collision integral $I(k, k')$, it has been shown [49,50] that the variational formula Eq. (53) can be written as

$$1/\tau_{tr} = \frac{\sum_{kk'}(v_{kx} - v_{k'x})^2 P_{kk'}}{2k_B T \sum_k v_{kx}^2(-\partial f/\partial\varepsilon_k)} \tag{66}$$

where $P_{kk'}$ is the rate of transitions from state k to state k' for the system in equilibrium. This is non-negative, and symmetric ($P_{kk'} = P_{k'k}$) which is the "principle of detailed balance." The factor $(v_{kx} - v_{k'x})^2$ becomes $(2v_F^2/3)(1 - \cos\theta_{kk'})$ in spherical symmetry.

Using the Hamiltonian Eq. (57), and writing $V_{ep}(kk'Q)$ as $M(kk')\delta(k' - k - Q)$, it is convenient to define a class of "electron–phonon spectral functions"

$$\alpha_w^2 F(\Omega) = N(0)\frac{\sum_{kk'}|M(kk')|^2 w(k,k')\delta(\Omega - \omega_{k-k'})\delta(\varepsilon_k)\delta(\varepsilon_{k'})}{\sum_{kk'} w(k,k')\delta(\varepsilon_k)\delta(\varepsilon_{k'})} \tag{67}$$

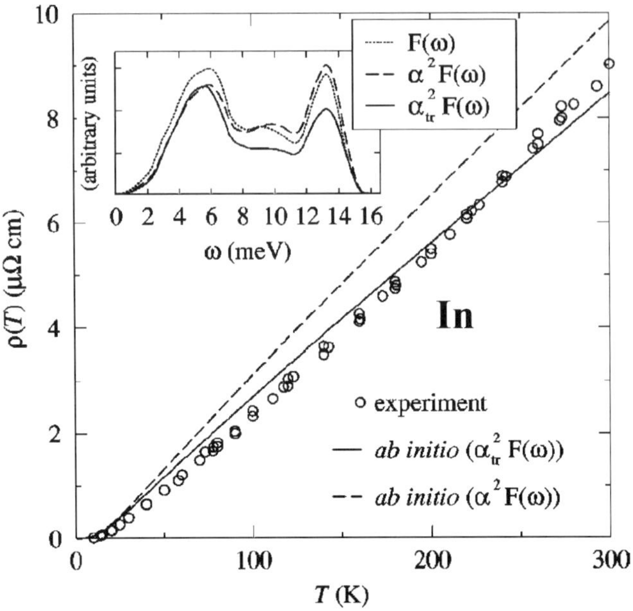

Fig. 6. Calculated resistivity of In versus T [63]. The two theoretical curves are computed using the spectral functions shown in the inset. The dashed line uses α^2F (corresponding coupling constant $\lambda = 0.88$), and the solid line uses $\alpha^2_{\mathrm{tr}}F$ (coupling constant $\lambda_{\mathrm{tr}} = 0.74$. The inset shows the similarity of these functions to the empirical phonon density of states.

The weight function $w(k, k')$ has various forms. When $w = 1$, the function called α^2F appears in the Migdal–Eliashberg theory of superconductivity [59–62]. When $w = (v_{kx} - v_{k'x})^2$, the function is called $\alpha^2_{\mathrm{tr}}F$. The weighted relaxation rate is defined correspondingly as

$$1/\tau_w = \frac{4\pi k_\mathrm{B}T}{\hbar}\int_0^\infty \frac{\mathrm{d}\Omega}{\Omega}\alpha^2_w F(\Omega)\left[\frac{\hbar\Omega/2k_\mathrm{B}T}{\sinh(\hbar\Omega/2k_\mathrm{B}T)}\right]^2 \tag{68}$$

When $w = 1$ this is the electron–phonon contribution to the Fermi surface average quasiparticle equilibration rate (except stripped of the renormalization factor $(1+\lambda)^{-1}$). When $w = (v_{kx} - v_{k'x})^2$, it gives the electron–phonon part of $1/\tau_{\mathrm{tr}}(T)$ which determines the resistivity in variational approximation. Numerical computations show that for elemental metals (see the inset of Fig. 6 for the case of In [63]; for Cu see ref. [58]) the various functions $\alpha^2_w F(\Omega)$ bear a close resemblance to the phonon density of states $F(\Omega)$. Finally, we define dimensionless coupling constants λ_w by the equation

$$\lambda_w = 2\int_0^\infty \frac{\mathrm{d}\Omega}{\Omega}\alpha^2_w F(\Omega) \tag{69}$$

The case $w = 1$ gives the coupling constant λ which determines the superconducting transition temperature of conventional metals [64,65]. At high T ($T > \Theta_D$)), relaxation rates $1/\tau_w \rightarrow 2\pi\lambda_w k_B T/\hbar$ are determined by the constants λ_w.

Quite a few calculations have been made of the superconducting function $\alpha^2 F(\Omega)$ and λ, but fewer of the corresponding transport function $\alpha_{tr}^2 F$. Shown in Fig. 7 are calculations [66] for four metals, done without adjustment of parameters. Agreement with experiment is at the 10% level. These calculations predate the development of efficient methods for calculating phonon dispersion ω_Q [67–69]. Therefore, ω_Q was taken from fits to experiment. The potential $\partial V/\partial u_\ell$ which determines the matrix element $M(kk')$ was not calculated self-consistently, but approximated as the rigid shift of the local "muffin-tin" potential. These approximations are appropriate for elemental metals only. Fully self-consistent calculations, using theoretical phonon curves, were reported for 8 metals in a monumental paper by Savrasov and Savrasov [70], for In in a careful study by

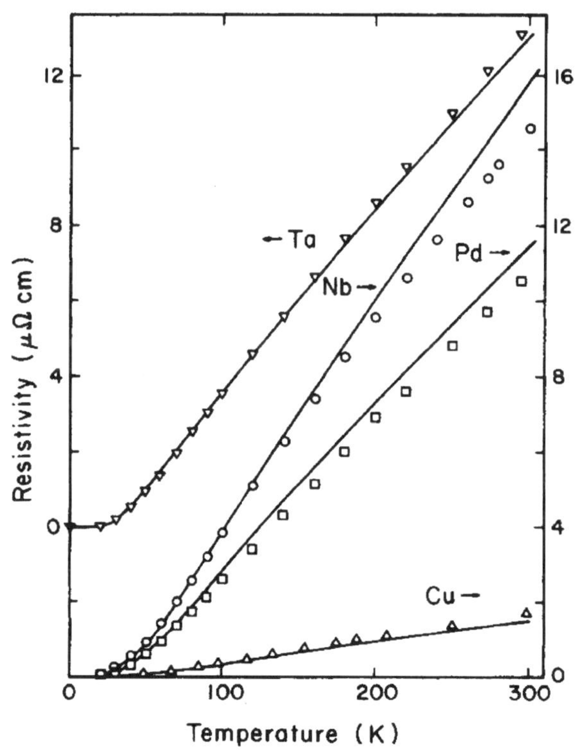

Fig. 7. Electrical resistivity of four transition metals versus temperature, showing experiment compared with computations. The calculations used experimental phonon dispersion curves and electronic band theory with no adjustable parameters as discussed in the text. See Allen et al. (1986) [66] and references therein.

Rudin et al. [63] shown in Fig. 6, and for 4 metals in an ambitious paper by Bauer et al. [71]. A result from Ref. [71] is shown later in Section 10.

Modern "first-principles" theory permits calculations for more complicated cases. The superconducting functions have been calculated for $SrCuO_2$ [72] and for MgB_2 [73–76a]. The latter material has very anisotropic electron–phonon scattering as k varies around the Fermi surface. Analogous to the known variation of the superconducting gap Δ_k is the variation of the shift τ_k Eq. (50) which fixes the non-equilibrium distribution function. One should expect significant k-dependence in τ_k for MgB_2, so the resistivity should deviate from the simplest variational approximation with constant τ_k. This in turn should give big deviations from Matthiessen's rule. The effect has been seen experimentally and explained by Mazin et al. [76b]. There are few serious attempts to calculate the electron-phonon resistivity of metals related to La_2CuO_4. There are two reasons: (1) metallic behavior is found only after doping which is an added challenge to theory; (2) there is doubt on the applicability of Landau Fermi liquid theory.

6.5. More Phenomenological Treatments

In Bloch's original work [4,77], an approximate formula emerged which Grüneisen [78] popularized. This "Bloch–Grüneisen" formula is just our variational result, evaluated for a spherical Fermi surface and a Debye phonon spectrum. It can be written as

$$\rho_{\mathrm{BG}} = \rho_0 + \frac{16\pi^2 \lambda_{\mathrm{tr}} \omega_{\mathrm{D}}}{4\pi (n/m)_{\mathrm{eff}} e^2} \left(\frac{2T}{\Theta_{\mathrm{D}}} \right)^5 \int_0^{\Theta_{\mathrm{D}}/2T} \mathrm{d}x \frac{x^5}{\sinh^2 x} \tag{70}$$

where the denominator of the prefactor, $\omega_{\mathrm{P}}^2 = 4\pi (n/m)_{\mathrm{eff}} e^2$ defines the "Drude" plasma frequency. For many metals this formula gives an excellent fit to the resistivity, with three adjustable parameters. The residual resistivity ρ_0 shifts the resistivity up and down. The Debye temperature Θ_{D} stretches the T axis. Finally, the strength of the electron–phonon part of the resistivity is fixed by the parameter $\lambda_{\mathrm{tr}}/\omega_{\mathrm{P}}^2$. It would be desirable to obtain separate values of λ_{tr} and ω_{P}^2. In principle, ac measurements (see Eq. (6)) should be able to give the necessary extra information. Unfortunately, this is rarely the case. The biggest problem is that interband transitions often overlap the Drude region very severely. The case of Cu shown in Fig. 1 is far more benign than most. Another difficulty is that infrared-active vibrational modes appear in the spectrum of most interesting compound metals, complicating the fitting. For quite a few metals it has been shown that the Drude plasma frequency ω_{P}^2 given by density functional theory (DFT) [79] agrees remarkably well with experiment. This allows determination of an empirical value of λ_{tr} [80] provided the resistivity fits well to the Bloch–Grüneisen formula, and provided the theoretical ω_{P}^2 is known.

Non-cubic metals have anisotropic resistivities ($\rho_{xx} \neq \rho_{zz}$) which, according to Eq. (54) can arise both from anisotropy in $(n/m)_{\mathrm{eff}}$ or in τ_{tr}, Eqs. (53) and (66). The similarity of λ_{tr} to λ shows that in many metals, there is little correlation between velocities v_{kx} and matrix elements $I_{k,k'}$. Therefore we expect τ_{tr} to be fairly isotropic, and this agrees with experiment for metallic elements [81].

6.6. Quasiparticles from Band Theory

It is well established that DFT with good exchange-correlation potentials does surprisingly well for ground state properties, including vibrational spectra, but fails to give band gaps for higher energy excitations. At higher $\hbar\omega$, $\sigma(\omega)$ probes band gaps. Not only at higher ω, but also in the dc limit, calculations of σ should use quasiparticle energy bands [82] rather than DFT bands. However, experience shows that low-lying quasiparticle excitations of metals are very similar to DFT eigenvalues, for no known reason. Shapes of Fermi surfaces were predicted accurately well before the modern era of fully self-consistent DFT bands, and continue to be well described by DFT theory. Presumably they are not extremely sensitive to the potential. Group velocities are critical to accurate transport calculations. The isotropic average Drude plasma frequency squared can be written as $(e^2/3\pi^2\hbar)\int \mathrm{d}A_k|\mathbf{v}_k|$, where $\mathrm{d}A_k$ is an element of Fermi-surface area. The ratio $\lambda_{\mathrm{tr}}/\Omega_p^2$ determines the magnitude of σ in a metal. Using $\lambda_{\mathrm{tr}} \approx \lambda$ for metals where λ is accurately known from superconducting tunneling spectroscopy, the fitted Ω_p^2 agrees well with the DFT value. We can conclude that $\mathbf{v}_{k,\mathrm{DFT}}$ agrees with $\mathbf{v}_{k,\mathrm{QP}}$ for these metals [80]. Therefore at present, DFT eigenstates seem sufficient for transport theory.

6.7. Resistivity of High T_c-Superconductors

Figure 8 shows the temperature-dependence of electrical resistivity of two of the most famous high T_c compounds. The samples are believed to be extremely good [83,84]. Theoretical understanding is limited; transport phenomena in 3d metals still has surprises. For pure $YBa_2Cu_3O_7$, the data, on carefully detwinned samples, show "metallic" resistivity in all three crystallographic directions. The large magnitude of ρ (somewhat bigger than most conventional metals) can be assigned to the small carrier density. There is a factor of 2 anisotropy between the a-axis (lower resistivity) and b-axis. This was predicted before the experiment by DFT calculations of the Drude plasma frequency tensor $\Omega_{p\alpha\beta}$ [85], using the assumption that τ_{tr} is isotropic. Along the b-axis, where the CuO chains run, DFT gives $\hbar\Omega_p \approx 4.4\,\mathrm{eV}$, while perpendicular to the chains, on the a-axis, it is 2.9 eV. The squared ratio agrees nicely with experiment. The c-axis anisotropy is not so well predicted. The calculated c axis plasma frequency of 1.1 eV predicts ρ_c/ρ_b to be 7, while experiment gives 33. One might conclude that band theory works surprisingly well, as is also seen in photoemission spectroscopy.

On the other hand, part (b) of Fig. 8 shows what happens when holes are removed from optimally Sr-doped La_2CuO_4. Similar results are seen in $YBa_2Cu_3O_7$ [84]. At higher T, the nominally insulating samples have "metallic" ($\mathrm{d}\rho/\mathrm{d}T > 0$) resistivities. When the carrier density is scaled out, giving the inverse mobility $ne\rho$,

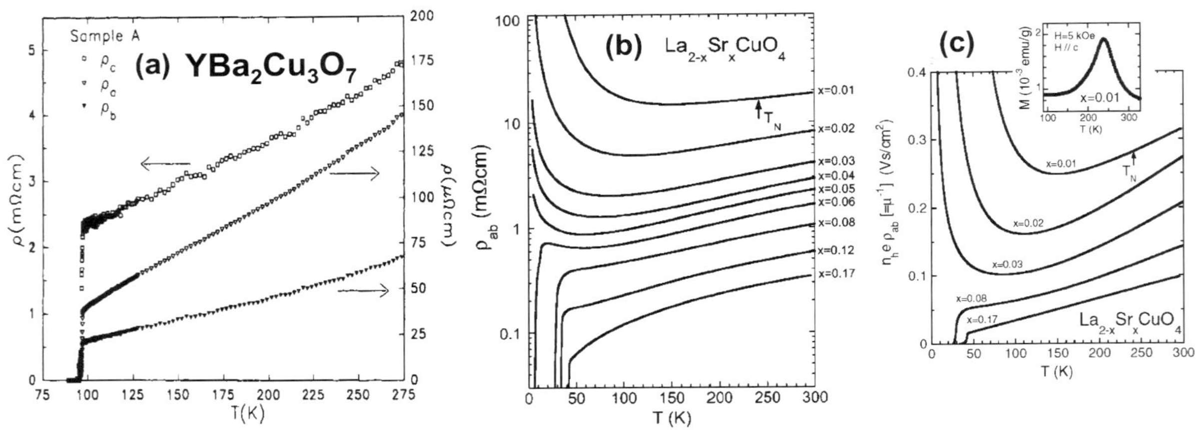

Fig. 8. Resistivity versus temperature of high T_c compounds. Part (a) shows the anisotropic resistivity of a detwinned single crystal of $YBa_2Cu_3O_7$ (from Friedmann et al., 1990 [83]). Part (b) shows the in-plane resistivity of a series of single crystals of $La_{2-x}Sr_xCuO_4$, with hole doping x going from a small x antiferromagnet to an optimum x superconductor (from Ando et al., 2001 [84]). (c) The same data (from Ando et al., 2001 [84]) on a linear scale after dividing out the nominal carrier density to get inverse mobility. The inset shows the Neel transition of the $x = 0.01$ sample detected as a peak in magnetization.

the mobilities of even antiferromagnetic "insulators," with 0.01 holes per unit cell, are only smaller than the mobilities of the optimum superconductor by a factor of 3. This strongly contradicts conventional notions that electrons in good metals behave ballistically, but dilute carriers in antiferromagnets have frustrated mobilities [86].

Whatever theory accounts for the optimum superconductor should therefore also account for the far under-doped antiferromagnet. The shape of $\rho(T)$ for the optimal superconductors is close to linear. A simple explanation is that this agrees with the Bloch–Grüneisen formula. The lower temperature region, where $\rho_{BG}(T)$ deviates from linear, is hidden from sight by superconductivity. However, this explanation does not work for under-doped materials. Also, the estimated mean-free path is too small at the more resistive end to justify quasiparticle transport [87]. The Hall coefficient has unusual temperature-dependence. The conclusion is that these data are still not understood.

7. KONDO EFFECT AND RESISTIVITY MINIMUM IN METALS

Figure 9 [88] shows resistivity of Au before and after implanting 60 ppm Fe impurities. The small resistivity upturn at low T has captured a huge amount of

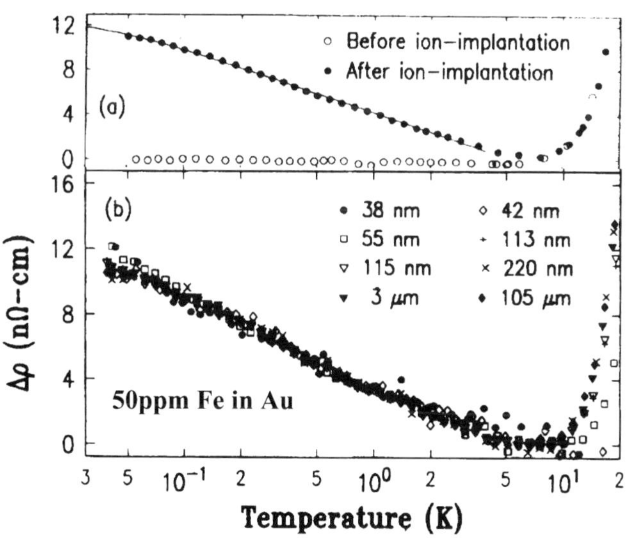

Fig. 9. Kondo resistivity [88]. The high-purity gold films used here are about 30 nm thick and have size-limited residual resistivities of order $\rho_0 \approx 2\,\mu\Omega\,\text{cm}$. At 300 K, phonon scattering contributes an additional $\rho - \rho_0 \approx 2\,\mu\Omega\,\text{cm}$. At temperatures shown here, the phonon scattering term is unobservable below 8 K, and its turn-on can be seen in the interval $8\,\text{K} < T < 20\,\text{K}$. After Fe implantation, a new T-dependent Kondo term $\Delta\rho \approx A - B\ln[1 + (T/\Theta)^2]$ is seen, whose magnitude is about 0.005 times ρ_0 and is independent of wire width (after subtracting an impurity-enhanced electron–electron contribution).

attention. The minimum in $\rho(T)$ at around 8 K is the crossover between phonon scattering causing $\rho(T)$ to increase with T at high T, and magnetic impurity scattering causing $\rho(T)$ to decrease with T at low T.

Anderson [89] showed how a transition metal impurity can either retain or lose its local magnetic moment when dissolved in a metal. Kondo [90,91] first glimpsed the complexity of the behavior when the moment is retained. The relevant part of the Hamiltonian is

$$H_{\text{Kondo}} = -\sum_{i,\ell} J(\mathbf{r}_i - \mathbf{R}_\ell)\sigma_i \cdot \mathbf{S}_\ell \tag{71}$$

where the ith electron of spin σ_i has a spin-flipping interaction with the ℓ'th magnetic impurity with spin $\mathbf{S}_\ell$. Perturbation theory for spin-flipping interactions differs from ordinary potential scattering from non-magnetic impurities in having a time-dependent impurity spin. For any time-dependent perturbation in a metal, the sharpness of the Fermi distribution causes logarithmic singularities in integrals. These diminish with T at least as fast as $\log(\varepsilon_F/k_B T)$ because of blurring of the Fermi distribution. An exact solution was found by Wilson [92,93] using the renormalization group, and by Andrei [94] using the Bethe Ansatz. These solutions were given a physical interpretation by Nozieres [95,96]. The subject is by no means closed. In particular, newer experimental tools applied to nanosystems permit more detailed exploration [97]. Finite biases in tunnel junctions allow the "Kondo resonance" to be explored by inelastic spectroscopy [98].

8. DIRTY FERMI LIQUIDS AND INTRINSICALLY DIFFUSIVE STATES

The Bloch–Boltzmann equation works beyond the naive expectation that weak interactions are required. However, the mean-free path of electrons in metals can often be reduced below 10 Å, where the definition of the wavevector of a quasi-particle is fuzzy. Metallic dirty alloys, liquids, and glasses are in this category. The other requirements for Fermi liquid status may still apply – there are complicated single-particle states of charge $\pm e$, spin $\pm\frac{1}{2}$, and energies sharp on a scale of ε_F. Because there is no wavevector, spectroscopies such as photoemission or Fermi-surface resonant techniques are not available to prove the value of this picture, but disorder does not automatically destroy the single-particle picture.

Dirty alloys are simplest, since the locations of atoms are known. Figure 10 shows an example [99]. In the region of (x, T) with $\rho < 125\,\mu\Omega\,\text{cm}$, $\rho(T)$ curves merge if shifted vertically – Matthiessen's rule is obeyed. When $\rho > 125\,\mu\Omega\,\text{cm}$, it is violated, but Boltzmann theory can no longer possibly be valid, because quasiparticles have had their k-vectors destroyed by either thermal disorder (in the pure V metal) or alloy disorder. In this regime, resistivity varies more weakly with (x, T) than when resistivity is smaller and quasiparticles exist. This phenomenon is called "resistivity saturation" after the paper by Fisk and Webb [100], and has been recently reviewed by Gunnarsson et al. [101–104].

First let us focus on $T = 0$. Theory should give reliable alloy disorder resistivity, because this is not a many-body problem. There are two limits. If the disorder is extremely large, states at the Fermi level may be Anderson-localized. Then the material will be insulating, meaning that as $T \to 0$, $\rho(T) \to \infty$. This is especially important in 2D disordered films, outside the scope of this article. In $d = 3$, localization is harder to achieve. Section 11 will discuss the metal-insulator transition in 3d-doped semiconductors, where Anderson localization does occur. Since the localized option occurs in $d = 3$, and since localization cannot be found by perturbation theory starting with delocalized basis functions, it is good to use a non-perturbative approach for resistivity of dirty alloys, such as exact diagonalization of finite subsamples. However, Brown et al. [105] showed that the "coherent potential approximation," a self-consistent perturbation theory, agrees with non-perturbative methods for at least one very dirty alloy.

For dirty 3d metals, as in Fig. 10, $\rho(T)$ shows reduced T-dependence. The dirtiest alloys ($V_{0.67}Al_{0.33}$) have a small negative $d\rho/dT$. There is no well-accepted explanation. Since $\rho(T)$ does not diverge at low T, true (or "strong") localization has not set in; the alloy is still a metal. Relatively few $d = 3$ metallic systems can be driven to the localized non-metallic state. Examples are $Ge_{1-x}Au_x$ [106,107] for small Au concentration $x \leqslant 0.12$, in both polycrystalline and amorphous films, and similarly

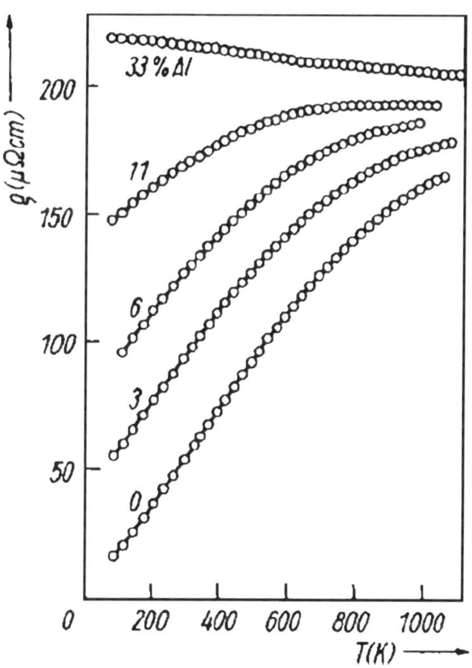

Fig. 10. Electrical resistivity of $Ti_{1-x}Al_x$ alloys versus temperature [99]. Matthiessen's rule (the total resistivity is intrinsic plus a constant upward shift from impurity scattering) is well obeyed whenever $\rho < 125\,\mu\Omega\,\text{cm}$. At larger resistivities, "saturation" is seen.

$Si_{1-x}Nb_x$ [108] for $x \leqslant 0.115$. It is necessary to make a heavy dilution of a metal like gold in a non-metal like Ge or Si, in order to get an insulator. At low T (below 20 K), a small upturn of $\rho(T)$ with decreasing T is often found, caused by effects beyond Boltzmann theory. If the effect is enhanced by disorder, then it is not a Kondo effect, but is denoted "weak localization" or "quantum corrections." Their origin is similar to the origin of Anderson or "strong" localization, but the resistivity remains well under 1000 $\mu\Omega$ cm, and the net change between 20 and 0 K is a few percent or less. These effects are discussed in Section 9.

There is an incorrect belief that $k_F\ell \approx 1$ or $\ell \approx a$ is the criterion for localization in 3d. Experiment clearly shows that this is not true, and that instead, metallic resistivity $100\,\mu\Omega\,\text{cm} < \rho < 1000\,\mu\Omega\,\text{cm}$ (indicating $k_F\ell \approx 1$) occurs commonly with no sign of a true insulator. What is the mechanism of transport? Here is a thought experiment which could be computationally implemented on a large computer. For good crystalline metals, the propagating nature of Bloch states is proved by constructing Gaussian wavepackets out of **k**-states centered on particular wavevectors $\mathbf{k}_0$. The Schrödinger equation evolves the state in time, the center of the wavepacket moving with the group velocity $\mathbf{v}_{\mathbf{k}0}$. This excitation transports charge and spin ballistically. Because of inevitable impurities, after a sufficiently long time ($t > \tau_k$) the wavepacket degrades and the center of charge stops moving. The squared width of the wavepacket continues to spread as $\langle r^2 \rangle \propto 3Dt$, where D is the diffusion constant, $D = v_k^2 \tau_k / 3$. Diffusion continues until the sample boundary is reached. A $d = 1$ computer experiment has been published for the case of weak disorder [109]. One dimension has the extra feature that the diffusion does not last forever, but evolves finally into localization with $\langle r^2 \rangle$ constant. In $d = 2$ the same effect should occur at extremely long times and distances, but lies beyond the power of computer experiment for weak disorder.

Consider the same construction in $d = 3$. The wavepacket should be built from eigenstates in a narrow energy window of the *delocalized* part of the spectrum of a very dirty metal, but the phases should be adjusted so that the resulting packet is spatially localized at $t = 0$. The time evolution is then computed, and it is found that $\langle r^2 \rangle \propto 3Dt$ starts immediately and holds to the boundaries, with $D \approx \hbar a^2/W$ where W is the band-width of the metallic energy band ($\hbar/W$ is the time to hop to a nearest neighbor). One should experiment with the phases of the different component eigenstates trying to create a propagating packet. The effort will fail; no packet can be made that shows ballistic propagation; The states of the dirty metal are "intrinsically diffusive" and do not propagate ballistically farther than an interatomic distance. Therefore one cannot define a mean free path, but if forced to make a choice, one would have to say $\ell \approx a$. Such states are not teetering on the border of localization. They are generic in the spectrum of dirty metals. As you move in the one-electron spectrum toward a band gap, there is a "mobility edge" where $D \to 0$ and localization sets in. The localized states are a small minority and are far from the Fermi level in ordinary metals.

Exact calculations of ρ for dirty metals can be done in one-electron approximation from Eq. (25) if the eigenstates $|n\rangle$ are all known. If the states m, n are localized, then this formula will correctly give $\sigma = 0$, because whenever two states are nearly degenerate, they will be separated spatially, with vanishing current matrix elements $\langle i|j_x|j\rangle = 0$. Successful calculations for alloys were reported by various groups [33–35,105,110,111].

Now return to $T > 0$. At high T, many relatively conventional crystalline metals show resistivities like those in Fig. 10 that deviate from the linear T-dependence predicted by Boltzmann theory. The reason is that the mean-free path ℓ has gotten too short (less than 10 Å). To see how electron–phonon interactions can give short mean-free paths, consider the lifetime broadening $\hbar/\tau = 2\pi\lambda k_B T$. Since λ is often of order 1, and $2\pi k_B T$ is 0.16 eV at room temperature, the levels are not necessarily narrower than the separation of adjacent bands. Consider Nb_3Sn, with $\lambda > 1$ and 8 atoms in a cubic unit cell. The total band width of the 4d levels is about 10 eV, and there are 30 d states in this band, for a mean level separation of 0.3 eV. Thus individual quasiparticle levels are not sharply defined. The resistivity [100] deviates strongly from the Bloch–Grüneisen form at room temperature. Although we understand why the theory fails, a useful theory to fix it [101–104,112] is not easily constructed.

9. WEAK LOCALIZATION AND QUANTUM CORRECTIONS

A huge range of fascinating low T transport effects goes under the various names "weak localization", or "mesoscopic fluctuations", or "interaction effects" or "quantum corrections." These effects show up as small corrections when resistivities are, large, but can be more significant when samples are small, especially in $d = 1$ or 2. An example is shown in Fig. 11 [113]. Quantum coherence is not just a property of well-organized propagating Bloch states where coherence is easily predictable. All solutions of t-independent Schrödinger equations are coherent. Components of a wavefunction interfere with other components of a superposition state. The coherence is only destroyed by t-dependent environmental perturbations such as scattering by phonons. Let $\hbar/\tau_{\text{inel}}$ be the lifetime broadening of a single-particle state caused by an environmental inelastic process. At low T, the scattering rate gets very small. Electrons therefore remain coherent for a long time, and may diffuse coherently over distances $L_{\text{coh}} = \sqrt{D\tau_{\text{inel}}}$ where the diffusion constant D is determined by elastic processes. In weakly disordered material, $D \approx v_F^2\tau_{\text{elast}}$ while in strongly disodered systems the diffusion constant is a^2/τ_{hop} where τ_{hop} is the time to hop to a nearest neighbor a distance a away. This time is of order $\hbar/W$ where W is the band width. When the sample is smaller than L_{coh}, large "mesoscopic" fluctuations can be expected. The same wavefunction coherence is required for a single-particle state to become Anderson-localized. This is why quantum coherency corrections are called "weak localization" even though the system may be very far from true localization. Electron–electron Coulomb interactions also become enhanced at low T by the effects of disorder. For perfect Bloch states, Coulomb interactions are suppressed by Fermi degeneracy, $\hbar/\tau_C \approx \varepsilon_F(k_B T/\varepsilon_F)^2$. However, if the propagation is diffusive, two electron states, close enough to interact with each other, see the same pattern of disorder and tend to propagate similarly, giving an

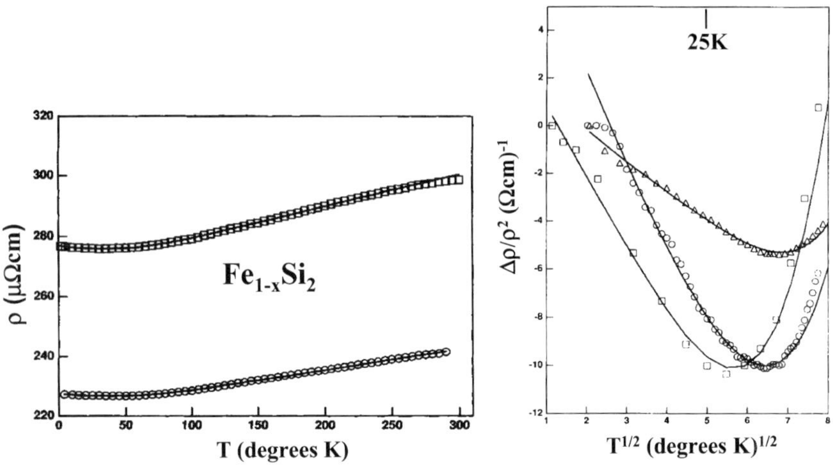

Fig. 11. Absolute (left) and relative (right) resistivity versus temperature for polycrystalline iron-deficient α-$FeSi_2$ films of thickness 100 nm (circles), 70 nm (squares), and 35 nm (triangles) [113]. The reduced and possibly saturated T-dependence characteristic of dirty metals is seen at $T > 100$ K. At $T < 50$ K, at the level of 1% of the total ρ, there is an interesting T-dependent upturn obeying approximately the law $A - BT^{1/2}$, characteristic of weak localization effects in $d = 3$.

enhancement of the Coulomb interaction. When samples are fairly clean, the corrections to ballistic propagation are weak, and perturbative theories predict leading corrections [114]. These theories go beyond conventional Fermi liquid theory, and have been confirmed in numerous experimental tests.

10. NEUTRON, PHOTOEMISSION, AND INFRARED SPECTROSCOPIES

Although these spectroscopies are not normally classified as transport, in fact they can measure otherwise inaccessible transport properties. Consider the lifetime-broadening of a phonon, as seen in infrared, Raman, or neutron scattering. The line shape formula is a spectral function analogous to the one defined for electrons in Eqs. (44), (45) and (48). On expects a broadened Lorentzian line shape, with broadening (full width at half maximum) Γ_Q of the normal mode Q given by 2 times the imaginary part of the corresponding phonon Green's function (see Fig. 12). This is a transport property. The phonon distribution obeys a Boltzmann equation written first by Peierls [115]. After linearizing, there is a rate of change due to collisions of the phonon distribution function

$$\left(\frac{\partial N_Q}{\partial t}\right)_{\text{coll}} = -\sum_{Q'} I(Q,Q')\delta N_{Q'} \tag{72}$$

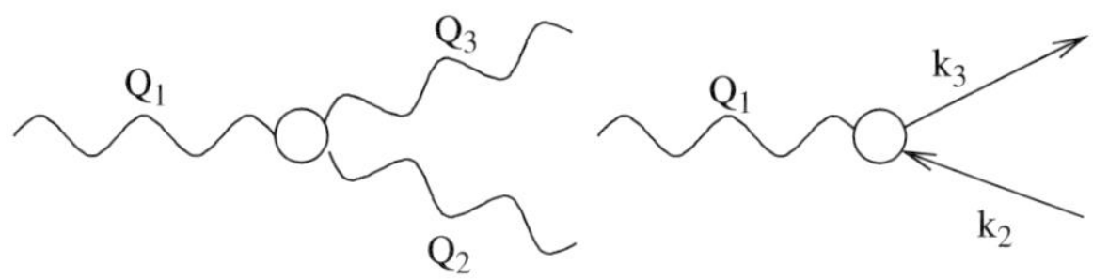

Fig. 12. The phonon (wiggly line) can decay into two phonons by the third-order anharmonic coupling V_{123} or into an electron-hole pair by the electron–phonon coupling M_{123}.

If the only normal mode out of equilibrium is the mode Q, then the right-hand side is just $-I(Q,Q)\delta N_Q = -\Gamma_Q \delta N_Q$ and the population N_Q returns to equilibrium $n_Q = [\exp(\hbar\omega_Q/k_B T) - 1]^{-1}$ exponentially with time constant $1/\Gamma_Q$. By the usual arguments of time-dependent perturbation theory used to construct scattering terms in Boltzmann equations, one finds the formulas

$$\Gamma_1^{\text{anh}} = \frac{\pi}{\hbar^2}\sum_{23} |V_{123}|^2[(n_2+n_3+1)\delta(\omega_1-\omega_2-\omega_3) + 2(n_2-n_3)\delta(\omega_1+\omega_2-\omega_3)] \tag{73}$$

$$\Gamma_1^{\text{ep}} = \frac{2\pi}{\hbar}\sum_{23} |M_{123}|^2(f_2 - f_3)\delta(\hbar\omega_1 + \varepsilon_2 - \varepsilon_3) \tag{74}$$

where $1, 2, \ldots$ are short for $Q_1, Q_2, \ldots$. Here f and n are the equilibrium Fermi-Dirac and Bose–Einstein distributions, $\hbar\omega$ and ε are the phonon and electron quasiparticle energies, and V_{123} and M_{123} are the matrix elements for phonon–phonon scattering (third-order anharmonicity) and phonon–electron scattering, each being restricted by crystal momentum conservation (Q_1 must have the same wavevector *modulo* a reciprocal lattice vector as $Q_2 + Q_3$ or $k_3 - k_2$). The anharmonic matrix element V_{123} involves $(\partial^3 V_N/\partial u_1 \partial u_2 \partial u_3)A_1 A_2 A_3$, where A_i is the amplitude factor $\sqrt{\hbar/2M\omega_i}$ of the ith normal mode, and V_N the total nuclear potential energy (the same order of magnitude as ε_F). The second derivative $\partial^2 V_N/\partial u_1 \partial u_2$ is of order $M\omega^2$ where M is the nuclear mass. By counting factors appearing in these equations, one can determine that the order of magnitudes are

$$\Gamma_1^{\text{anh}}/\omega_{\text{ph}} = 2\pi |V_{123}|^2 F(\omega_{\text{ph}}) n_{\text{ph}} \approx \frac{k_B T}{V_N} \tag{75}$$

$$\Gamma_1^{\text{ep}}/\omega_{\text{ph}} = 2\pi |M_{123}|^2 F(\omega_{\text{ph}}) n_{\text{pairs}} \approx \frac{\hbar\omega_{\text{ph}}}{\varepsilon_F} \tag{76}$$

where $F(\omega_{\text{ph}})$ is the average phonon density of states, approximately $1/\hbar\omega_{\text{ph}}$. These equations use the previously mentioned size estimate $M(kk') \approx \sqrt{\varepsilon_F \hbar\omega_{\text{ph}}}$ and the corresponding estimate $V_{123} \approx \sqrt{(\hbar\omega_{\text{ph}})^3/V_N}$. Thus, we see that the two decay modes for phonons are roughly the same size. An important difference is that Γ_{anh} increases linearly with T at higher temperature, while Γ_{ep} is roughly independent of temperature. Other details affect the magnitude quite a lot. The electron–phonon process depends on $n_{\text{pairs}} \approx (N(0)\hbar\omega)^2$, the number of electrons and holes within a

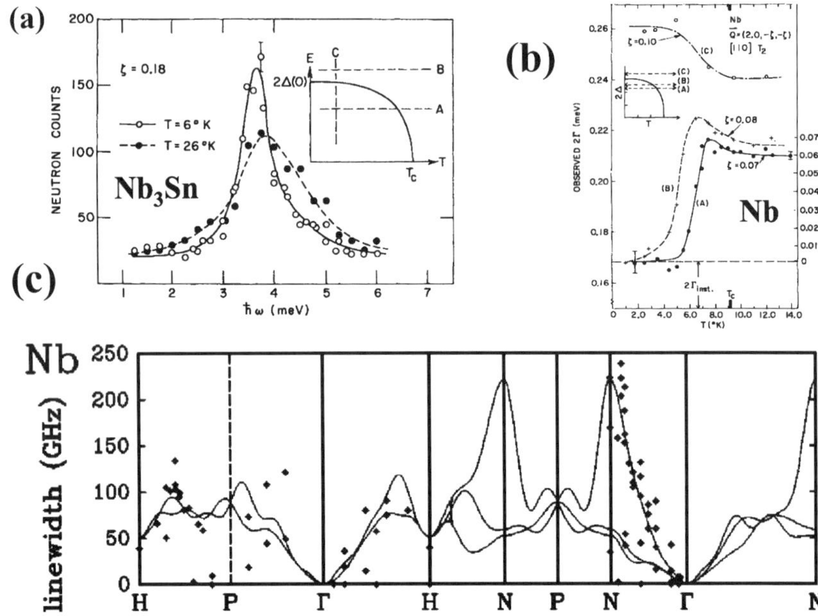

Fig. 13. (a) Change (for $T > T_c$ and $T < T_c$) of the line shape detected by neutrons for an acoustic phonon in the superconductor Nb_3Sn (from Axe and Shirane, [117]). (b) Measured line width as a function of T for various acoustic phonons in Nb (from Shapiro and Shirane, [116]). (c) Theoretical [71] and experimental [118] phonon line widths for Nb.

phonon energy ω of the Fermi energy. Different metals have quite different densities of states $N(0)$ at the Fermi energy – Pb and Nb differ by a factor of 3 (0.50 states/eV atom for Pb, 1.46 for Nb [7]) which appears squared in the phonon decay rate. Shapiro et al. [116–118] were able to see the extra broadening of Nb phonons caused by decay to electron-hole pairs, partly by exploiting the change in Γ when the superconducting transition occurs (see Fig. 13). Habicht et al. [119], using neutron echo techniques, saw no evidence for the electron-hole decay channel in Pb, consistent with a 10 times smaller expected electron decay relative to Nb, and a stronger anharmonic interaction.

Can we similarly measure electron equilibration rates $1/\tau_k$? Both photoemission and infrared spectroscopy provide partial measurements. In photoemission, electrons are ejected from a clean surface into vacuum where energy and wavevector are measured. Since the energy and wavevector of the incident photon are known, subtraction gives the energy and wavevector of the hole that was left behind, allowing mapping of energy bands. Two complications make the process less ideal. (1) The kosher theory of the process shows that a higher-order Green's function is needed [47,48]. (2) Since translational invariance is broken, crystal momentum k_z perpendicular to the surface is not conserved in the emission process – the sample can absorb arbitrary amounts of perpendicular momentum. This complicates the process of mapping bands and broadens the lineshapes. This second complication is

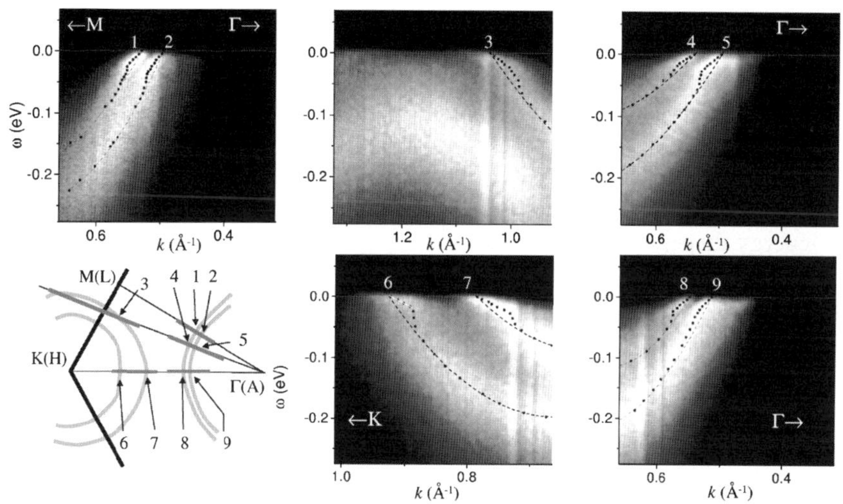

Fig. 14. Energy versus wavevector distributions of hole spectral weight (coded by color in the original [120]) seen in photoemission spectroscopy of the layered metal 2H-$NbSe_2$. The wavevector scans include nine Fermi surface crossings. Near each crossing the hole dispersion curve has a self-energy shift $-\omega\lambda_k(\omega)$, and the shift decreases at larger ω, leaving a kink in the dispersion curve which measures the size of λ_k. The values $\lambda_k \approx 0.85 \pm 0.15$ are deduced at all crossings except number 6 where the value 1.9 ± 0.2 is found.

eliminated if the hole lies in a surface state (**k** is 2d and has no z component) or lies in a quasi-2d band (ε_k depends very weakly on k_z). Figure 14 shows data [120] for 2H-$NbSe_2$, a quasi-2d metal with a strong electron–phonon interaction ($T_c = 7.2\,K$). The energy resolution is 4–6 meV and k-resolution is $0.0025\,\text{Å}^{-1}$. Therefore, the intrinsic fuzziness of line shapes is larger than the resolution and reflects actual broadening and shifting of bands in $A(k,\omega)$. The data clearly show both broadening associated with $\mathrm{Im}\,\Sigma$ and a rapidly varying shift associated with $\mathrm{Re}\Sigma$ from electron–phonon interactions.

The lifetime broadening $1/\tau_k = -2\mathrm{Im}\,\Sigma$ can be found from Boltzmann theory using the diagonal part $k = k'$ of the linearized collision integral, Eq. (49). The answer is

$$\hbar/\tau_k = 2\pi \sum_{k'Q} |M(kk')|^2[(1 - f_{k'} + n_Q) \times \delta(\varepsilon_k - \varepsilon_{k'} - \omega_Q) + (f_{k'} + n_Q)\delta(\varepsilon_k - \varepsilon_{k'} + \omega_Q)] \tag{77}$$

where $q = k - k'$. The alternate way to derive this is from the electron–phonon self-energy, for which, as shown by Migdal [121], perturbation theory behaves well. The self-energy to lowest order in the small parameter $\hbar\omega_{ph}/\varepsilon_{el}$ is

$$\Sigma(k,\omega) = \sum_{k'Q} |M(kk')|^2 \left(\frac{1 - f_{k'} + n_Q}{\omega + i\eta - \varepsilon_{k'} - \Omega_Q} + \frac{f_{k'} + n_Q}{\omega + i\eta - \varepsilon_{k'} + \Omega_Q} \right) \tag{78}$$

From the imaginary part evaluated at the band energy $\omega \to \varepsilon_k$, we get the same answer as in Boltzmann theory.

The consequences of the self-energy Eq. (78) have been seen experimentally in many metals, especially in superconducting tunneling experiments using planar junctions [122]. At low ω, Σ has the form $-\lambda_k\omega$. When λ_k is averaged over the Fermi surface, one gets the electron–phonon coupling constant λ which is of order 1 at low T. At higher excitation ω, λ goes to zero, which causes the kink seen in the near-Fermi-energy dispersion of Fig. 14.

Infrared spectroscopy is an alternate way to see the same physics as in planar tunneling junctions, and improvements in infrared sources and detectors make this method increasingly powerful. Holstein [123] argued that at infrared probing frequencies $\omega \sim \omega_{\text{ph}}$ there would be corrections to the Drude formula not contained in the ordinary low-frequency Boltzmann approximation. The starting point is Kubo's formula Eq. (16) for the conductivity,

$$\sigma(\omega) = \frac{i}{\omega}\left[r(\omega) + \frac{ne^2}{m}\right] \tag{79}$$

$$r(\omega) = i\int_0^{\infty} \mathrm{d}t e^{\mathrm{i}\omega t}\langle[j(t), j(0)]\rangle \tag{80}$$

This formula can be evaluated only for simple systems without interactions. To get perturbative expressions for systems with interactions, a Wick-ordered ($\hat{T}$) imaginary time ($0 \leqslant \sigma \leqslant \beta = 1/k_{\text{B}}T$) version of $r(\omega)$, is used, namely

$$r(\mathrm{i}\omega_\mu) = -\int_0^{\beta} \mathrm{d}\sigma e^{\mathrm{i}\omega_\mu\sigma}\overline{\langle\hat{T}j(\sigma)j(0)\rangle} \tag{81}$$

$$r(\mathrm{i}\omega_\mu) = -\frac{e^2}{\beta}\sum_{kk'\nu} v_{k'x}\Gamma(kk', \mathrm{i}\omega_\mu, \mathrm{i}\omega_\nu)G(k, \mathrm{i}\omega_\nu + \mathrm{i}\omega_\mu)G(k, \mathrm{i}\omega_\nu) \tag{82}$$

$$G(k, \mathrm{i}\omega_\nu) = \frac{1}{\mathrm{i}\omega_\nu - \varepsilon_k - \Sigma(k, \mathrm{i}\omega_\nu)} \tag{83}$$

When analytically continued from the imaginary (Matsubara) frequencies $\mathrm{i}\omega_\mu = 2\pi\mu/\beta$ and $\mathrm{i}\omega_\nu = 2\pi(\nu + 1/2)/\beta$, with μ and ν integers, to just above the real frequency axis $\omega + \mathrm{i}\eta$ (η is infinitesimal), these functions become $r(\omega)$ as in Eq. (80) and G and Σ as in Eq. (44). The vertex function Γ is related to the self-energy Σ by a Ward's identity.

For electron–phonon systems, Holstein derived [22] the integral equation (generalized Boltzmann equation) for Γ at the Migdal level of approximation. Allen [124] showed that the resulting conductivity had the form

$$\sigma(\omega) = \frac{ine^2}{m\omega}\int_{-\infty}^{\infty} \mathrm{d}\omega' \frac{f(\omega') - f(\omega' + \omega)}{\omega - \Sigma_{\text{ir}}(\omega' + \omega + \mathrm{i}\eta) + \Sigma^*_{\text{ir}}(\omega' + \mathrm{i}\eta)} \tag{84}$$

The self-energy Σ_{ir} is almost exactly Σ of Eq. (78). The difference is in a k-dependent weighting factor $w(k, k')$, Eq. (67) similar to $(1 - \cos\theta_{kk'})$, that appears in the

Fermi-surface integration, and can be omitted to reasonable accuracy over much of the spectral range. Infrared experiments [125,126] have seen Holstein's predicted deviations from simple Drude behavior.

Unfortunately, Eq. (84) lacks the simplicity of the Drude form Eq. (6). Götze and Wölfle [127] suggested a simplified way to compute optical response in metals using peturbation theory for a "memory function" $M(\omega)$ defined as

$$\sigma(\omega) = \frac{ine^2/m}{\omega + M(\omega)} \tag{85}$$

This is much closer to the Drude form, and in particular, the imaginary part of $M(\omega)$ is the generalization of the scattering rate $1/\tau$. Götze and Wölfle gave a closed formula for $M(\omega)$ at the lowest level of approximation. In the dc limit, their formula correctly reproduces the lowest-order variational solution of Boltzmann theory. Higher-order approximations for $M(\omega)$ are very messy, and the method is less reliable than Eq. (84). Infrared experiments, together with Kramers–Kronig analysis, can be used to extract $M(\omega)$, which is sometimes called an "optical single-particle self energy" [128]. An example is in Fig. 15, which shows interesting structure in $M(\omega)$ similar to what may be expected in a self-energy. Comparing

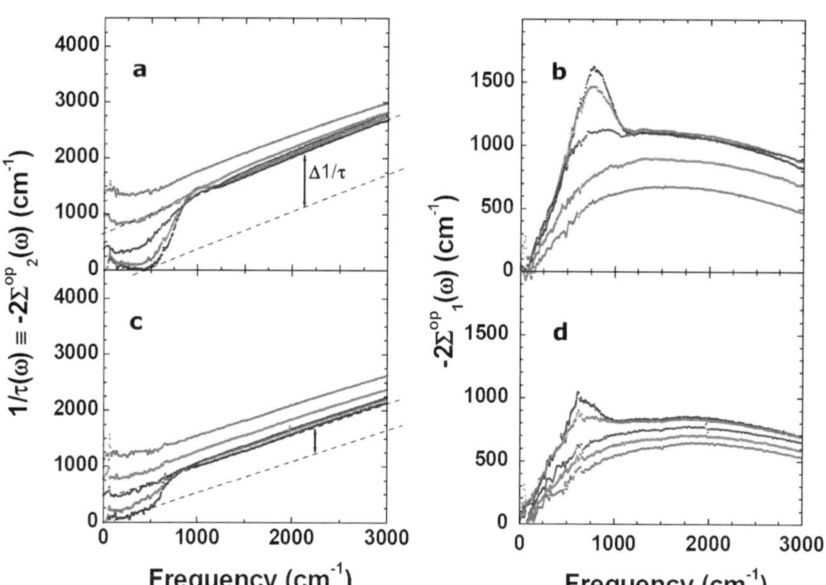

Fig. 15. Real (parts **b** and **d**) and imaginary (parts **a** and **c**) parts of the memory function $M = 2\Sigma^{op}$ measured by reflectance of untwinned single crystals of $Bi_2Sr_2CaCu_2O_{8+\delta}$, with optimal ($T_c = 96$ K; parts **a** and **b**) and somewhat overdoped ($T_c = 82$ K; parts **c** and **d**) oxygen concentrations [128]. Values are shown for five temperatures in each panel, namely, from bottom to top (left panels) and top to bottom (right panels), 27 K, 71 K, 101 K, 200 K, and 300 K.

Eqs. (84) and (85), one finds at low ω the relation

$$M(\omega) \sim -\omega \frac{\mathrm{dRe}\Sigma_{\mathrm{ir}}(\omega)}{\mathrm{d}\omega} - 2i\mathrm{Im}\,\Sigma_{\mathrm{ir}}(\omega) = \lambda_{\mathrm{ir}}\omega + i/\tau_{\mathrm{ir}} \tag{86}$$

In the dc limit $M \to i/\tau_{\mathrm{ir}}$, the mass renormalization λ_{ir} disappears and the correct dc result is $ne^2\tau_{\mathrm{dc}}/m$, where τ_{dc} is the dc limit of τ_{ir}. Thus, in agreement with the semiclassical Boltzmann approach, the mass renormalization does not enter dc transport properties.

11. SEMICONDUCTORS AND THE METAL/INSULATOR TRANSITION

In semiconductors, carriers of electrical current are much more dilute than in metals. They are thermally activated out of filled bands, or injected by light, charged particles, or through tunnel barriers, or produced by intentional doping. Transport theory in semiconductors therefore differs from theory for metals. One main difference is that theory is often motivated by device applications [129], involving junctions and high fields, which takes us outside the linear ohmic regime [130]. Another difference is that dilute carriers and low temperatures opens up the fascinating insulator to metal transition. And a third difference is that hopping provides an alternate mechanism to band transport.

In a pure semiconductor, electrical transport occurs *via* thermally activated electron and hole carriers $n_{\mathrm{e}} = n_{\mathrm{h}} \propto \exp(-E_{\mathrm{g}}/2k_{\mathrm{B}}T)$, with E_{g} the gap between occupied valence and empty conduction bands. The conductivity σ is written as $n_{\mathrm{e}}e\mu_{\mathrm{e}} + n_{\mathrm{h}}e\mu_{h}$. The mobilities μ_{e} and μ_{h} of electron and hole carriers are typically quite large. Hall measurements provide values for the carrier density, and mobilities can be measured by drift velocity methods. In Si at 300 K, electrons have $\mu_{\mathrm{e}} \approx 1.4 \times 10^3\,\mathrm{cm}^2/\mathrm{V\,s}$, and holes $\mu_{\mathrm{h}} \approx 4.5 \times 10^2\,\mathrm{cm}^2/\mathrm{V\,s}$ [131]. Experimental values of the temperature-dependent electron mobility are shown in Fig. 16 [132]. To explain these results [133] in detail requires solution of Boltzmann's equation, as discussed in a classic paper by Herring and Vogt [134], and performed by many authors using often Monte-Carlo procedures [132,135–137]. Silicon has electron carriers in 6 equivalent pockets, and holes in a heavy, a light, and a split-off hole band. One needs the scattering matrix elements for both intervalley and intravalley scattering by acoustic and optic phonons, making the total picture rather complex.

In doped semiconductors, the number of carriers no longer obeys $n_{\mathrm{e}} = n_{\mathrm{h}}$, but is determined by the temperature and the Fermi level which is fixed near the binding energy of the dominant impurity type. At higher temperatures, carriers of the majority type (electron or hole depending on whether doping is n- or p-type) are activated out of the impurity levels into band states which carry current by normal quasiparticle propagation. Scattering by ionized impurities now enters and often dominates.

At low temperatures, very interesting things happen to the transport properties of doped semiconductors. First consider lightly doped silicon. Fig. 17 shows the low T

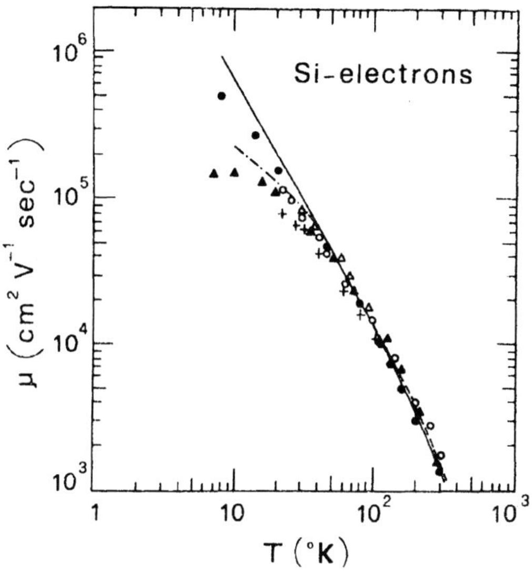

Fig. 16. Mobility versus temperature [132] for electrons in Si measured by time-of flight. The closed circles from Canali et al. used a very pure sample (donor and acceptor densities $\approx 10^{12}\,\text{cm}^{-3}$), while earlier data shown by other symbols have impurity densities larger by up to 100, causing ionized impurity scattering to dominate at lower T. The power law $\mu \propto T^{3/2}$ from acoustic phonon scattering [133] is in rough accord, but the data do not have any simple power law.

conductivity of Si with 2.7×10^{17} phosphorus donor atoms per cm^3, and 0.8×10^{15} boron acceptors [138]. Notice how small the conductivity is. The value of σ closely follows $\exp(-12\,\text{meV}/2k_BT)$ until the temperature is lowered to a frequency-dependent point where the T-dependence becomes much weaker. Taking the dc limit, one sees that the conductivity below 2 K bends away from the activated line, to a weaker T-dependence. The mechanism for this tiny residual conductivity is hopping between localized impurity states. Most of the donor electrons are bound in hydrogenic localized orbitals with Bohr radius $\approx 14\,\text{Å}$, about 10 times smaller than the typical donor atom separation. However, because of the 300 times smaller concentration of boron acceptors, about 0.3% of the donor states become empty, the bound phosphorus electron recombining with a bound boron hole, leaving a P^+ and a B^- ion. The P^+ sites offer a place for a bound electron (on a neutral P) to hop to. This requires some non-zero thermal energy, because the donor P sites do not all have the same donor binding energy. Their energies are perturbed by Coulomb fields $e^2/\varepsilon R$ of the neighboring P^+ and B^- ions. At the dopant density of Fig. 17, this spatially fluctuating potential has characteristic size of several meV. Therefore there are empty higher energy sites and filled lower energy sites, separated by a Fermi level, and zero conductivity at $T = 0$. At finite temperature, random thermal fluctuations occur. The dominant fluctuation is hopping of isolated electrons back and forth between two nearby sites whose site energies happen to lie on either side of the Fermi level. These fluctuations couple to an oscillatory E field, giving an ac

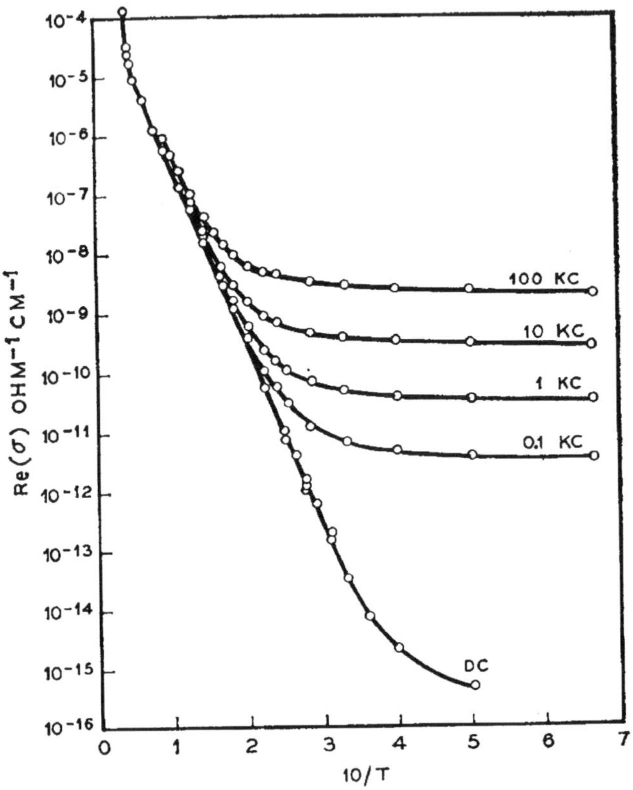

Fig. 17. Conductivity versus $10\,\mathrm{K}/T$ for Si with carrier density 10 times smaller than critical [138]. The linear slope comes from thermal activation out of the bound impurity states, and the strongly frequency-dependent low T limit comes from hopping.

conductance like a collection of random capacitors coupled by resistors. At a lower probability, there are longer range paths of electron propagation, giving a weakly activated hopping conduction which goes to zero as T goes to zero.

Mott [139] predicted successfully the form of the weakly activated hopping conduction, $\sigma_{\mathrm{VRH}} \propto \exp[-(T_0/T)^{1/4}]$, seen in many systems with localized charge carriers. The idea of "variable-range hopping" (VRH) is that the thermal energy $k_B T$ available to promote a hop may be too small to give a decent rate for nearby hops where wavefunctions $\psi \propto \exp(-\alpha r)$ have large overlap. Especially at low T, the hop may have to go to a farther neighbor with smaller overlap. A compromise is reached between the overlap $\exp(-2\alpha R)$ and the probability of available energy $\exp(-\Delta E(R)/k_B T)$ where $\Delta E(R)$ is the likely minimum energy hop available within a radius R of the starting point. This energy scales as $1/N(\varepsilon_F)R^3$. The optimum distance is found by minimizing the product by R, and gives a likely range $R_0 \propto [\alpha N(\varepsilon_F)k_B T]^{-1/4}$. Weakly activated hopping with $T^{1/4}$ is the result. Mott's arguments not only agree with many experiments [140] but have also been confirmed theoretically by various methods [141–143].

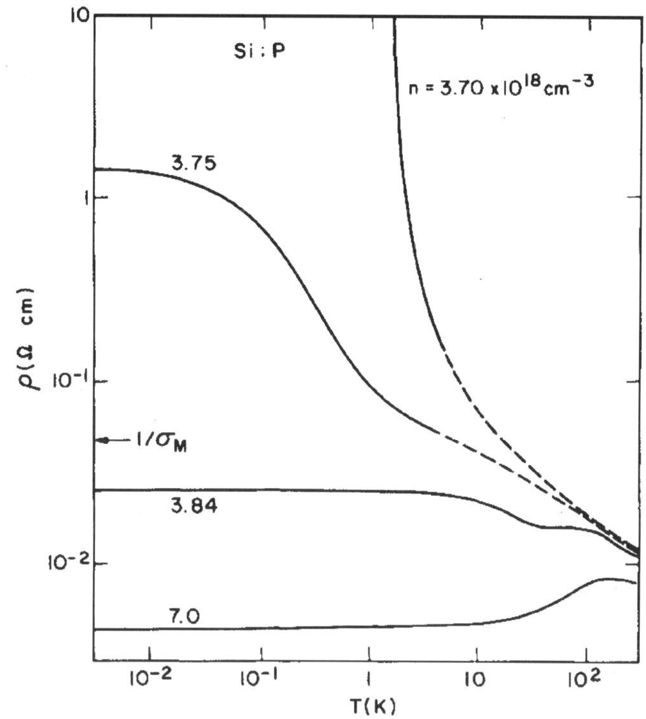

Fig. 18. Resistivity versus T of uncompensated Si, for four phosphorus concentrations near the critical concentration $\approx 3.75 \times 10^{18}\,\mathrm{cm}^{-3}$ [144].

Contrast this weak conductance at low doping with the measured resistivity shown in Fig. 18 [144] at phosphorus doping higher by 10. At a sharp critical donor concentration $n_c \approx 4 \times 10^{18}\,\mathrm{cm}^{-3}$ [145,146], where the average separation of localized impurity states drops to about 4 times the Bohr radius of these states, a continuous transition begins. Electron states at the Fermi level delocalize, and the $T = 0$ conductivity is no longer 0. The material is no longer insulating, and so must be called a "metal." This "Anderson transition" [147] is now known to occur in 3d but not in 1d where *all* states are localized no matter how weak is the disorder. The intermediate case of 2d is marginal and still controversial [148].

The transition from localized to delocalized is somewhat subtle, in that it does not show up in the single-particle density of states or in the single-particle Green's function averaged over a macroscopic system. Figure 19 [149] shows infrared spectra for three different doping levels, all below the critical concentration. At light doping, the hydrogenic impurity levels show up as lines in the infrared as expected. At heavier doping, it is perhaps not surprising that the lines are broadened by impurity overlap, which eliminates any sign of discrete levels at doping n 3 times below n_c. One might guess that the spectrum at the highest doping shown would indicate delocalized states, but dc conductivity shows otherwise – it goes to zero as T goes to zero, although at any achievable temperature it is far higher than the

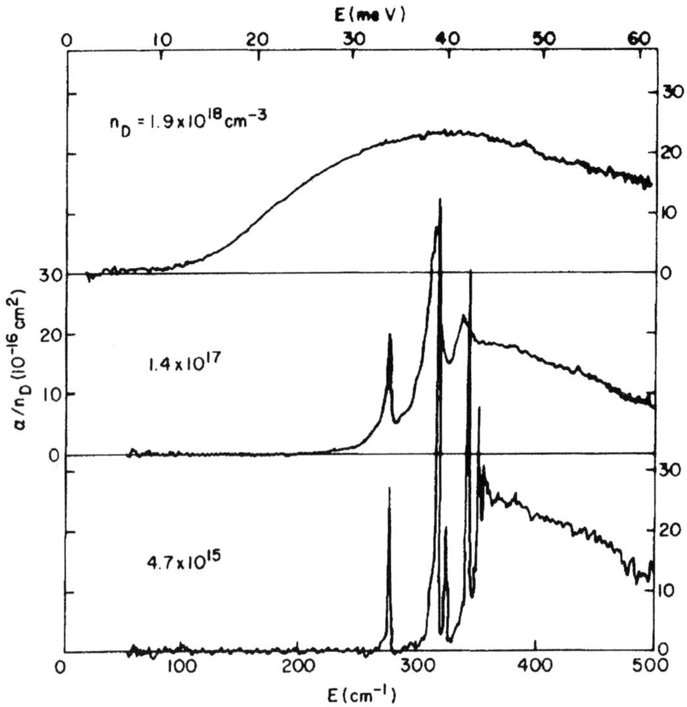

Fig. 19. Absorption coefficient α (in cm^{-1}) per donor (in cm^{-3}) measured at $T = 2\,K$ for uncompensated P-doped Si versus infrared photon energy [149]. Bound to bound impurity transitions are seen at the lowest doping, and broaden due to overlap of neighboring impurity states at heavier doping.

conductivity in Fig. 17 at 1.2 K. Before Anderson's paper, it was assumed that delocalization always occurred to some degree. After Anderson's paper was understood and generalized by Mott and others [150], a different picture emerged. The impurity levels are always localized when they are very dilute. As the concentration increases, overlapping pairs and clusters occur. The excited states of these dopant atoms overlap even more, and perhaps become delocalized, and probably merge with the unoccupied conduction band. However, the n_{imp} ground levels, which are distributed in energy by random perturbations, have the ability to resist delocalization, at least in the lower energy part of their spectrum. Somewhere higher in the spectrum there is necessarily a sharply defined energy ε_c, called the "mobility edge," which separates localized from delocalized states. The insulator to metal transition occurs when the mobility edge coincides with the Fermi level. The measured resistivity [151] at very low T is plotted in Fig. 20 for a series of samples across this transition.

It is also known [152,153] that Coulomb interactions between electrons have important consequences for the metal/insulator transition. Many of the experimental studies have been on "uncompensated" samples where only one species of

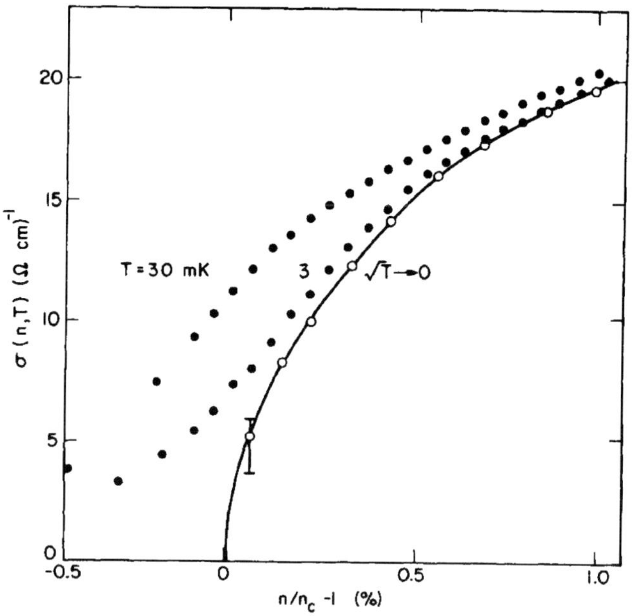

Fig. 20. Conductivity versus dopant density at two temperatures (30 and 3 mK) plus (open circles) the extrapolation to $T = 0$ assuming a $\sqrt{T}$ law [151]. The curve is a fit with critical exponent $\nu = 1/2$. The measurements are all on the same sample, with dopant density changes simulated by varying the applied stress.

dopant occurs. ("Compensated" means that both n and p type dopants occur.) At low and uncompensated doping and $T = 0$, all dopant levels are singly occupied. There are no ionized impurity atoms, and no doubly occupied impurity levels because of the significant on-site repulsion, comparable to the binding energy. As doping increases and dopant levels start to overlap, the transition to metallic conduction may be more like a Mott transition [154,155], dominated by correlations, instead of an Anderson transition, dominated by disorder. This is still a fascinating and controversial subject. The data of Fig. 20 show a critical exponent $\sigma(T = 0) \propto (n/n_c - 1)^\nu$ with $\nu \approx 1/2$. The theory of the pure Anderson transition (with no Coulombic electron–electron interactions) predicts ν to be close to 1. When samples are intentionally compensated, most experiments seem to show an exponent closer to 1, as is also seen in diluted metals like Nb_xSi_{1-x} [108]. However, the data are not necessarily good close enough to the transition to measure the critical exponent [156].

12. COULOMB BLOCKADE

Coulomb interactions alter transport properties in many ways. In homogeneous 3d metals, little influence is seen in low-frequency transport, apart from screening all interactions and thus affecting the single-particle spectrum. Electron–phonon or impurity scattering overpower Coulomb scattering as a relaxation mechanism, owing to the Pauli principle which suppresses Coulomb scattering by $(k_B T/\varepsilon_F)^2$ as shown in Section 6.3.

More dramatic Coulomb effects are seen at low T in special situations, of which the simplest is the "Coulomb blockade" of electron tunneling. Consider a case such as Fig. 4 or 21c, with electrons tunneling through a small conducting island separated from both leads by tunnel barriers. The addition of a single electron to the island raises the electrostatic energy of the island by $\Delta U_I = e^2/2C_{eff}$ where C_{eff} is the total island capacitance. A gate electrode coupled to the island capacitively with capacitance C_g and gate potential V_g, can raise and lower the energy of this added electron by $\Delta U_g = e(C_g/C_{eff})V_g$. If (1) source-drain bias V_{SD}, (2) temperature $k_B T$, and (3) single-particle electron level spacing $\Delta\varepsilon_I$ on the island, are all small compared to ΔU_I, conductance through the island is suppressed. The suppression is periodically modulated by the gate. Whenever the gate voltage is tuned so that the energy for adding one electron aligns with the source and drain electrode Fermi level, the conductance peaks. This happens periodically with spacing $\Delta V_g = e/C_g$, and corresponds to successive increases of the island's net electron charge. These effects in the single-particle tunneling regime were predicted by Averin and Likharev [157] and seen first by Fulton and Dolan [158]. The device is called a "single-electron transistor," and an "orthodox theory" [159] gives accurate fits to data. When the island level spacing $\Delta\varepsilon_I$ is small compared with $k_B T$, a simple theoretical expression due to Kulik and Shekhter [160] applies,

$$G = G_{max} \frac{\Delta U/k_B T}{\sinh(\Delta U/k_B T)} \tag{87}$$

where $\Delta U = \Delta U_g - \Delta U_I$ and $1/G_{max} = 1/G_1 + 1/G_2$ is the peak conductance, with G_1 and G_2 the conductances of the tunnel barriers to the source and drain electrodes.

Figure 21(a and b) shows the conductance versus gate voltage for a junction of metallic Al electrodes and an Al island (size $\approx (40\,\text{nm})^3$) separated by an aluminum oxide tunnel barrier [161]. The conductance peak G_{max} is smaller by 100 than the quantum unit $G_0 = 2e^2/h$ seen in the ballistic point contact, Fig. 3. This indicates that tunneling rather than ballistic conductance is occurring. The level spacing of Al valence states in such an island is $\approx 3\,\text{mK}$, less than the temperatures used (400, 200, and 6 mK in panel (a), and 300, 200, 100, and 50 mK in panel (b)). The data on panel (b) fit very well to the theory of Averin et al. [162] which extends the formula of Kulik and Shekhter to include "cotunneling," a higher-order process where tunneling across both barriers is coordinated, leaving an electron-hole pair on the island. This process increases with T quadratically as the phase space for the electron-hole pairs increases. Figure 21d, for a smaller Al island [163], shows the more complete story for source drain voltages which are no longer small compared with the other energies, and when finite level spacing due to island size quantization starts to set in.

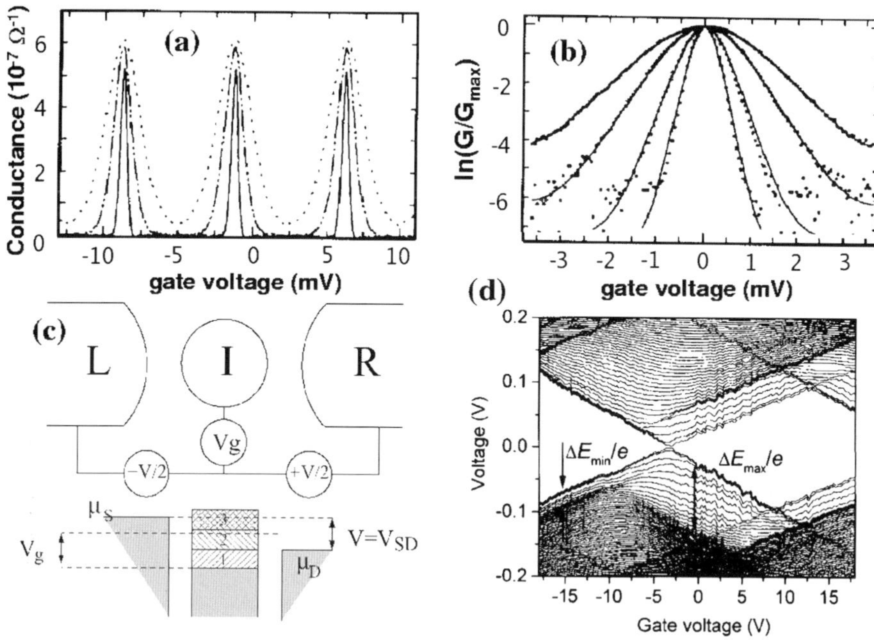

Fig. 21. Coulomb blockade of electron tunneling through a metallic Al island separated by oxide from Al electrodes [161,163]. Panels (a) and (b), for small source-drain voltage $V_{SD} = V$, show conductance versus gate voltage at various temperatures: from top to bottom, T = 400 mK, 200 mK, and 6 mK in (a), and T = 300 mK, 200 mK, 100 mK, and 50 mK in (b). The linear conductance peaks whenever the gate voltage is tuned to a point where one more electron can hop onto the island without increase of energy. This is indicated in the schematic diagram (c) where the energy levels of the island are denoted as continua with different discrete island Fermi levels corresponding to additional electrons 1, 2, 3 on the island. For infinitesimal V_{SD}, conductance is shut off unless the Fermi level of the island for some change state n aligns with the source and drain chemical potentials μ_S, μ_D. Panel (d) shows data at 4.2 K for a very small island. The lines are contours of constant current in increments of 50 pA. Data of panels a and b (on bigger islands) correspond to the horizontal line $V = V_{SD} = 0$. At constant but non-zero V_{SD}, current can flow for a non-zero interval of gate voltage. The width of this interval increases with V_{SD} giving diamond-shaped openings of blocked current. Fine structure is caused by energy level discreteness on the island.

13. COULOMB GAP

Important but more subtle effects of the Coulomb interaction are seen in systems near a metal to insulator transition, and appear as a suppression of electron density of states $N(\varepsilon)$ for small $|\varepsilon - \mu|$ where μ is the Fermi level. Efros and Shklovskii [152] found that $N(\varepsilon)$ vanishes as $|\varepsilon - \mu|^p$ with $p \approx 2$ when the Fermi level lies on the insulating side of the transition. Altshuler and Aronov [153] and McMillan [164] found a cusp-like suppression, $N(\varepsilon) = N(\varepsilon_F)[1 + (|\varepsilon - \mu|/\delta)^{1/2}]$ when ε_F lies on the metallic side. Tunneling conductance of boron-doped silicon near the metal

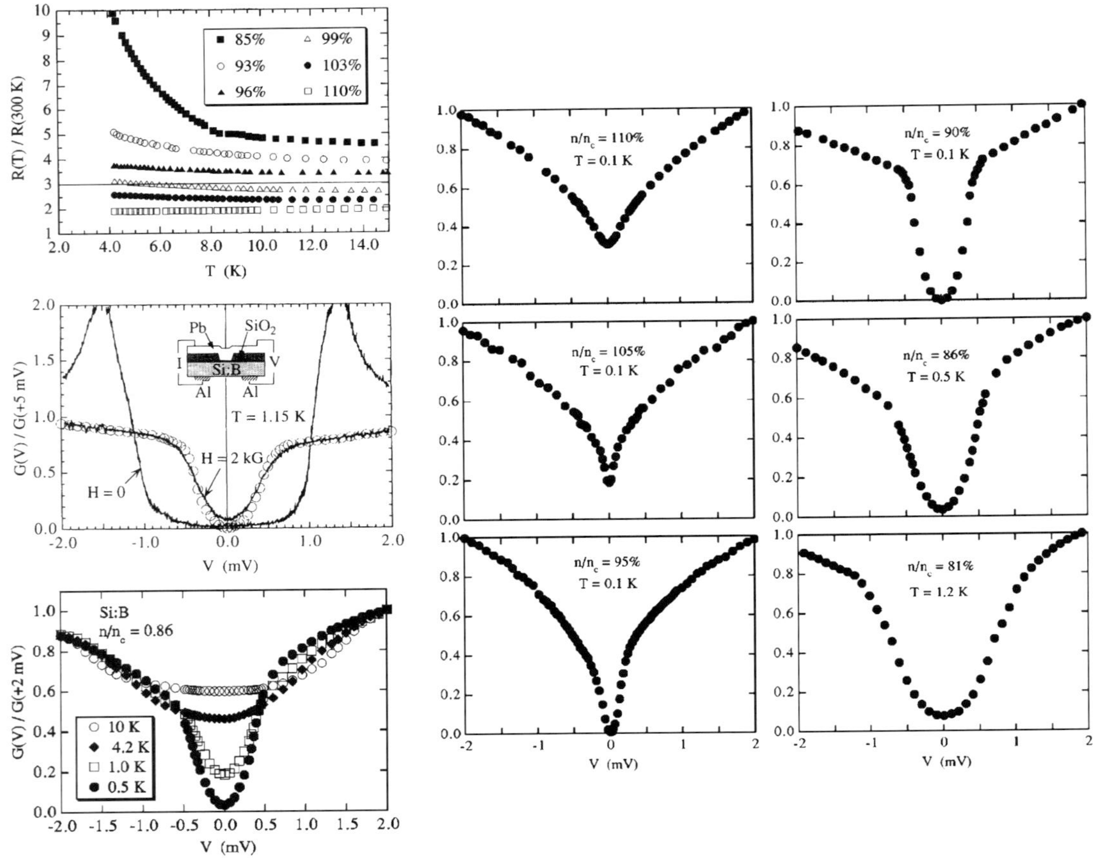

Fig. 22. Coulomb gap in boron-doped silicon at concentrations near the metal to insulator transition [165–167]. Upper left: Resistance versus T for six samples with various dopings n/n_c with $n_c = 4 \times 10^{18}\ \text{cm}^{-3}$ the critical concentration. Middle left: $H = 0$ conductance shows the gap of the Pb electrodes, verifying that junctions do exhibit tunneling. Six panels on right: Conductance, intepreted as thermally smeared density of states with a cusp in the metallic samples and a soft Coulomb gap in the insulating samples. Lower left: By 10 K, the anomaly is gone and all samples look rather similar.

insulator transition at $n_c = 4 \times 10^{18}\,\text{cm}^{-3}$ measured by Lee et al. [165–167], shown in Fig. 22, confirm these ideas. In the tunneling regime, with a barrier height large compared to bias voltage, the tunneling conductance $G(V)$ is proportional to the thermally broadened density of states $\int d\varepsilon N(\varepsilon)\partial f(\varepsilon - eV)/\partial V$. Very similar effects were seen for amorphous Nb_xSi_{1-x} near the metal to insulator transition at $x = 0.115$ by Hertel et al. [108].

On the insulating side, the nearly complete depletion for ε near μ is called a "soft" Coulomb gap. The explanation by Efros and Shlovskii is very simple. Suppose there are localized electron states at random positions $\mathbf{R}_i$, with a random distribution of energies h_i, before adding the Coulomb repulsion between the electrons which will occupy some of these sites. For each occupied pair, there is a repulsive energy $v_{ij} = e^2/\kappa R_{ij}$ where κ is the dielectric constant and R_{ij} the distance. The problem is to find the stable occupancy assuming fewer electrons than sites. The energy to remove one electron, leaving the rest fixed, is $\xi_i = h_i + \sum_j v_{ij} n_j$ where n_j is the occupancy of site j. The energy to add one more electron in a previously empty state is $\xi_k = h_k + \sum_j v_{kj} n_j$. The density of states is

$$N(\varepsilon) = -\frac{1}{\pi}\sum_i \text{Im}G(i,\varepsilon) = \sum_i \delta(\varepsilon - \xi_i) \tag{88}$$

where the single particle removal energy is used for $\xi < \mu$ and the single particle addition energy for $\xi > \mu$, as is usual for the spectral density of states defined from the Green's function, and measured in the tunneling experiment. Efros and Shklovski point out that the energy difference Δ between the N particle ground state and the excited state with state i removed and state j added is $\xi_j - \xi_i - v_{ij}$. Although the first part $\xi_j - \xi_i$ is positive, the second part is negative, and it is not obvious that the sum is positive for small $|\xi_j - \xi_i|$. In fact, the only way to guarantee that Δ is always positive is for the ground state occupancy to be organized such that the states ξ_i are depleted near μ.

The Altshuler–Aronov argument for the cusp on the metal side is quite different, and relies on being able to compute perturbatively around the metallic state with good Fermi-liquid behavior and no disorder. Thus, there is no evident reason why the theoretically predicted anomalies on the two sides of the transition should be related. Experiment shows forcefully the unity of the phenomena. For $T \geqslant 10\,\text{K}$ (shown in panel (a) of Fig. 22) the resistivity does not distinguish insulator from metal. Tunneling conductances measured at low T show $N(\varepsilon)$ recovering rapidly from the low ε anomaly and behaving similarly for metal and insulator. It is reminiscent of the resistivity of high T_c superconductors in Fig. 8, which at higher T look similar for metal and insulator [84], defying theory.

ACKNOWLEDGMENTS

I thank D. V. Averin, Y. Gilman, and K. K. Likharev for help in preparing this review. The work was supported in part by NSF Grant ATM-0426757.

REFERENCES

[1] P. Drude, Zur Elektronentheorie der Metalle, *Ann. Phys. (Leipzig) Ser.* **4**(1), 566–613 (1900).

[2] A. Sommerfeld, Zur elektronentheorie der metalle auf grund der fermischen statistik, *Zeits. and Physik* **47**, 1–32 and 43–60 (1928).

[3] A. Sommerfeld and H. Bethe, Elektronentheorie der Metalle, *Handbook Phys.* **24/2**, 1–290 (1932).

[4] F. Bloch, Über die quantenmechanik der elektronen in kristallgittern, *Zeits. Phys.* **52**, 555–600 (1928).

[5] L.D. Landau, Oscillations in a fermi liquid, *J. Exp. Theor. Phys.* **30**, 1058 (1956), engl. transl. in *Men of Physics: L. D. Landau I*, edited by D. ter Haar, (Pergamon, Oxford, 1965), pp. 105–118.

[6] P.B. Johnson and R.W. Christy, Optical constants of copper and nickel as a function of temperature, *Phys. Rev. B* **11**, 1315–1323 (1975).

[7] D.A. Papaconstantopoulos, *Handbook of the Band Structure of Elemental Solids* (Plenum, New York, 1986).

[8] J.C. Maxwell, *A Treatise on Electricity and Magnetism*, Dover, New York.

[9] Y.V. Sharvin, A possible method for studying Fermi surfaces, *Sov. Phys. JETP* **21**, 655 (1965).

[10] G. Wexler, Size effect and non-local Boltzmann transport equation in orifice and disk geometry, *Proc. Phys. Soc. London* **89**, 927 (1966).

[11] B. Nikolic and P.B. Allen, Electron transport through a circular constriction, *Phys. Rev. B* **60**, 3963–3966 (1999).

[12] B.J. van Wees, H. van Houten, C.W.J. Beenakker, J.G. Williamson, L.P. Kouwenhoven, D. van der Marel and C.T. Foxon, Quantized conductance of point contacts in a two-dimensional electron gas, *Phys. Rev. Lett.* **60**, 848–850 (1988).

[13] B.J. van Wees, L.P. Kouwenhoven, E.M.M. Willems, C.J.P.M. Harmans, J.E. Mooij, H. van Houten, C.W.J. Beenakker, J.G. Williamson and C.T. Foxon, Quantum ballistic and adiabatic electron transport studied with quantum point contacts, *Phys. Rev. B* **43**, 12431–12453 (1991).

[14] D.A. Wharam, T.J. Thornton, R. Newbury, M. Pepper, H. Ahmed, J.E.F. Frost, D.G. Hasko, D.C. Peacock, D.A. Ritchie and G.A.C. Jones, One-dimensional transport and the quantisation of the ballistic resistance, *J. Phys. C* **21**, L209–L214 (1988).

[15] R. de Picciotto, H.L. Störmer, L.N. Pfeiffer, K.W. Baldwin and K.W. West, Four-terminal resistance of a ballistic quantum wire, *Nature* **411**, 51–54 (2001).

[16] M.A. Topinka, B.J. LeRoy, S.E.J. Shaw, E.J. Heller, R.M. Westervelt, K.D. Maranowski and A.C. Gossard, Imaging coherent electron flow from a quantum point contact, *Science* **289**, 2323–2326 (2000).

[17] R. Kubo, Statistical mechanical theory of irreversible processes. I. General theory and simple applications to magnetic and conduction problems, *J. Phys. Soc. Jpn.* **12**, 570–586 (1957).

[18] G. Rickayzen, *Green's Functions and Condensed Matter* (Academic Press, London, 1980).

[19] G.D. Mahan, *Many Particle Physics (Physics of Solids and Liquids)*, 3rd ed. (Kluwer Academic, 2000).

[20] G.M. Eliashberg, Transport equation for a degenerate system of Fermi particles, *Sov. Phys. JETP* **14**, 886–892 (1962).

[21] R.E. Prange and L.P. Kadanoff, Transport theory for electron–phonon interactions in metals, *Phys. Rev.* **134**, A566–A580 (1964).

[22] T.D. Holstein, Theory of transport phenomena in an electron–phonon gas, *Ann. Phys.* **29**, 410–535 (1964).

[23] J.M. Ziman, *Models of Disorder* (Cambridge University Press, Cambridge, 1979).

[24] D.A. Greenwood, The Boltzmann equation in the theory of electrical conduction in metals, *Proc. Phys. Soc.* **71**, 585–591 (1958).

[25] R. Landauer, Spatial variation of currents and fields due to localized scatterers in metallic conduction, *IBM J. Res. Dev.* **1**, 223–331 (1957), **32**, 306–316 (1988).

[26] D.S. Fisher and P.A. Lee, Relation between conductivity and transmission matrix, *Phys. Rev. B* **23**, 6851–6854 (1981).

[27] S. Datta, *Electronic Transport in Mesoscopic Systems* (Cambridge University Press, Cambridge, 1995).
[28] H. Haug and A.-P. Jauho, *Quantum Kinetics in Transport and Optics of Semiconductors* (Springer, 1996).
[29] Y. Imry and R. Landauer, Conductance viewed as transmission, *Rev. Mod. Phys.* **71**, S306–S312 (1999).
[30] Y. Meir and N.S. Wingreen, Landauer formula for the current through an interacting electron region, *Phys. Rev. Lett.* **68**, 2512–2515 (1992).
[31] C. Caroli, R. Combescot, P. Nozieres and D. Saint-James, Direct calculation of the tunneling current, *J. Phys. C: Sol. State Phys.* **4**, 916–929 (1971).
[32] D. Kalkstein and P. Soven, A Green's function theory of surface states, *Surf. Sci.* **26**, 85–99 (1971).
[33] T.N. Todorov, Calculation of the residual resistivity of three-dimensional quantum wires, *Phys. Rev. B* **54**, 5801–5813 (1996).
[34] B.K. Nikolic and P.B. Allen, Resistivity of a metal between the Boltzmann transport regime and the Anderson transition, *Phys. Rev. B* **63**, 020201:1–4 (2001).
[35] Y. Gilman, J. Tahir-Kheli, P.B. Allen and W.A. Goddard III, Numerical study of resistivity of model disordered three-dimensional metals, *Phys. Rev. B* **70**, 224201:1–3 (2004).
[36] S. Datta, Nanoscale device modeling: The Green's function method, *Superlatt. Microstr.* **28**, 253–278 (2000).
[37] N.D. Lang and P. Avouris, Electrical conductance of individual molecules, *Phys. Rev. B* **64**, 125323:1–7 (2001).
[38] D.M. Adams, L. Brus, C.E.D. Chidsey, S. Creager, C. Creutz, C.R. Kagan, P.V. Kamat, M. Lieberman, S. Lindsay, R.A. Marcus, R.M. Metzger, M.E. Michel-Beyerle, J.R. Miller, M.D. Newton, D.R. Rolison, O. Sankey, K. Schanze, J. Yardley and X. Zhu, Charge transfer on the nanoscale: Current status, *J. Phys. Chem B* **107**, 6668–6697 (2003).
[39] A. Nitzan and M.A. Ratner, Electron transport in molecular wire junctions, *Science* **300**, 1384–1389 (2003).
[40] M. Brandbyge, J.-L. Mozos, P. Ordejn, J. Taylor and K. Stokbro, Density-functional method for nonequilibrium electron transport, *Phys. Rev. B* **65**, 165401:1–17 (2002).
[41] A.F. Andreev, The thermal conductivity of the intermediate state in superconductors, *Sov. Phys. JETP* **19**, 1228–1231 (1964).
[42] R.J. Soulen Jr., J.M. Byers, M.S. Osofsky, B. Nadgorny, T. Ambrose, S.F. Cheng, P.R. Broussard, C.T. Tanaka, J. Nowak, J.S. Moodera, A. Barry and J.M.D. Coey, Measuring the spin polarization of a metal with a superconducting point contact, *Science* **282**, 85–88 (1998).
[43] G.E. Blonder, M. Tinkham and T.M. Klapwijk, Transition from metallic to tunneling regimes in superconducting microconstrictions: Excess current, charge imbalance, and supercurrent conversion, *Phys. Rev. B* **25**, 4512–4532 (1982).
[44] Report of the investigation committee on the possibility of scientific misconduct in the work of Hendrik Schön and coauthors, Tech. rep., http://publish.aps.org/reports/lucentrep.pdf (2002).
[45] J. Yamashita and S. Asano, Electrical resistivity of transition metals I, *Prog. Theor. Phys.* **51**, 317–326 (1974).
[46] F.J. Pinski, P.B. Allen and W.H. Butler, Calculated electrical and thermal resistivities of Nb and Pd, *Phys. Rev. B* **23**, 5080–5096 (1981).
[47] W.L. Schaich and N.W. Ashcroft, Model calculations in the theory of photoemission, *Phys. Rev. B* **2**, 2452–2465 (1971).
[48] G.D. Mahan, Theory of photoemission in simple metals, *Phys. Rev. B* **2**, 4334–4350 (1970).
[49] J.M. Ziman, *Electrons and Phonons* (Oxford University Press, London, 1960), chap. VII.
[50] P.B. Allen, Boltzmann Theory and Resistivity of Metals, *in Quantum Theory of Real Materials* (Kluwer, Boston, 1996), chap. 17.
[51] L.D. Landau and E.M. Lifshitz, *Statistical Physics*, 3rd ed. part I (Pergamon Press, Oxford, 1980), sect. 55.
[52] P.B. Allen, Fermi surface harmonics: A general method for non-spherical problems. Application to Boltzmann and Eliashberg equations, *Phys. Rev. B* **13**, 1416–1427 (1976).

[53] P.B. Allen, New method for solving Boltzmann's equation for electrons in metals, *Phys. Rev. B* **17**, 3725–3734 (1978).
[54] P.B. Allen, Empirical electron–phonon λ values from resistivity of cubic metallic elements, *Phys. Rev. B* **36**, 2920–2923 (1987).
[55] V.F. Gantmakher and Y.B. Levinson, *Carrier Scattering in Metals and Semiconductors* (North-Holland, 1987).
[56] B. Raquet, M. Viret, E. Sondergard, O. Cespedes and R. Mamy, Electron–magnon scattering and magnetic resistivity in 3*d* ferromagnets, *Phys. Rev. B* **66**, 024433 (2002).
[57] J. Bass, Deviations from Matthiessen's rule, *Adv. Phys.* **21**, 431 (1972).
[58] T.P. Beaulac, P.B. Allen and F.J. Pinski, Electron–phonon effects in copper. II. Electrical and thermal resistivities and Hall coefficient, *Phys. Rev. B* **26**, 1549–1558 (1982).
[59] G.M. Eliashberg, Interactions between electrons and lattice vibrations in a superconductor, *Sov. Phys. JETP* **11**, 696–702 (1960).
[60] D.J. Scalapino, The electron–phonon interaction and strong-coupling superconductors, *Superconductivity* (M. Dekker, New York, 1969) chap. 10.
[61] P.B. Allen and B. Mitrović, Theory of superconducting T_c, *Solid State Physics* (Academic, New York, 1982) **37**, pp. 1–92.
[62] J.P. Carbotte, Properties of boson-exchange superconductors, *Rev. Mod. Phys.* **62**, 1027–1157 (1990).
[63] S.P. Rudin, R. Bauer, A.Y. Liu and J.K. Freericks, Reevaluating electron–phonon coupling strengths: Indium as a test case for *ab initio* and many-body theory methods, *Phys. Rev. B* **58**, 14511–14517 (1998).
[64] W. McMillan, Transition temperature of strong-coupled superconductors, *Phys. Rev.* **167**, 331–344 (1968).
[65] P.B. Allen and R.C. Dynes, Transition temperature of strong-coupled superconductors reanalyzed, *Phys. Rev. B* **12**, 905–922 (1975).
[66] P.B. Allen, T. Beaulac, F. Khan, W. Butler, F. Pinski and J. Swihart, dc transport in metals, *Phys. Rev. B* **34**, 4331–4333 (1986).
[67] S. Baroni, P. Giannozzi and A. Testa, Green-function approach to linear response in solids, *Phys. Rev. Lett.* **58**, 1861–1864 (1987).
[68] P. Giannozzi, S. de Gironcoli, P. Pavone and S. Baroni, *Ab initio* calculation of phonon dispersions in semiconductors, *Phys. Rev. B* **43**, 7231–7242 (1991).
[69] S. Baroni, S. de Gironcoli, A.D. Corso and P. Giannozzi, Phonons and related crystal properties from density-functional perturbation theory, *Rev. Mod. Phys.* **73**, 515–562 (2001).
[70] S.Y. Savrasov and D.Y. Savrasov, Electron–phonon interactions and related physical properties of metals from linear-response theory, *Phys. Rev. B* **54**, 16487–16501 (1996).
[71] R. Bauer, A. Schmid, P. Pavone and D. Strauch, Electron–phonon coupling in the metallic elements Al, Au, Na, and Nb: A first-principles study, *Phys. Rev. B* **57**, 11276–11282 (1998).
[72] S.Y. Savrasov and O.K. Andersen, Linear-response calculation of the electron–phonon coupling in doped $CaCuO_2$, *Phys. Rev. Lett.* **77**, 4430–4433 (1996).
[73] H.J. Choi, D. Roundy, H. Sun, M.L. Cohen and S.G. Louie, The origin of the anomalous superconducting properties of MgB_2, *Nature* **418**, 758–760 (2002).
[74] K.P. Bohnen, R. Heid and B. Renker, Phonon dispersion and electron–phonon coupling in MgB_2 and AlB_2, *Phys. Rev. Lett.* **86**, 5771–5774 (2001).
[75] J. Kortus, I.I. Mazin, K.D. Belashchenko, V.P. Antropov and L.L. Boyer, Superconductivity of metallic boron in MgB_2, *Phys. Rev. Lett.* **86**, 4656–4659 (2001).
[76] (a) T. Yildirim, O. Gulseren, J.W. Lynn, C.M. Brown, T.J. Udovic, Q. Huang, N. Rogado, K.A. Regan, M.A. Hayward, J.S. Slusky, T. He, M.K. Haas, P. Khalifah, K. Inumaru and R.J. Cava, Giant anharmonicity and nonlinear electron–phonon coupling in MgB_2 A combined first-principles calculation and neutron scattering study, *Phys. Rev. Lett.* **87**, 037001 (2001).
(b) I.I. Mazin, O.K. Andersen, O. Jepsen, O.V. Dolgov, J. Kortus, A.A. Golubov, A.B. Kuz'menko and D. van der Marel, Superconductivity in MgB_2: Clean or Dirty? *Phys. Rev. Lett.* **89**, 107002 (2002).

[77] F. Bloch, Zum elektrischen Widerstandsgesetz bei tiefen Temperaturen, *Z. Phys.* **59**, 208–214 (1930).

[78] E. Grüneisen, Die Abhängigkeit des elektrischen Widerstandes reiner Metalle von der Temperatur, *Ann. Phys. (Leipzig)* **4**, 530–540 (1933).

[79] W. Kohn and L.J. Sham, Self-consistent equations including exchange and correlation effects, *Phys. Rev* **140**, A1133–A1138 (1965).

[80] P.B. Allen, Electron–Phonon coupling constants, *Handbook of Superconductivity* (Academic Press, San Diego, 2000) pp. 478–489.

[81] B.A. Sanborn, P.B. Allen and D.A. Papaconstantopoulos, Empirical electron–phonon coupling constants and anisotropic electrical resistivity in hcp metals, *Phys. Rev. B* **40**, 6037–6044 (1989).

[82] L.J. Sham and W. Kohn, One-particle properties of an inhomogeneous interacting electron gas, *Phys. Rev.* **145**, 561–567 (1966).

[83] T.A. Friedmann, M.W. Rabin, J. Giapintzakis, J.P. Rice and D.M. Ginsberg, Direct measurement of the anisotropy of the resistivity in the a-b plane of twin-free, single-crystal, superconducting $YBa_2Cu_3O_{7-\delta}$, *Phys. Rev. B* **42**, 6217–6221 (1990).

[84] Y. Ando, A.N. Lavrov, S. Komiya, K. Segawa and X.F. Sun, Mobility of the doped holes and the antiferromagnetic correlations in underdoped high-T_c cuprates, *Phys. Rev. Lett.* **87**, 017001:1–4 (2001).

[85] P.B. Allen, W.E. Pickett and H. Krakauer, Anisotropic normal-state transport properties predicted and analyzed for high-T_c oxide superconductors, *Phys. Rev. B* **37**, 7482–7490 (1988).

[86] W.F. Brinkman and T.M. Rice, Single-particle excitations in magnetic insulators, *Phys. Rev. B* **2**, 1324–1338 (1970).

[87] M. Gurvitch and A.T. Fiory, Resistivity of $La_{1.825}Sr_{0.175}CuO_4$ and $YBa_2Cu_3O_7$ to 1100 K: absence of saturation and its implications, *Phys. Rev. Lett.* **59**, 1337–1340 (1987).

[88] V. Chandrasekhar, P. Santhanam, N.A. Penebre, R.A. Webb, H. Vloeberghs, C.V. Haesendonck and Y. Bruynseraede, Absence of size dependence of the Kondo resistivity, *Phys. Rev. Lett.* **72**, 2053–2056 (1994).

[89] P.W. Anderson, Localized magnetic states in metals, *Phys. Rev.* **124**, 41–53 (1961).

[90] J. Kondo, Resistance minimum in dilute magnetic alloys, *Progr. Theoret. Phys. (Kyoto)* **32**, 37–69 (1964).

[91] G. Grüner and A. Zawadowski, Magnetic impurities in non-magnetic metals, *Rep. Prog. Phys.* **37**, 1497 (1974).

[92] K.G. Wilson, The renormalization group: Critical phenomena and the Kondo problem, *Rev. Mod. Phys.* **47**, 773–840 (1975).

[93] H.R. Krishna-murthy, J.W. Wilkins and K.G. Wilson, Renormalization-group approach to the Anderson model of dilute magnetic alloys. I. Static properties for the symmetric case, *Phys. Rev. B* **21**, 1003–1043 (1980).

[94] N. Andrei, K. Furiya and J.H. Lowenstein, Solution of the Kondo problem, *Rev. Mod. Phys.* **55**, 331–402 (1983).

[95] P. Nozieres, A 'fermi-liquid' description of the Kondo problem at low temperatures, *J. Low. Temp. Phys.* **17**, 31–42 (1974).

[96] A.C. Hewson, *The Kondo problem to Heavy Fermions* (Cambridge University Press, Cambridge, 1997).

[97] K. Nagaoka, T. Jamneala, M. Grobis and M.F. Crommie, Temperature dependence of a single Kondo impurity, *Phys. Rev. Lett.* **88**, 077205 (2002).

[98] D. Goldhaber-Gordon, H. Shtrikman, D. Mahalu, D. Ambusch-Magder, U. Meirav and M.A. Kastner, Kondo effect in a single-electron transistor, *Nature* **391**, 156–159 (1998).

[99] J.H. Mooij, Electrical conduction in concentrated disordered transition-metal alloys, *Phys. Stat. Sol.* **a17**, 521–530 (1973).

[100] Z. Fisk and G.W. Webb, Saturation of the high-temperature. Normal-state electrical resistivity of superconductors, *Phys. Rev. Lett.* **36**, 1084–1086 (1976).

[101] O. Gunnarsson, M. Calandra and J.E. Han, Colloquium: Saturation of electrical resistivity, *Rev. Mod. Phys.* **75**, 1085–1099 (2003).

[102] M. Calandra and O. Gunnarsson, Electrical resistivity at large temperatures: Saturation and lack thereof, *Phys. Rev. B* **66**, 205105:1–20 (2002).
[103] M. Calandra and O. Gunnarsson, Violation of Ioffe-Regel condition but saturation of resistivity of the high-T_c cuprates, *Europhys. Lett.* **61**, 88–94 (2003).
[104] O. Gunnarsson and J.E. Han, The mean free path for electron conduction in metallic fullerenes, *Nature* **405**, 1027–1030 (2000).
[105] R.H. Brown, P.B. Allen, D.M. Nicholson and W.H. Butler, Resistivity of strong-scattering alloys: absence of localization and success of coherent-potential approximation confirmed by exact supercell calculations in $V_{1-x}Al_x$, *Phys. Rev. Lett.* **62**, 661–664 (1989).
[106] B.W. Dodson, W.L. McMillan, J.M. Mochel and R.C. Dynes, Metal-insulator transition in disordered Germanium-Gold alloys, *Phys. Rev. Lett.* **46**, 46–49 (1981).
[107] W.L. McMillan and J. Mochel, Electron tunneling experiments on amorphous $Ge_{1-x}Au_x$, *Phys. Rev. Lett.* **46**, 556–557 (1981).
[108] G. Hertel, D.J. Bishop, E.G. Spencer, J.M. Rowell and R.C. Dynes, Tunneling and transport measurements at the metal-insulator transition of amorphous Nb: Si, *Phys. Rev. Lett.* **50**, 743–746 (1983).
[109] P.B. Allen and J. Kelner, Evolution of a vibrational wavepacket on a disordered chain, *Am. J. Phys.* **66**, 497–506 (1998).
[110] R. Kahnt, The calculation of the resistivity of liquid and amorphous transition metals *via* the Landauer formula, *J. Phys. C: Condens. Matter* **7**, 1543–1556 (1995).
[111] R. Arnold and H. Solbrig, Disorder-induced resistivity of liquid and amorphous transition metals calculated within the scattered-wave supercell concept, *J. Non-Cryst. Solids* **205–207**, 861–865 (1996).
[112] P.B. Allen and B. Chakraborty, Infrared and d.c. conductivity in metals with strong scattering: non-classical behavior from a generalized Boltzmann equation containing band mixing effects, *Phys. Rev. B* **23**, 4815–4827 (1981).
[113] K.K. Larsen, M.V. Hove, A. Lauwers, R.A. Donaton, K. Maex and M.V. Rossum, Electronic transport in metallic iron disilicide, *Phys. Rev. B* **50**, 14200–14211 (1994).
[114] B.L. Altshuler, P.A. Lee and R.A. Webb (Eds), *Mesoscopic Phenomena in Solids* (North-Holland, Amsterdam, 1991).
[115] R.E. Peierls, On the kinetic theory of thermal conduction in crystals, *Ann. Phys. (Leipzig) Ser.* **5**(3), 1055–1101 (1929).
[116] S.M. Shapiro, G. Shirane and J.D. Axe, Measurements of the electron–phonon interaction in Nb by inelastic neutron scattering, *Phys. Rev. B* **12**, 4899–4908 (1975).
[117] J.D. Axe and G. Shirane, Influence of the superconducting energy gap on phonon linewidths in Nb_3Sn, *Phys. Rev. Lett.* **30**, 214–216 (1973).
[118] N. Wakabayashi, Phonon anomalies and linewidths in Nb at 10 K, *Phys. Rev. B* **33**, 6771–6774 (1986).
[119] K. Habicht, R. Golub, F. Mezei, B. Keimer and T. Keller, Temperature-dependent phonon lifetimes in lead investigated with neutron-resonance spin-echo spectroscopy, *Phys. Rev. B* **69**, 104301:1–8 (2004).
[120] T. Valla, A.V. Fedorov, P.D. Johnson, P.-A. Glans, C. McGuinness, K.E. Smith, E.Y. Andrei and H. Berger, Quasiparticle spectra, charge-density waves, superconductivity, and electron–phonon coupling in 2H-$NbSe_2$, *Phys. Rev. Lett.* **92**, 086401:1:4 (2004).
[121] A.B. Migdal, Interaction between electrons and lattice vibrations in a normal metal, *Sov. Phys. JETP* **7**, 996–1001 (1958).
[122] W.L. McMillan and J.M. Rowell, Tunelling and Strong-Coupling Superconductivity, *Superconductivity* (M. Dekker, New York, 1969), chap. 11.
[123] T. Holstein, Optical and infrared volume absorptivity of metals, *Phys. Rev.* **96**, 535–536 (1954).
[124] P.B. Allen, Electron–phonon effects in the infrared properties of metals, *Phys. Rev. B* **3**, 305–320 (1971).
[125] R.R. Joyce and P.L. Richards, Phonon contribution to far-infrared absorptivity of superconducting and normal lead, *Phys. Rev. Lett.* **24**, 1007–1010 (1970).

[126] B. Farnworth and T. Timusk, Phonon density of states of superconducting lead, *Phys. Rev. B* **14**, 5119–5120 (1976).
[127] W. Götze and P. Wölfle, Homogeneous dynamical conductivity of metals, *Phys. Rev. B* **6**, 1226 (1972).
[128] J. Huang, T. Timusk and G. Gu, High transition temperature superconductivity in the absence of the magnetic resonance mode, *Nature* **427**, 714–717 (2004).
[129] S.M. Sze, *Physics of Semiconductor Devices*, 2nd ed. (Wiley, New York, 1981).
[130] E.M. Conwell, *High Field Transport in Semiconductors* (Academic, New York, 1967).
[131] See the electronic archive New Semiconductor Materials maintained at the Ioffe Institute: ⟨http://www.ioffe.rssi.ru/SVA/NSM/Semicond/⟩.
[132] C. Canali, C. Jacoboni, F. Nava, G. Ottaviani and A.A. Quaranta, Electron drift velocity in silicon, *Phys. Rev. B* **12**, 2265–2284 (1975).
[133] F. Seitz, On the mobility of electrons in pure non-polar insulators, *Phys. Rev.* **73**, 549–564 (1948).
[134] C. Herring and E. Vogt, Transport and deformation-potential theory for many-valley semiconductors with anisotropic scattering, *Phys. Rev.* **101**, 944 (1956).
[135] P. Norton, T. Braggins and H. Levinstein, Impurity and lattice scattering parameters as determined from Hall and mobility analysis in n-type silicon, *Phys. Rev. B* **8**, 5632–5646 (1973).
[136] L.R. Logan, H.H.K. Tang and G.R. Srinivasan, Analytic solutions to the Boltzmann equation for electron transport in silicon, *Phys. Rev. B* **43**, 6581 (1991).
[137] B.A. Sanborn, P.B. Allen and G.D. Mahan, Theory of screening and electron mobility: Application to n-type silicon, *Phys. Rev. B* **46**, 15123 (1992).
[138] M. Pollak and T.H. Geballe, Low-frequency conductivity due to hopping processes in silicon, *Phys. Rev.* **122**, 1742–1753 (1961).
[139] N.F. Mott, Conduction in non-crystalline materials: 3. localized states in a pseudogap and near extremities of conduction and valence bands, *Philos. Mag.* **19**, 835 (1969).
[140] M. Pollak and B. Shklovskii (Eds), Hopping Transport in Solids (North-Holland, New York, 1991).
[141] V. Ambegaokar, B.I. Halperin and J.S. Langer, Hopping conductivity in disordered systems, *Phys. Rev. B* **4**, 2612–2620 (1971).
[142] W. Brenig, G. Döhler and P. Wölfle, Theory of thermally assisted electron hopping in amorphous solids, *Z. Phys.* **246**, 1–12 (1971).
[143] M. Pollak, A percolation treatment of dc hopping conduction, *J. Non-Cryst. Solids* **11**, 1024 (1972).
[144] T.F. Rosenbaum, R.F. Milligan, M.A. Paalanen, G.A. Thomas, R.N. Bhatt and W. Lin, Metal–insulator transition in a doped semiconductor, *Phys. Rev. B* **27**, 7509–7523 (1983).
[145] U. Thomanschefsky and D.F. Holcomb, Metal–insulator transition in the compensated semiconductor Si: (P,B), *Phys. Rev. B* **45**, 13356–13362 (1992).
[146] P. Dai, Y. Zhang and M.P. Sarachik, Critical conductivity exponent for Si: B, *Phys. Rev. Lett.* **66**, 1914–1917 (1991).
[147] P.W. Anderson, Absence of diffusion in certain random lattices, *Phys. Rev.* **109**, 1492–1505 (1958).
[148] S.V. Kravchenko, G.V. Kravchenko, J.E. Furneaux, V.M. Pudalov and M. D'Iorio, Possible metal-insulator transition at B = 0 in two dimensions, *Phys. Rev. B* **50**, 8039–8042 (1994).
[149] G.A. Thomas, M. Capizzi, F. DeRosa, R.N. Bhatt and T.M. Rice, Optical study of interacting donors in semiconductors, *Phys. Rev. B* **23**, 5472 (1981).
[150] N.F. Mott, *Metal-Insulator Transitions* (Taylor and Francis, London, 1974).
[151] G.A. Thomas, M. Paalanen and T.F. Rosenbaum, Measurements of conductivity near the metal-insulator critical point, *Phys. Rev. B* **27**, 3897–3900 (1983).
[152] A.L. Efros and B.I. Shklovskii, Coulomb gap and low temperature conductivity of disordered systems, *J. Phys. C: Solid State Phys.* **8**, L49–L51 (1975).
[153] B.L. Altshuler and A.G. Aronov, Zero bias anomaly in tunnel resistance and electron–electron interaction, *Solid State Commun.* **30**, 115–117 (1979).
[154] D. Belitz and T.R. Kirkpatrick, The Anderson–Mott transition, *Rev. Mod. Phys.* **66**, 261–380 (1994).

[155] M. Imada, A. Fjimori and Y. Tokura, Metal-insulator transitions, *Rev. Mod. Phys.* **70**, 1039–1263 (1998).

[156] H.v. Löhneisen, The metal-insulator transition in Si: P, *Festkörperprobleme* **30**, 95–111 (1990).

[157] D.V. Averin and K.K. Likharev, Coulomb blockade of single-electron tunneling, and coherent oscillations in small tunnel junctions, *J. Low. Temp. Phys.* **62**, 345–372 (1986).

[158] T.A. Fulton and G.J. Dolan, Observation of single-electron charging effects in small tunnel junctions, *Phys. Rev. Lett.* **59**, 109–112 (1987).

[159] D.V. Averin and K.K. Likharev, Single-electronics: a correlated transfer of single electrons and Cooper pairs in systems of small tunnel junctions, in *Mesoscopic Phenomena in Solids*, edited by B.L. Altshuler, P.A. Lee, and R.A. Webb (North-Holland, 1991), chap. 6.

[160] I.O. Kulik and R.I. Shekhter, Kinetic phenomena and charge discreteness effects in granulated media, *Sov. Phys. JETP* **41**, 308–321 (1975).

[161] L.J. Geerligs, M. Matters and J.E. Mooij, Coulomb oscillations in double metal tunnel junctions, *Physica B* **194–196**, 1267–1268 (1994).

[162] D.V. Averin, Periodic conductance oscillations in the single-electron tunneling transistor, *Physica B* **194–196**, 979–980 (1994).

[163] Y. Pashkin, Y. Nakamura and J. Tsai, Room-temperature Al single-electron transistor made by electron-beam lithography, *Appl. Phys. Lett.* **76**, 2256–2258 (2000).

[164] W.L. McMillan, Scaling theory of the metal-insulator transition in amorphous materials, *Phys. Rev. B* **24**, 2439–2443 (1981).

[165] J.G. Massey and M. Lee, Direct observation of the coulomb correlation gap in a nonmetallic semiconductor, Si:B, *Phys. Rev. Lett.* **75**, 4266–4269 (1995).

[166] J.G. Massey and M. Lee, Electron tunneling study of coulomb correlations across the metal-insulator transition in Si:B, *Phys. Rev. Lett.* **77**, 3399–3402 (1996).

[167] M. Lee, J.G. Massey, V.L. Nguyen and B.I. Shklovskii, Coulomb gap in a doped semiconductor near the metal-insulator transition: Tunneling experiment and scaling ansatz, *Phys. Rev. B* **60**, 1582–1591 (1999).

AUTHOR INDEX

SUBJECT INDEX